A History of Irish Forestry

Born in Cork in 1927 and educated at Newbridge College, Eoin Neeson served in the Irish Air Corps before turning to writing and journalism. During the 1950s he was editor of *The Kerryman* and *The Munster Tribune* and had two thrillers published.

In 1961 he joined Radio Telefís Éireann, scripting plays and films, and editing news programmes. From 1968 to 1972 he was head of the Government Information Bureau, and thereafter became Director of Special Projects in the Forestry Division of the Department of Energy.

He is author of several works on Irish mythology and history, which include his pioneering account, *The Civil War 1922-23* (1966, 1989) and *The Life and Death of Michael Collins* (1968). He also translated *Poems from the Irish* (1966, 1985) and has written a series of historical novels under the pseudonym Donal O'Neill. Those published to date are *Crucible* (1986), *Of Gods and Men* (1987) and *Sons of Death* (1988).

'Cad a dhéanfhaimíd feasta gan adhmad

Tá deireadh na gcoillte ar lár…?'

CORRECTION

The attribution in footnote 51 of chapter 17, page 367, to the speech on the Forest Estimate by the Minister for Forestry, Michael Smith TD, in 1988 as the source of 'stated forest policy' on page 277, is not correct.

The quotation attributed to Dr Niall O'Carroll in the footnote related to this incorrect statement and not to the published contents of the Minister's speech. Dr O'Carroll's observation, therefore, is correct.

The author and publishers regret the error and apologize to Dr O'Carroll for any possible inference that Dr O'Carroll was not aware of government policy. Any such inference is wholly without foundation.

A History of Irish Forestry

Eoin Neeson

THE LILLIPUT PRESS
in association with
THE DEPARTMENT OF ENERGY

Copyright © The Department of Energy

All rights reserved. No part of this publication
may be reproduced in any form or by any means
without the prior permission of the publisher.

First published in 1991 by
THE LILLIPUT PRESS LTD
4 Rosemount Terrace, Arbour Hill,
Dublin 7, Ireland.

A CIP catalogue record
for this title is available
from The British Library.

ISBN 0 946640 70 X
ISBN 0 946640 71 8 (pbk)

Jacket and inset illustrations
by Wendy Walsh
Set in 10.5 on 11pt Garamond Original by
Diskon Technical Services Ltd of Dublin
and printed by Colour Books Ltd.

CONTENTS

	Acknowledgments	vii
	Foreword by Lord Killanin	ix
	Introduction	1

PART ONE   THE HISTORICAL BACKGROUND

ONE	Primeval Forest and Early Man	9
TWO	Woodlands, Land Title and Tenure in Celtic Ireland	24
THREE	The Normans and Early Medieval Ireland	33
FOUR	From the Later Middle Ages to the Tudor Conquest	46
FIVE	Tudor Plantations and After: The Effect on Irish Forests	57
SIX	Ships and Shipbuilding	79
SEVEN	Estate Forestry in the Eighteenth and Nineteenth Centuries	91

PART TWO   FORESTRY IN MODERN IRELAND

EIGHT	State Forestry: The New Idea	117
NINE	The First World War	137
TEN	Saorstát Éireann: The First Decade (1922-32)	150
ELEVEN	The Thirties and the Emergency Years (1933-46)	170
TWELVE	A Policy Defined (1944-58)	184
THIRTEEN	A Critical Decade (1956-66)	200
FOURTEEN	The Rising Tide (1968-76)	215
FIFTEEN	A Decade of Decision (1976-86)	228
SIXTEEN	Timber and Species	247
SEVENTEEN	Private Forestry	258
EIGHTEEN	Northern Ireland; Biomass	278

APPENDICES

1	Disafforestation: Glendalough, Co. Wicklow	295
2	General Deforestation in the Seventeenth Century	297
3	Some Irish Statutes before 1800	301
4	Note on the Measures Adopted by the (Royal) Dublin Society to encourage the planting of forest trees in the eighteenth century, and the results attained	309
5	Land Ownership	319
6	Irish Statutes since 1920 Relating to Forestry	322
7	'A National Scheme of Afforestation'	323
8	Some Senior Officials Responsible for Forestry since 1904	327
9	Ministers Responsible for Forestry since 1904	329
	Notes	333
	Select Bibliography	369
	Index	375

ACKNOWLEDGMENTS

This book covers a period of approximately 2000 years, from Celtic Ireland to the 1990s. There is an introductory reference to the post-glacial period during which the ecological foundations were laid, but the book is primarily concerned with the history and development of forestry during the historical period. It is hoped that the reader will find it a useful guide both to earlier times and as a record of the growth and development of afforestation in the twentieth century.

During the writing of the book a great deal of help, encouragement and information came from many sources. My first thanks must go to the Advisory Group which I established at the outset of the project to guide me through virgin territory and who attended numerous meetings and waded through notes and drafts at no little inconvenience to themselves. I was privileged to be able to draw on the collective experience, freely and generously offered, of the members – particularly the late Mr Tim McEvoy, former chief inspector, FWS, whose inital encouragement was unstinting and compelling; Mr Patrick Howard, former principal officer, FWS, who was a constant source of advice and a rock of common sense; Professor Tom Clear, former professor of forestry, UCD and Professor Padraic Joyce, professor of forestry UCD, whose combined knowledge and experience in the field was made freely available; and Dr Niall O'Carroll, chief inspector, professional unit, Forest Service, Department of Energy.

The late Seán MacBride was, as always, exceptionally courteous and generous of his time and knowledge, as was Dr T. K. Whitaker, Chancellor of the NUI. Thanks are also due to Mr Dan McGlynn, former assistant chief inspector of the FWS, and to Mr H. M. Fitzpatrick, doyen of Irish foresters, for their time, extensive comments and insights; Mr John Tyrrell, who was very helpful about ship building and Wicklow timber sources; Mr Darach Connolly, who made available the papers of the late Joseph Connolly; Dr Ronan Fanning; Mr Patrick Whooley, former secretary of the Department of Fisheries and Forestry; the late Mr Thomas Rea; Mr R. M. Keogh; Mr Brendan Halligan, who clarified an important aspect of Bord na Móna activities in relation to forestry; the staff of the Forestry School at Avondale whose kindness and attention were of great assistance and to the many officers of the Forest Service who offered encouragement and advice and those who, even when unaware of being so, were a stimulus.

Thanks and acknowledgment are due the ministers under whose authority this work progressed – Liam Kavanagh, TD; Brendan Daly, TD; Ray Burke, TD; Michael Smith, TD; Robert Molloy, TD – and to the secretaries of the Departments – the late Mr Seamus de Paor,

ACKNOWLEDGMENTS

whose response to the idea was enthusiastic from the outset; Mr Patrick Whooley and Mr John Loughrey – who were unfailing in their support throughout. My thanks are due to the staff of the National Library of Ireland, the library of Trinity College Dublin and also the libraries of the Departments of Energy, Agriculture and Coillte.

Finally I owe a particular debt of gratitude to Helen Litton, whose painstaking editing contributed in no small way to the end product. Whatever mistakes, errors or omissions that remain are my own.

**

The views and opinions expressed in the book, except where otherwise stated, are mine alone and do not necessarily, or at all, represent the official view of the Minister for Energy or of the Forest Service of the Department of Energy.

Eoin Neeson

FOREWORD

When invited by Antony Farrell of The Lilliput Press to write the Foreword to *A History of Irish Forestry* it appeared to me strange, as I come from a rocky area where stone walls replace wooden fencing and trees are at a minimum. In Dublin I often stroll beneath the trees of the resuscitated Phoenix Park (OPW) and Temple Gardens (Corporation), or in the west of Ireland among the few trees clipped by the south-west gales and bending to the north-east. After reading the proofs I learned what part Irish forestry had played in the history of Ireland, politically, socially and economically.

Having had a lifelong interest in our heritage, and thanks to the greening of the Emerald Isle and the new appoach to our countryside, this book, the subject of which has never yet been covered so fully, on pages themselves derived from forestry, is most opportune. It is a comprehensive work, starting from the earliest times to the modern day.

A new, generous understanding of the past now extends to trees, whether planted in the wide streets of eighteenth-century or modern Dublin or in forests and landlords' demesnes. Our earliest trackways across the vanishing bogs of Ireland, now being protected by those conscious of the environment, were made of timber, and throughout history wood has played a vital part in Irish life, as seen in medieval castles such as those restored at Aughnanure and Drimnagh or in the great Georgian houses in the country and city. These buildings are unique not only in their façades or interior stucco work but in their fenestration and wood furnishings, which include the magnificent doors under fanlights (many of them in Dublin spoilt by too many name-plates and unfortunate modern door furnishings), and the best of furniture which our forests were able to provide for the craft cabinet-maker.

I have always been anxious about some aspects of tree farming which encourages quick-growing conifers that can be harvested in the shortest time. Successive governments did not take into consideration a longer-term investment in our countryside of broad leaf or hardwoods and ignored landscaping with the harsh, unlandscaped tree lines. When Coillte, the State-owned company for forestry, was formed it came under the Department of Energy, and 'energy' infers the burning of wood rather than the growing of wood for other commercial purposes. But the Department, which is represented with the Department of Finance on Coillte, has to be business-like and now has an Environmental Officer and a policy of what it plants and where, avoiding areas of scientific interest or National Parks. The National Heritage Council under my chairmanship made several representations on this subject, and I am glad to say these were heeded, and that long-term planning should ensure forestry does not end as have our bogs.

A History of Irish Forestry appears to me of inestimable worth to those commercially involved and to others with a social and historical interest in forestry. How lucky Eoin Neeson is to have an Irish publisher. Except for religious and school publishing, they were thin on the ground when I was an ambitious, aspiring writer over fifty years ago. The appendices, sources of information and index will be invaluable to the reader who wishes to study further.

Killanin
Dublin and Spiddal, 1991

INTRODUCTION

During the last two hundred years in Ireland the topic of forestry has given rise to much controversy and comment. But, hitherto, no comprehensive assessment of what looks like becoming one of our major industrial enterprises within the next fifty years has been published.

In order to attempt a broad perspective, a progression must be traced, and periods of emphasis categorized. The most effective method is historical, since it includes species-dominance and life-cycle as well as references to the contemporary legal status of forests and woodlands. Except for a very general introduction, the geologic primeval period is excluded.

The assessment may be divided into four periods. Firstly we have what may be called the 'Gaelic period', which dates from the proto-historic period to the end of the Tudor conquest. It embraces the definitive change of dominant species during that time; rights of title and ownership; obligations and penalties under Brehon Law tracts and derivative claims of ownership; Norman claims to absolute ownership of land and consequentially assumed rights, and the decline, revival and destruction of the Gaelic Order before and during the Tudor conquest.

The second period runs from the Tudor conquest to the Act of Union, when, so far as may be judged, the greatest exploitation and decline of Irish natural forests occurred. During what I call this, the 'Period of Foreign Exploitation', some species (and whole forests) virtually vanished from the landscape. What remained was of little value. This period also produced developments in timber industries, and the beginning of extensive timber imports. It ended with the Act of Union when, for the first time, the forest laws of Great Britain applied to the forests of this country. One result of this legal 'regularization' was the development of a climate of public opinion, principally among farmers, hostile to forests and forestry, the residual effects of which still exist.

During this period, from about the middle of the eighteenth century, another development began which, for want of a better name, is called 'Estate Forestry', to distinguish it from later 'State Forestry', of which it was an important precursor. It exemplifies a trend amongst landowners in England, Scotland and Ireland, partly fashionable and scientific, towards enlightened self-interest. Such people, both the dilettante and the genuinely committed, were in a position to undertake long-term economic investment while at the same time entertaining an interest in 'scientific' forestry. 'Estate Forestry' and those engaged in it were, for 180 years, to have profound and formative effects on the whole subsequent course of Irish forestry.

The third period covered is the nineteenth century, ending in 1899, when the Department of Agriculture and Technical Instruction (DATI), which went on to form a committee for forestry, was established. During the nineteenth century far-reaching changes occurred in the course of Irish forestry. Principles of forest management, economics and species changed, not only in Ireland, but also, if for widely differing reasons, in Britain and continental Europe. In Ireland great social and political changes also occurred. Some of these, in respect of land and tenure, greatly affected forestry and attitudes towards it.

Finally, the twentieth century saw the emergence, growth and development of an Irish forest policy which was to provide the foundation for the national afforestation programme undertaken after independence in 1922. Forest policy was largely promoted, encouraged and, to some extent, controlled by landowners. They were the inheritors of 'Estate Forestry'. They were also (the new tenantry apart) those most affected by the Land Acts that were altering the whole social order in rural Ireland, with all the resulting consequences, legal and otherwise, for forestry and land title. The movements for national expression and self-determination, and the First World War, also influenced forest policy-makers.

Although the idea had existed since 1884, the beginning of the State Forestry programme effectively coincided with the period when the foundations of the State itself were being laid. Clearly, having regard to the times, events, individuals and interests involved, the possibility of conflict already lurked, so to speak, on the forest floor when the State came into being. The interests and motives of those who knew most about forestry, namely the landowners, were, for a number of reasons, suspect. Almost all were representative of a minority traditionally regarded as repressive and with an alien tradition. Since they represented a resented elitism their very existence fuelled hostility among rural smallholders, even though the Land Acts offered tenants certain land purchase rights. In frequently creating many holdings from one, this had a direct bearing on the later problem of fractured holdings and piecemeal acquisition.

Most landlords belonged to a class which regretted the passing of the ruptured link with Britain. Since such forestry as existed was (and for many years remained) 'Estate Forestry'; and because those who had managed government forest policy under the DATI tended to be either members of the landed class or of the English tradition (for the very good reason that they knew most about the subject), forestry was generally associated with this elitism and was resented in the Irish rural community. This affected the questions of tenure, land for planting and any conceivable land purchase or land-use policy. 'The landlord was regarded as an alien, both by birth and religion, possessing by right of conquest what the farmers considered to be theirs by hereditary right' (Dardis).

On the other hand, many enthusiastic nationalists, at a time when it was believed that natural resources hardly existed in the country, and regardless of their own lack of experience or expertise, held views on forestry coloured by a romanticism which envisaged it as some sort of boundless natural panacea of limitless economic and social potential. They tended to join in unlikely alliance with the landowners to promote programmes of forestry and reafforestation, but did not always agree among themselves and, for a considerable period after independence, some might say to the present day, there were almost as many forest policies as there were forest policy makers.

There were other important factors. The natural political polarization which is at the heart of representative democracy, while it examines legislation in the making on any issue with care and caution, also tends to a short-term rather than a long-term view of economic planning, particularly when the national purse is lean and its strings tightly drawn. Short-term planning, however, is impractical when forestry is being undertaken virtually *de novo*, without established forests of any significance and in the absence of any social forestry conscience, tradition or system of established management. In such circumstances a capital-intensive undertaking such as new forestry is costly, but has a low political priority. Their high level of professionalism notwithstanding, it strains the limits of common sense to suppose that the key English or Scottish forest experts introduced to assist the new State's forest programme were all free of partisanship. Inevitably they came into conflict both with landowners and enthusiasts; with, at a more active level, politicians and smallholders and, sometimes, with their own staffs; and, of course, with each other.

Once the State programme got under way the influence of 'Estate Forestry' began to decline. This was not foreseen, perhaps, and had some curious consequences, the most important being the metamorphosis of State input to the planned National Programme from a minority one to one approaching 80 per cent. Circumstances also led to existing estate forests being drastically reduced during the Second World War, weakening the landowners of whom remarked Mr H. M. FitzPatrick, 'they knew more about forestry than all the rest of us put together'. During the early years proposals for forestry programmes and bases for policy emerged at regular intervals from a variety of sources. They were sometimes so self-cancelling that advocates of one policy would neither speak to, nor give credence to, the advocates of another.

These were some of the problems that beset the infant Irish National Forest Programme. In spite of this Irish forestry has grown from less than 101,173.6 (South and North combined) hectares in 1922, to more than 404,695 hectares (1,000,000 acres in the State) today. This considerable progress, the type of woodlands created, their location, economic viability, ancillary industries and potential are described in Part Two, while Part One is concerned with the historical development up to 1900.

The question of the role and function of a bureaucracy in relation to the initiation, administration and continuing management of a productive forestry marketing programme on such a vast scale requires consideration. The National Forest Programme was undertaken largely from scratch; such managed forests as were in the country were in private hands. Forests had to be created from nothing, with limited tradition and experience, and little capital outlay. The obstacles were great, but there were also opportunities – to avoid the limitations of imposed outmoded traditions and methods, for instance. But there is a point between the development of a State forestry programme, *de novo*, and the successful marketing of its product, where the management dynamics radically alter.

Forestry means more than simply the natural or managed activity of dense tree growth on a piece of land. We take the use of woods, of timber and of wood products by man from beyond the dawn of history so much for granted that we overlook the fact that the relationship between these two living organisms, man and tree, is no less than the relationship between man and domestic animals. One may speak of 'domestic forests' in much the same sense and for similar reasons.

In the Preface to his book *Trees*, Rushford writes:

If it were not for the spread of the human race there would be vastly more trees and greater areas of forests on the face of the earth, and their appearance would often be entirely different. Mankind has brought about great changes in his environment both by reckless destruction of forests and the pursuit of rapid financial gain, and by significant alteration of forest types. In the context of the geological timescale it was physical factors – climate, altitude, soil-type and so on – that determined whether a region could support a forest of some kind, but man's influence has now made itself felt with relative suddenness and left its mark on every part of the world.

In areas where the equable climate has favoured the development of human civilization, the exploitation of forest resources has caused damage that is irreversible.

INTRODUCTION

Without some appreciation of its past or sense of continuity, any account of Irish forestry must be valueless. This book has attempted to identify the trends and influences that have led to the forests of Ireland today, and to pursue a chronology faithful to that development. The research and collation of material has been a pioneering task, aspects of which will no doubt prove controversial.

During the research a number of issues, other than that of the general and historically superior one of the disappearance of vast forest areas between 1600 and 1800, stood out as worthy of note.

First was the exceptional decline of hazel over a twenty-to-thirty-year period at the turn of the eighteenth century. The inference I have drawn here points clearly to a new and perhaps significant historical conclusion.

Secondly, the concept of a national forestry programme envisaged in the seminal 1908 Committee Report underwent an extraordinary metamorphosis. The original proposal was virtually stood on its head – and, paradoxically, achieved the target set by the 1908 Committee almost to the year, if by a route and means totally unforeseen by that committee.

Thirdly, at a time when the co-operative movement was well established among the rural population, co-operative methods – until recently – failed to make any impact on the development of forestry.

In general, the history of Irish forests and forestry has paralleled the political history of the country. Given the social and economic importance of woodlands – ever-changing though these aspects were from century to century – that is less surprising than it might seem. What does surprise is how little attention has been given to it.

PART ONE

The Historical Background

CHAPTER ONE

Primeval Forest and Early Man

'And out of the ground made the Lord God to grow every tree that is pleasant to the sight, and good for food' – Genesis.

THE TOPOGRAPHY OF IRELAND was determined about 2,000,000 years ago when an increasingly cold climate led to dramatic alterations in the landscape. Intense cold replaced wet, tropical conditions in which dense woods and forests, consisting largely of trees now either extinct or exotic in this country, flourished on very deep soils of up to thirty metres, compared with an average of about one metre today. The subsequent Ice Age covered much of the country with ice and glaciers.

When, after some 1,990,000 years, the ice withdrew, retreating glaciers scoured the country, creating new valleys and ravines. The hitherto hidden rocky skeleton of the land was exposed. The resulting much altered landscape is basically what we know today. The low-lying central plain, floored essentially by carboniferous limestone, is surrounded by a rim of mountains, in a general saucer shape. The mountains (some more than 300,000,000 years old), eroded and rounded, are far older than the lowlands.

Between 10,000 and 8000 years ago land bridges connecting Ireland with Britain and the Continent were inundated. The forests and woods of that time, established since the ice withdrawal, consisted of species that endured, in some cases, until today. This was the foundation from which all subsequent forests and forest lands developed.[1]

When man came to Ireland between 9000 and 8000 years ago, so far as we know, he found an island with a climate somewhat warmer than today. It has remained much the same size; 84,000 sq. km. (32,000 sq. miles), about 486 km. (302 miles) north to south and 275 km. (171 miles) east to west. It had, as now, many modest-sized lakes but fewer turf bogs, some of them still undeveloped, filling ancient lake basins with the detritus of the ice withdrawal.

This mountain-rimmed island west of the European mainland, dominated by the Atlantic Ocean, has a 'typical west maritime climate... It is modified by the Gulf Stream flowing north-eastwards... from the warm regions of the Carribean', producing a climate with an 'absence of extremes favourable to the growth... of tree and shrub species... and the climate is especially suitable for forest trees... The method by which these so-called primeval forests were destroyed is not clearly understood, owing to the superimposition of climatic change and the activities of early settlers – both forces for vegetation change. However there is considerable evidence that the deterioration of climate which began about 3000 BC stimulated bog development at the expense of forests in many regions.'[2]

The author goes on to state:

Soil is the most important side factor in influencing tree growth. There are considerable variations in forest soils in Ireland. This is a result of a wide range of origins. It is further complicated by climatic, ecologic and topographic factors; an excess of precipitation over evapotranspiration in large areas of the country results in a tendency towards podzolization – intensely leached mineral soil – where drainage is free; gleying or peat formation where it is impeded.

When the first mesolithic people arrived, the forests were mainly established woods of lowland oak, elm, ash and pine, and hazel, alder and birch on poorer soils. These forests may already have been somewhat in decline, and man's activities after his arrival presumably hastened the process. The currently available evidence is scanty, but suggests that the earliest inhabitants were of the later mesolithic period. They lived here more or less undisturbed and unchanging for about 3000 years, when they were joined by a very different people, neolithic agriculturalists.

Pollen counts indicate that for some 2000 years after man's arrival, that is until about 7000 years ago, the carpet of forest layering the country was virtually undisturbed. At that time the first of two factors to affect the blanket forest adversely became increasingly evident. This was the growth of raised bog, a process that overlapped the arrival of the neolithic farmers, who introduced the second factor, forest clearance to create arable.

The terms paleolithic, mesolithic and neolithic, generally speaking, refer to periods of prehistoric time. It is not necessarily the case that a level of attainment in a later period was more advanced than that of an earlier one, or that one culture did not overlap another. For instance the Maori and aboriginal people of New Zealand and Australia in the early part of the last century were culturally paleolithic/mesolithic. The Bushmen of the Kalahari Desert today are mesolithic. The term *paleolithic* refers to the earliest period of the stone age, when stone weapons and tools were first developed by hunter/fisher peoples. The *mesolithic* period indicates a hunter/fisher pastoral, transitional period of the Stone Age, and the *neolithic* period means the later stone-age, food-producing period, characterized by polished stone implements, some of which were used for agricultural purposes.

In the historic meaning, paleolithic and mesolithic refer to a period lasting over a million years, when people lived a largely nomadic existence hunting and collecting wild plants and berries. Such *food-gathering* societies were gradually converted to or replaced by neolithic *food-producing* societies. This change also laid the foundations of civilization. 'Hitherto, communities had been restricted by the number of game animals and the amount of edible plants available: now it was possible to plant more seed, till more land and breed more animals as the population increased... the word "neolithic" in fact simply implies food production based on crops and domesticated stock, without metals...'[3]

There is evidence of early human activity at Mount Sandel near Coleraine in Co. Down, from where artefacts have been carbon-dated to 8650 years ago, + or - 50 years. This site has provided flint tools, mainly small flint points called microliths, and small hand-axes made of flint flakes, typical of a hunter/fisher community that did not engage in farming. There is evidence of similar late mesolithic communities in Denmark and of early mesolithic in both Denmark and England. No evidence has so far been discovered indicating the presence of early mesolithic or earlier paleolithic peoples in Ireland.

The mesolithic peoples of Mount Sandel and the Larne (the Larnian people, so-called because the stone implements they used are common in the raised beach gravels at Larne, Co. Antrim) appear to have had little knowledge of agriculture. The country was smothered in dense forest and there were no plains on which they could hunt. They existed as hunter-fishers roaming the edges of forests, coasts, the shores of lakes and rivers, and probably operated from semi-permanent or seasonal camps, as the Amazonian Indians do today. In a primitive way they may have cultivated certain naturally regenerative crops, such as bracken and nettles.

A comparison can be made with the seasonal migrations of the Maori at the time when they first came in contact with white settlers about 200 years ago:

(They) had a wider range of implements than the Larnians, but were, like them, essentially food-collectors... For one such group the following yearly pattern in search of protein has been recorded – September and October, up to Tuturau for lampreys at the Mataura Falls; November on to the Wainea plains to get eels; December, back to Tuturau to dry food; January, to the coast to catch fish and collect sea-weed; February, making sea-weed bags; March, to the offshore islands; April, catching and smoking sea-birds; May, return from the islands; June, bringing presents of smoked sea-birds to friends and relatives; July and August, catching forest birds.[4]

Notably two months – February and April – are devoted to manufacturing. Some similar cyclic pattern may have operated amongst the mesolithic hunter/fishers of Ireland. Since much of the available evidence about them consists of worked or partially worked flint implements it seems reasonable to conclude that they devoted considerable time to manufacturing these.

The mesolithic people probably cured meat and fish by salting, smoking and drying on racks as is still done in Iceland, Greenland and elsewhere. They certainly had boats of some kind, and were clearly capable wood-workers. Perhaps part of the attraction of Ireland for these nomadic people was the abundance of available timber. Such dense woods, besides being a multiple food source, provided the raw materials for fuel, fishing and building, a means of making containers (of bark; they had no pottery), and, fundamental to north European nomadic hunters, the means of making vessels to penetrate the otherwise inaccessible country via its rivers and lakes. They probably used some form of dug-out or burnt-out canoes. From Toome Bay on Lough Neagh, as Mitchell[5] demonstrates, comes evidence of wood which had been worked for a special purpose. By the time these mesolithic people were absorbed or superseded by the neolithic farmers, they had inhabited the country for some 3000 years, about 400 years more than the period separating us, today, from the first coming of the Celts. And after them the neolithic people inhabited it for as long before the Celts arrived.

The decline of forest cover accelerated with the arrival of these neolithic farmers between 5000 and 6000 years ago. One of their first concerns would have been to clear land for tillage and grazing, a task in many ways simpler and less demanding in forest land than in open plains or savannah, which require the use of ploughs to break up the tangled grass roots.

It is hard for us to picture the majesty and silence of those primeval woods, that stretched from Ireland far across northern Europe. We are accustomed to an almost treeless countryside, and if we can find anywhere some scraps of 'native' woodland, we are disappointed by the quality of the trees. For thousands of years man has been roving the Irish woodlands seeking for 'good' timber for houses, ships and other uses. As a result all the well-grown 'good' trees have long since disappeared, and what are left are the progeny of 'bad' trees rejected by earlier carpenters. If we visit the National Museum we can see wooden shields 1m in diameter, worked from a slice taken from the trunk of a well-grown forest alder. We could not find in Ireland to-day a single alder tree capable of supplying a blank for such a shield. In some remote parts of Europe scraps of upland valleys have escaped the loggers' attentions, and there we can recapture something of the vanished dignity of the Irish forests.[6]

From the arrival of the neolithic farmers, through the early and late Bronze Ages and the coming of the Celts about 800 BC to the beginning of the Iron Age, was a period of great change and activity in the Irish landscape, especially in relation to woodlands. Given the time-scale involved, the change was never immediately apparent, unlike the man-made environmental changes dramatically taking place today.

About 5000 BC Ireland would have presented a mosaic of different woodland types composed of a few species, depending upon local soil and climatic conditions. The lowlands and sheltered valleys had a covering of mixed broadleaf forest, chiefly of oak; pines and birch were dominant on poorer sites, especially in the west. Alder and willows formed local scrub woodlands on marshy sites near lakes and rivers. The remains of these forests can still be seen in cutaway bogs at altitudes ranging from sea level to 600 metres (1800 ft) at Turlough Hill in Co. Wicklow.[7] Why these early forests disappeared is not fully understood. But there is evidence to suggest, among other reasons, that climatic deterioration, which began about 3000 BC, stimulated bog development at the expense of forests. Extensive areas of forest were cleared by felling, by fire and by grazing stock. These clearances, and the prevailing climate, encouraged bogland to expand, downwards from the mountains and upwards from wet hollows, providing the foundations on which the later enormous bogland increases occurred. These recurring vast boggy areas, not without reason, have been called 'wet deserts'.

The stone monuments from these remote times are impressive, but they give the impression that stone was the most commonly used material for building purposes. That was not the case. Wooden buildings were the universal type in Europe at that time and there is no reason to think that Ireland was any different in this respect. The scarcity of significant remains is due to the climate, in which timber, especially light, treated timber such as wattling, perishes. 'In 1967 a rectangular house, 7 × 6m. and walled on two sides with thin planks of oak, was found at Ballynagilly, Co. Tyrone in Northern Ireland. Associated pottery suggests a date of about 3000 BC...'[8] Irish building in wood continued up to the seventeenth century, maybe later. The ancient stone constructions that survive, including Newgrange, circular forts, the remains of monasteries and so on would have had substantial timber buildings both within and without the walls. Many later monasteries, of which only some stone fragments remain, were teaching establishments with up to 4000 people, all of whom required to be catered for in their everyday needs. Hence their attraction for the Norse invaders.

Since the early migratory settlers to this island came by sea, it follows that they were both boat-builders and craftsmen of some skill.

> With space for the crew and two dogs, and bedding for the stock, its [a frame boat covered with hides] load for a sea-passage might have been two adult cows and two calves, or about six pigs or ten sheep or goats. Because of the difficulty of watering stock on a voyage, such trips must of necessity have been short... Case thinks that stock, if well watered beforehand might hold out for two days... it is clear that sea-voyages would have to be planned with some care, and that undue holdups had to be avoided.[9]

Domestic animals were very much smaller at that time.

A people capable of constructing vessels to carry them across miles of open sea were not only accomplished wood-workers, but also tool-makers and planners. Craft construction required worked timber of some kind, whether it was frames covered in hides, dug- or burnt-out canoes, rafts or timber boats. Boat-building is a derived skill, and it is not reasonable to assume that it was the primary skill of these people, or that their planning and manufacturing skills were confined to it and to seafaring. Accordingly we may infer that the peoples who first inhabited this country demonstrated from the outset some skill in the use and working of timber.

The techniques they used in building their boats were very likely adapted from techniques of domestic building, farming – clearance, tillage, pasturage – and, of course, hunting and fishing. They also brought with them considerable farming knowledge, skills in the manufacture and in the use of tools, and the organizational ability demanded by communal migration overseas. Such skills and co-operation would not manifest themselves spontaneously for this purpose alone.

H. J. Case concluded that 'lightness and manoeuvrability were more important than size' and that a boat similar to the currachs and naomhogs of the west coast, about thirty feet long, with about eight oarsmen and able to carry about three tons or up to forty people, could have been the optimum.[10] In 1976-7 Tim Severin with several companions crossed the Atlantic in a similar boat made in traditional manner of lengths of timber, rods, pegs of wood, binding and hides.

No one knows where these early settlers came from or where they first landed. However, the coasts of Antrim and Down were clearly visible from Britain. People of advanced neolithic culture, roughly equivalent to nineteenth-century Australian aboriginals, were capable of coming from anywhere on the northern coast of the continental mainland, and of making landfall on the southern and western coasts of Ireland as well as in the north-east.

There would have been considerable advance scouting before major community migrations took place. While a migratory group would include men, women, children and infants, together with breeding and milking livestock and seed-corn, it is unlikely that an entire community would embark simultaneously, and not impossible that there was more than one migration from different sources, at roughly the same time. Such voyages could only be made at certain times of the year, probably between July and October, when the work of harvesting crops was over and there was still grass and leaf-fodder in the new destination.

Such early neolithic settlers built plank as well as clay and wattle houses, as did their European counterparts. The roofs, usually of thatch, were supported by interior posts. They surrounded their settlements and farms with stockades for protection; they built storage and refuse pits and sowed crops. To the present day their descendants have used the native timber for precisely similar purposes; to construct homes, make boats, tools and utensils.

The fitting of wooden handles to his stone axes was of great importance to all neolithic man's activities. It converted them from hand-held chopping implements of limited potential, to axes, mattocks, picks, adzes and so on which could be wielded with great power, enabling the user to fell trees, work large timbers, and erect substantial buildings. This effective conjunction of wooden handle and blade has remained one of the single most important technical developments in the history of mankind, second only, perhaps, to the discovery and effective use of the wheel. The weapon technology of the neolithic people would have been similarly progressive and both arrowheads and scrapers of flint, useful for making stems for basket-work and making and arming arrows, have been found. Though they have not been found in Ireland, 'bows of yew wood have been found in Somerset'.[11]

Coincidentally with man's arrival in Ireland some forest trees, notably and quite dramatically the elm, went into decline. Hazel scrub was universal. Pollen counts of the period 5500 to 5200 years ago clearly demonstrate this. Mitchell writes:

I picture that quite a small group of men could sweep through an elm-wood quickly, ring-barking the trees, and so indeed putting an end to the production of elm pollen. In any case the soil of the tillage patch created was going to become exhausted quite quickly and it would have been stupid to put into land that was only going to have a short life the immense amount of effort necessary to produce neat fields, free of tree trunks and stumps.[12]

This point has been evidenced more recently in Brazil where the Amazonian forest clearances provided cleared land for settlers, but the experiment was of mixed success. For every plot of good land with vital soil, there were several of land of poor quality which – when it grew anything – produced only stunted crops or brush.

The American geographer, Carl Sauer, wrote:

Primitive agriculture is located in woodlands. Even the pioneer American farmer hardly invaded the grasslands until the second quarter of the past century. His fields were clearings won by deadening, usually by girdling (ring-barking), the trees. *The larger the trees, the easier the task:* brush required grubbing and cutting; sod stopped his advance until he had plows capable of ripping through matted grass roots. The forest litter he cleared up by occasional burning; the dead trunks hardly interfered with his planting. *The American pioneer learned and followed Indian practices.* It is curious that scholars, because they carried into their thinking the tidy fields of the European plowman and the felling of trees by ax, have so often thought that forests repelled agriculture and that open lands invited it[13] (author's italics).

The evidence, from the remains of the forests found in cutaway bogs, is that the early settlers in Ireland followed a similar practice to that described above.

The disposition of trees in those early times tended to be mainly oak on the lowlands and pine on the highlands, but this was by no means exclusively the case. The native Scots pine, possibly for climatic reasons, suffered a serious decline at an early stage and by the twelfth or thirteenth century AD was altogether extinct. Work by Bord na Móna on some bogs, including lowland bogs, has uncovered what have come to be called 'forest graveyards'; great cemeteries of Scots pine tree-stumps, some of them 5000 years old. Many are blackened by fire rings and some appear to display the clean, flat planes of skilful axe-work, thus exhibiting precisely the sort of tell-tale evidence one would expect from the residue of forest cleared in the manner described.

The efficacy of the stone axes employed by neolithic man is surprising. Experiments were conducted in Denmark in the 1950s:

A genuine neolithic blade was fitted to an ash-wood haft, copied from an original haft dating from neolithic times recovered from Sigersslev Bog. It was found that the blade broke if fitted too tightly; it had to be left free to vibrate slightly in the haft. The axe was most effective when swung only from the elbow, with short, sharp cuts rather than swinging blows from the shoulder, which is the most effective way of using a metal blade.

Three men managed to clear 600 square yards of silver birch forest in four hours. More than 100 trees were felled with one axe-head, which had not been sharpened for 4000 years.[14]

Forest regions such as those referred to above were cleared and palisades and stockades were erected for protection – mainly against natural predators like the bear, the wolf, the lynx and the fox, all of which abounded and for whom the availability of domestic animals was a gastronomic bonanza.

Charcoal (*c.* 4500 years old) indicates that hazel, ash, hawthorn and holly were the most frequently used firewoods. 'The hazel at all times found conditions in Ireland very favourable... In its first expansion

...its contribution to the ... pollen-rain seems to have been greater than elsewhere in Europe... (When) a woodland area which had been cleared by farmers was abandoned, hazel immediately expanded into it... The pollen count shows that ... from the first woodland disturbance more than 5000 years ago, until the Tudor clearance in the late sixteenth century, there must have been very extensive hazel scrub in Ireland.'[15]

There is abundant evidence of timber use. Much of it has been revealed from accidental discoveries in the bogs and fenlands which preserve fibrous artefacts that might otherwise have decayed. We possess dug-out canoes a thousand years old, plates, implements, utensils and house frames. In some cases the very walls and flooring of the houses themselves, constructed of woven wicker-work and wattle, often with traces of the mud, lime and dung covering still adhering to it, have been preserved. There is also clear evidence of the deliberate cultivation of timber for specific purposes. The Dal gCais of Thomond deliberately cultivated an ash and holly wood for weapon and domestic building purposes. Hazel, the principal ingredient of wattle building, was cultivated for this purpose from time immemorial.

By the year AD 300 forest regeneration was no longer capable of swallowing up the habitation sites that must have come and gone countless times over the preceding 3000 years. In the intervening period appeared monuments such as those in the Boyne valley. Others, less enduring, were laid down and their foundations established; for instance roads.

Many of the roads of ancient Ireland were what came to be called, in America, 'corduroy roads' – that is a surface of planks or of tree trunks laid parallel and often used where the soil was soft or boggy, or within the habitat of a community. Such classes of road are recorded in the Brehon Laws (the law tracts of ancient Ireland), referred to collectively as *conai cai* (any kind of) road. The seven classes, in order of size and merit, were: 1, *slíghe*, such as one of five great highways from Tara; 2, *rannt*, a major road; 3, *bóthar*, a good road; 4, *rot*, clearly of later usage and of foreign derivation; 5, *tuagrota*, and 6, *lam rota* of which the same might be said. In addition there were two specific types of road relevant to forestry – a *bealach*, meaning a pass (or cleft) through a forest which required to be maintained (Bealach Mughna, Ballaghmoon in Carlow, is one example), and a *togher*, a pathway.

The use of logs and wattles as a community surface is an obvious practical development, as the speed with which forest mud can be churned up in the Irish climate and the time it takes to harden again (if ever) when a site is in continual use, indicate. Besides *bealachs* and *toghers*, bridges and fords of timber were subject to laws, many of which had to do with maintenance and upkeep.

Clearly, from the very earliest times trees were an important part of the Irish landscape. The continuous and fundamental interaction between trees and men is often given recognition and acknowledgement in man's mysterious propitiatory religious rites. In a story concerning the mysterious Celtic division of the country into four parts, each of which is (apparently paradoxically and incorrectly) called a fifth, a tree is an important symbol of each.

Fintan of Munster, reviewing the history of Ireland at Tara, told 'of a strange personage called Trefuilngid Tre-eochair who suddenly appeared at a gathering of the men of Ireland on the day when Christ was crucified... In his left hand he carried stone tablets and in his right a branch with three fruits – nuts, apples and acorns'.[16] He confirmed that Ireland consisted of four quarters and a centre. These were the provinces of Leinster, Connacht, Ulster, Munster and the centre, which was Tara. Before he left, Trefuilngid gave Fintan some berries from his branch with an injunction to plant them in places appropriate to these divisions, which Fintan duly did. From them grew five trees: – *Bile Tortan*, the Ash of Tortu; *Eo Rossa*, the Bole of Ross (a yew); *Eo Mugna*, the Oak of Mugna; *Craebh Daithi*, the Bough of Daithi (also ash) and *Bile Uisnig*, the Ash of Uisneach. 'Though the location of most of these five places is uncertain there can be no doubt that the underlying idea is that the trees symbolize the four quarters around the centre.'[17]

In ancient Ireland lines without breadth symbolized the supernatural in the realm of space. For instance, 'Irish poets believed that the brink of water was always a place where *eicse* – "wisdom", "poetry", "knowledge" – was revealed. And the mystical fifth fifth of the five fifths of Ireland is held to be some such centre without dimension.'[18] This has been expressed in terms relating to trees, 'between the bark and the tree'.[19]

There is an established connection between trees, woodlands and druidism. There are some (possibly prejudiced) observations about druidism in Britain and on the Continent from Julius Caesar and others and some of these may refer to local variations not relevant to Ireland. Not enough is known about Irish druidism to indicate how or if it differed from the form practised in Britain and on the Continent. There appears to be sufficient common ground in other social customs among the Celtic peoples generally to warrant the assumption that trees, even if they were not worshipped here as in Gaul, were treated with considerable importance.

However, no tradition similar to that of Britain and the Continent survives in this country about the use of oak-trees and of mistletoe, which is not native to Ireland. The common view that Irish druids held religious meetings and performed their rituals under the shade of a sacred oak is based on the assumption that druidism in Ireland was identical with that of Britain and Gaul, which is doubtful. Sacred groves involving holly may have featured in Irish druidic worship, but even that is uncertain. That is not to suggest that Irish druidical practices were any more wholesome than those of their counterparts elsewhere, but they do appear to have been different in some respects, including the attitude to trees. It is possible that Irish druidism was laundered by Christian recorders and historians to eradicate traces of the pagan past. But so far as we know, the dominant Irish idol, Crom Cruach, was a permanent figure of stone, not of wood, and this fundamental fact may have resulted from a form of druidism in Ireland as distinctive as the language.

In Gaul and in Britain mistletoe featured in druidic rituals. One theory is that this was because its pale berries resemble the moon and druidism was a moon cult (the Celts counted the passage of time by nights rather than by days). Mistletoe is not native here and has no traditional significance in Ireland. The custom of kissing under mistletoe may, therefore, be imported. In view of its established sacred use elsewhere, the dearth of Irish tradition may be connected with a 'clean-up' more vigorous than in some other instances to eradicate a powerful druidic symbol and, perhaps, some more 'exuberant' customs.

Details of the druidic religion are very scanty, but we may confidently assume that it imposed certain laws, rituals and prohibitions which affected the people in general. As it would be fallacious to suppose that people are other than children of their age, we may presume that they reacted to such influences much as people today raised in a similar environment might do. It is also fairly certain that, as the inheritors of traditions predominantly Indo-European going back for thousands of years, they practised a religion containing elements already very ancient. We know that in Celtic Ireland stones and wells were objects of worship; but though certain kinds of tree, the ash and the yew for instance, were to some extent venerated, there is no evidence to suggest that trees were worshipped as they were in Gaul and Britain.[20]

While Irish records do not relate that Irish druids (*drui* or *dli*) shared the Gaulish veneration for the oak, it seems to have occasionally figured in rites and may have had some significance, perhaps accounting for the name '*doire*', oak-wood or grove, which seems to have had a special appeal. Other trees were also accorded a special place, in particular the yew, used frequently to mark the bounds of sanctuary land around a church and, therefore, associated with the word '*fidnemed*' and with the complicated laws of sanctuary in Ireland. When, for instance, Mael Mordha of Leinster sought refuge in a yew tree after the battle of Glen Mama in AD 999, he was probably seeking recognized sanctuary as much as concealment.

Fortified dwelling places, strategically sited as they frequently were, on the highest spot of the local terrain, were particularly liable to be struck [by lightning] and, even if the reason was completely obscure to the people of the period, it may have been common knowledge that a tall tree near the house would attract the flash and save the house itself from damage. This may, conceivably, be the reason for the custom of planting trees in the immediate precincts of dwelling places and the protective tree would, naturally, command the regard and the reverence of the inhabitants. With the coming of Christianity it would follow that its churches and cells would also be provided with guardian trees. When, therefore, the annals record that Ciaran's yew at Clonmacnoise and the *bile* of Swords were struck by lightning we ought, perhaps, to read between the lines that the adjacent buildings were saved from damage by the presence of these sacred lightning conductors and, further, that, however ignorant they may have been of the nature of electricity, the monks of both institutions were conscious of their protective function.[21]

The word *fidnemed* is connected etymologically with the Gaulish word for a sacred grove, *nemeton*. But in Ireland the yew (or the ash) took the place of the Continental and British oak. Giraldus Cambrensis wrote: 'Yews, with their bitter sap, are more frequently to be found in (Ireland) than in any other place I have visited; but you will see them principally in old cemeteries and sacred places where they were planted in ancient times by the hands of holy men to give them what ornament and beauty they could.'[22] The name of Newry in Co. Down is an anglicization of the Irish word, *iubher* – yew. The word for a yew wood is Eochaill, from which the town of Youghal in east Cork derives its name.

Where more than one community vies for the same territory, or where ideologies vie for control of a community, the dominant beliefs (or perhaps the beliefs of the dominant) tend to endure. Aspects of subordinate faiths or rituals can survive in adapted form; thus, from being part of a living ritual or worship, becoming a source of minor tradition. This process occurred with the adaptation of heathen places of veneration, often trees and/or wells, which were absorbed in large numbers by Christianity. Some, however, presumably because of their character, had their pagan significance altogether obliterated. The word *bile* is commonly found in placenames all over the country, indicating not only that the 'cult' of the sacred tree was a universal phenomenon, but also its tenaciousness – e.g. Knocknavilla (Galway, Mayo, Tipperary, Wexford), Gortavilla (Cork), Gortvilly (Tyrone), etc.

A gloss on the *Martyrology of Oengus*, compiled, according to tradition, at Tallaght suggest that a *bile* was to be found growing at every church. It refers to *bile na cille*, 'the *bile* of the church', as if on the assumption that it formed part of the view of every church.

From the story of Fintan it will be recalled that two of the five trees that grew from the berries Trefuilngid gave to Fintan were *billeanna*, the *Bile Uisnig* and the *Bile Tortan*, that neither of them were oaks and that three of them were ash. One of these, the *Bile Tortan*, features again in both the *Book of Armagh* and the *Tripartite Life of St Patrick*, which tells us that Patrick 'went thereafter to the *Bile Tortan* and near the *Bile Tortan* he built a church for Justinian the Presbyter which now belongs to the community of Ard Brecain' (Co. Meath).[23] One wonders if the *Bile Tortan* did not represent the mystic centre without dimension, and if it was not this fact that attached the Patrician legend to it. Inauguration trees and the trees outside a king's rath were also called *bile*. Most were ash trees.

If a claim for special veneration of any tree can be made it would seem to be the ash. Apart from its frequent occurrence as a *bile* or sacred tree, it is also associated with holy wells in large numbers, i.e., 75 ash, 7 oak, from a random sample of 210 such sites:

...the very frequency of ash trees at holy wells is, in itself, a testimony of the sacred character of the trees growing there, for nothing else could have saved them from use as fuel or timber in a countryside as starved for wood as was the greater part of Ireland during the eighteenth and nineteenth centuries, when the dearth of timber erected the winning of the semi-fossil timber from the bogs into a major rural industry.[24]

The word *craebh*, though having the literal meaning 'branch', was interchangeable with *bile*, and had the same sacred meaning – as used in respect of the *Bile* (or *Craebh*) *Uisnig*. The *Craebh Daithi* was similarly called the *Bile Daithi*. There seems little room for doubting that the ash was the tree most venerated by Irish Celts. A. T. Lucas cites the tenaciousness of this tradition: 'In 1834 there was still growing in Tombrickane, in Borrisokane parish, Co. Tipperary, a large ash, twenty-two feet in circumference at the base, which O'Donovan says was called "bellow-tree" in English... (or) "Big Bell tree" ... there can be no doubt that (these) are versions of the Irish *bile*.'[25] Even today folklore throughout the country maintains that the ash will be the first tree to be hit by lightning.

Stories associated with inauguration trees (called *bile* and also invested with the mystic qualities of *fidnemed* or 'sanctuary'), are common. One recalls the destruction of the Dal gCais *bile* at Magh Adhair in 982 by King Malachy, who carried it off to roof his castle; the destruction of the '*biledha*' at the O'Neill inauguration site at Tullaghogue in 1111; and that of the Uí Fiachrach Aidhne of Connacht by the O'Briens and Mac Carthys in 1129.

Eoin MacNeill attributed some early personal names to trees, suggesting that they derive from tree-worship. Mac Cuill (son of hazel) was so-called because 'hazel was a god to Mac Cuill', and he cites as other similar examples 'Mac Cairthin, son of rowan', 'Mac Ibair, son of yew' and 'Mac Cuilin, son of holly'.

But it is in the adaptation of the word *'fidnemed'* to Christian use that we see the true absorption of one tradition by another. *Fid* meaning 'wood', and *neimed* meaning 'sacred place', became absorbed by Christianity and associated almost exclusively with the numerous churches that were established – one must conclude by deliberate policy – on the sites of heathen sacred places. The point is further made that 'the very beginning of Armagh as an ecclesiastical site was due to the presence of this sacred grove, the original church having been founded near it as part of a policy to Christianize pagan cult centres by diverting the popular attachment to them into Christian channels'.[26]

Certain other species of tree retain a mysterious significance which may account for the stature they occupy in the legal hierarchy of trees in Celtic Ireland. One of these, the rowan (*sorbus*), quicken tree or mountain ash (a misnomer since it is not a member of the ash family) had a tutelary function, especially in the dairy, observed in parts of the country within living memory. In spite of its mysterious power the rowan (*caorthann*) does not seem to have belonged to any group of ancient sacred trees. Nonetheless it had a considerable role in popular magic. Besides its power to protect dairies and dairy produce, it was hung in the house to offset fire-raising and to keep the dead from rising, and would increase the speed of a hound if it were tied to its collar. Branches of rowan were placed over the doors of houses and byres to keep away witches and fairies alike.

It has been argued that the association between the rowan and milk derives from the possible use of its bark as winter feed for cattle. Furze, which also had defensive magical properties over milk and butter, was widely used in Ireland as cattle-fodder.[27] Alternatively it has been suggested[28] that the Irish may have adopted the magical virtues of the rowan from the Norse, who are known to have used its bark as cattle fodder and 'being, as they were, a people preoccupied almost beyond modern comprehension with cows and milk, they (the Irish) would have been predisposed to adopt with alacrity any new magical practices for the protection of both...'. An old term for the rowan was *fid na ndruad*, the 'wood' or tree of the druids.

In the historic tale of the siege of Knocklong, Co. Limerick, when the northern and southern armies confronted each other, druids on both sides made immense fires of quicken (rowan) boughs. These were all cut according to ritualistic formalities and the fires were lighted with solemn incantations. The purpose of these fires was to exercise a sinister influence on the opposing army, and from the movements of the smoke and flames the druids forecast the outcome of battle. On some occasions witches, druids or malignant phantoms cooked flesh, sometimes the flesh of dogs or horses, on rowan spits as part of a rite for the destruction of someone obnoxious to them. Many such superstitions have survived. Bring a rowan walking stick out with you at night and fairies will take care to give you a wide berth. Its efficacy is especially strong in respect of the dairy and, when a housewife is churning, if she puts a ring made of a strake from this tree on the handle of the churn, no evil-minded neighbour can rob her of her butter.

As recently as the Famine of 1847 bark, which was certainly used for food in Ireland for centuries, was again employed, and bark bread was common in Scandinavia late into the nineteenth century. There are several references in the lives of the saints to eating bark. The birch tree has associations in the Celtic world with love. The lovers' bower was, more often than not, beneath a birch tree or in a birch bush and wreaths of birch were used as love-tokens. And, of course, at the traditional source – or sources – of the seven great rivers of Ireland grew the nine hazels of wisdom. This is particularly recalled in the Well of Sergeis, Connla's Well, which was the source of inspiration and knowledge. The hazel nuts dropped into the well and caused bubbles of mystic inspiration to form on the streams that issued from it. They were also eaten by the salmon and whoever ate the nuts or the salmon which had done so, obtained the gift of the wisdom of the seer and the poet. The popular story of Fionn Mac Cumhail (mac Cuil – Son of Hazel) and the Salmon of Knowledge illustrates this tradition precisely.

CHAPTER TWO

Woodlands, Land Title and Tenure in Celtic Ireland

'This is the forest primeval' – Longfellow.

Gradually man controlled and began to utilize the natural forest and woodland resources in Ireland for pasturage and agriculture, destroying forest land to make room for fields, flocks and herds. Forests were also sources of food and provided the raw materials for building, cooking and weapons. In some cases trees – hazel, holly, ash – were rudely propagated for these purposes:

> In spite of this multiple use of forests, however, their vastness and apparent general indestructibility prevented man from conceiving ideas of possession until after he had cut most of them down and converted the land to tillage, pasture or wastes. The forests themselves remained a gift of nature for all men to use alike as they wished and were subject to unregulated multiple use by means of uncontrolled selective cutting.[1]

This notion of 'conceiving of possession' is important to any appreciation of the conflict between cultures and its consequences in Ireland over many centuries, a conflict that was significant for Irish forests.

It was otherwise in England, where the notion of forests as areas where special laws applied was introduced by the Normans. Forest Law followed the Norman Conquest and was rigorously enforced over much of England. Except in isolated instances (the legendary Robin Hood is an example) it prevented the forests from becoming refuges and fastnesses for those opposed to the new regime, as later happened in Tudor Ireland.

Forest Law in England, we are told (but on authority held by scholars to be dubious) had a precursor in the *Constitutiones de Foresta* attributed to the Danish King, Cnut (Canute) who, having effectively conquered England in 1012, is alleged to have enacted the following statute (11, Cnut, Sec. 80): 'I will like that every man be worthy of his hunting in wood and field on his own estates. And let every man abstain from my hunting: Look, wherever I will, that it should be freed, under full penalty.'

This ordinance is interesting not least because Cnut found it necessary to proclaim it at all. It was directed in the first instance at landowners and not at commoners or freemen. While it was evidently a law of the forest, its purpose was to preserve hunting rights and it had nothing directly to do with trees or woodland *per se*, or with rebels. And it may well be this that coloured the traditional and (at least in the early years) rather doubtful motive attributed to Norman Forest Law of being primarily to preserve the chase.

The availability and abundance that enabled man to use the forests and woodlands as an available resource gave way to the realization that, if forests were continuously exploited to meet increasing demand, a point would be reached where supply dried up unless effective control and management were introduced. Such control was introduced to England from Normandy, ostensibly for the same purpose as Cnut's law – the protection of game.

When most people use the word 'forest' today they mean a large area covered in trees. The word is of Old French (der. Latin) origin, anglicized, and in England the meaning was not always so restricted. Following the Norman Conquest in 1066 'forest' meant any area of land, with or without trees, that was subject to Forest Law and law enforcement, which was quite distinct from the Common Law. Such areas could include woodland, arable land, common land, pasturage, even villages and towns. At one period in the thirteenth century more than one quarter of all England was Royal Forest and subject to Forest Law.[2]

The use of the word in Ireland derives from and was influenced by English attitudes. Accordingly an understanding of English meaning and usage will help to illuminate the situation in Ireland. Thomas Hinde writes:

Forest Law was first defined by a statute of 1184 known as the Assize of Woodstock, but was almost certainly functioning as a separate system within its own coverts fifty years before in Henry I's reign, and had probably been established soon after the Conquest. The Assize of Woodstock, the basis of Forest law, is doubly interesting in that it pre-dates Magna Carta, the great Charter of Common Law, by eighty-one years.

It was complete and varied over the centuries. Broadly speaking it not only forbade hunting in royal forests, but ... limited what those who possessed land in a forest could do with their own property. In particular it forbade them to build on their land, to cut timber or wood except for their own use and to assart their land – turn it from woodland into agricultural land.[3]

Such interdicts referred, *inter alia*, to village, farm and common land.

J. G. Turner, the nineteenth-century English historian, defined a forest as follows: 'a definite tract of land within which a particular body of law was enforced... held by the king or held of the king by someone else, usually a noble or a religious institution'. The old tradition that William the Conqueror destroyed numerous villages and thirty-six churches when creating the New Forest in Hampshire may well refer to alteration in legal governance rather than to demolition.

A. C. Forbes, never one to temper strong opinions, wrote:

[This] old tradition ... requires a great deal more evidence to support it than is forthcoming. Those who know the class of country occupied by the New Forest know well enough that anything like intensive cultivation over the greater part of it is, and was, out of the question, and it was probably for that reason that it was brought under the forest laws, which existed long before the Norman Invasion.[4]

This statement is possibly closer to the truth than intended. It may well be precisely because of the rugged terrain in question and the refuge it offered those opposed to the Norman Conquest of England that the New Forest was put under Forest Law.

In the historical sense, therefore, the word forest has a meaning very different from that commonly used today. From the Irish point of view the historical meaning never had any great significance, the word *coille* or *fásach coille* generally being taken to mean dense, natural woodland.

Pollens have a long life when trapped in layers of turf (peat), lake mud or acid soil. By dating these layers the proliferation of trees can be identified and quantified with some accuracy. Obviously some precautions are necessary; for example against the popular notion that great oak forests covered the country in prehistoric times must be set the fact that oak pollen is wind-distributed, which might give the impression that oaks were more numerous than in fact they were. Limes, on the other hand, are insect pollinated and the distribution of their pollen could suggest fewer than actually existed.

Pollen counts have shown that considerable fluctuation in tree densities occurred from the time of the neolithic agriculturalists down to about AD 300. Primeval forest had virtually blanketed the country with oak, elm, hazel and pine from about 7000 BC to 3500 BC, when the neolithic period began. Within a relatively short period (geologically speaking) the elm, the Scots pine and ash declined drastically. 'About AD 300 elm and ash are swept away with a thoroughness from which they never recover, and the destruction-phase (a term of convenient reference here used) in which more advanced farming ultimately destroys the woodlands, opens here.'[5] Bogland also increased and remained prolific until the eighteenth century.

Gradually the high forest gave way to open grassland, much of it, at times, probably savannah-type resulting from a drier climate and the excessive use of the ard-plough (a simple plough with a wooden share, usually of oak, to make a shallow furrow; spades were often more efficient in heavy or stony soil). Heath and bog development as well as recurrent forest regeneration also took place. Wood as a working material was plentiful. By the twelfth century, when cultivation had continued for thousands of years, the greater part of the country was still clothed in trees.

The question of forest ownership in Ireland should be considered. In Irish law the family unit was the root *fine* (group) which was also a legal unit. This could be *gel-fine*, of three generations, *derb-fine*, of four, *iar-fine*, of five or *ind-fhine*, of six. The *derb-fine* was the usual elective legal group so far as succession, inheritance and disposal of goods and property were concerned. If an entitled *derb-fine* failed within three generations to have a member elected to the kingship, the right was lost to the whole line; for example if an eligible great-grandson failed to be elected to an office that had been held by his great-grandfather, and no closer kin succeeded, that line failed and was relegated to a lower order, a situation emphasized in the saying: 'Cúig gluine ó rígh go rámhainn, five generations from king to spade'. Since kings had certain powers of grant, as well as other powers, this was a vital legal matter as well as being of economic and social importance.

Mac Neill writes:

This Irish idea of dominion over land very much facilitated the Norman Conquest (*sic*!). The Irish lords thought they were submitting under duress to a new political authority. The feudal invaders believed that they were acquiring a rigid, complete and perpetual ownership of the 'land' from the zenith to the uttermost depths – an ownership more complete than that of any chattel – an ownership that they imagined to be self-existing even when the person in whom it should be 'vested' was, for the time being, unknown and unascertained.[6]

What, then, was the Irish legal situation in respect of forests and woodlands over which the Normans sought to establish possessory title?

Woodlands were generally common land and, presumably deriving from this, there were rights in common in each tuath (the people/territory ruled by a minor king) – 'the full property of every tuath, belonging in equal right to every condition (of persons)'.[7] The eighth-century *Bretha Comaithchesa*, that portion of the Brehon Law translated as 'the Laws of Neighbourhood' (or the Community), lists twenty-eight trees and shrubs in four classes. In fact this is twenty-eight plus one, in accordance with Celtic ritualistic numeral observance. Laws governed each class and species of tree and varied according to class. Penalties related to the nature of the offence. So far as we can judge from the Brehon Law and from other traditional sources, title in respect of the timber, as distinct from the land on which it stood, was vested in the sept, and more particularly in the *derb-fine*. In general timber was a source of common or communal ownership with some specific rights and regulations as to usage.

Among the things listed in the *Ancient Laws of Ireland*[8] that are 'the full property of every *Tuath*, belonging in equal right to every condition (of person)' are the following:

The night's supply of kindling from every wood.
The cooking material of every wood.
The nutgathering of every wood.
The frame-work of every vehicle, yoke and plough.
Timber of a carriage for a corpse.
The shaft fit for a spear.
A supple hand implement for a stable (an *echlach*, or horse-rod, used for guiding horses).
The tapering wood of the three parts of a spancel.
The makings of hoops (for barrels).
The makings of a churnstaff.
Every wood not subject to *treiniugud* (which MacNeill glosses: 'some sort of appropriation – "triple division" is the literal and official rendering').

The Brehon legal tract codifying penalties for unlawfully interfering with trees (and certain bushes) purports to give the fines in respect of each type of offence, from two and one half milch-cows for felling an important tree, to a sheep for destroying a shrub. Trees and shrubs were classified in four degrees paralleling the social order. These were *Airig Fed* – literally nobles or chieftains of woods or trees; *Aithig Fedo* – commoners or common trees; *Fodla Fedo* – the lower orders, and *Losa Fedo* – bushes, non-persons or slaves.[9]

There were seven species in each class. Nobles: oak, hazel, holly, yew, ash, pine, apple. Commoners: alder, willow, hawthorn, rowan, birch, elm and another – possibly the wild cherry – which is not known from its Irish name *idha* (variously spelled *idath*, *idadh*, *hidha*, *fidhat* and *fidhout*).[10] Lower orders: blackthorn or sloe-bush, elder or bore-tree, white hazel, spindle tree, aspen, arbutus and *crann fir*, possibly juniper. Bushes or slave trees: bracken (used for soap making, bedding and for providing potash for bleaching linen), bog-myrtle, furze or whin, brambles, heather, broom, gooseberry – or wild rose, opinion is divided – and ivy.[11]

The classification into groups of seven reflects the mystical qualities which the Celts attributed to numbers such as three, five, seven and nine. Here we have an extra *losa fedo* – ivy – making eight in that category, so that the grand total becomes a multiple of seven, plus one. Celtic fighting units, for instance, commonly consisted of eight men plus a leader (nine) banded together into a larger unit of three of these plus another leader, twenty-eight, divisible by seven.

Trees and shrubs were classified in accordance with their economic importance, evidently determined by their size, quality of timber and the use to which the timber or fruit might be put. The yew and the apple, for instance, though relatively modest in size, were accorded noble status and a higher class than larger species because of the fruit of the apple (probably wild and uncultivated for a considerable period) and the versatility of yew wood.[12] The apple was also esteemed for its bark, but what use was made of it is not known (it may have been for making a sort of bread). Yew wood was used for decorative purposes, ornamental furniture, household vessels, building and weapons. Giraldus Cambrensis opined that the poisonous juices and the exhalations of the yew-tree (*taxus baccata*) seriously checked the increase in bees. He also tells us that several troops of archers were quartered in the town of Finglas, Co. Dublin, where:

> The illustrious abbot Kenach and other holy men in succession, through whose fervent piety the place became celebrated, had formerly planted with their own hands ash trees and yews and various other kinds of trees, round the cemetery for the ornament of the church. The soldiers, short of fuel, cut down these trees but in retribution for their impiety they were smitten by 'a sudden and singular pestilence'.

Of the 'noble' trees the oak was highly regarded because of its size and appearance, for its acorns, which were very important as food for swine, and for its bark which was used for tanning. Hazel was regarded for its nuts, which were an important food source (hazel-mead was one form of this popular drink), and hazel rods were cultivated (coppiced in order to encourage staves for wattling, perhaps one of the few instances of deliberate management in early Irish forests). Holly was highly regarded because of its use for chariot and spear shafts. The upper branches were also used for fodder.[13] Ash, too, was used for spear shafts and for other weapons and 'for supporting the king', which may be an indication that its timber was used to build his throne, that it was used for the weapons of his troops, or that it was the *bile* we have already discussed. Within living memory ash was commonly used for making stools and chairs in rural Ireland. The pine was well regarded because of the many uses to which it could be put. Its timber was used for making roof and other puncheons, for bed-posts, the ridge-poles of houses and the numerous other uses to which strong, straight timber may be put; in some cases its branches may have been used for bedding (there is a reference in the twelfth-century poem 'Suibhne Geilt') and its resin was made into pitch for caulking boats and other weathering. The indefatigable, if often inaccurate, Giraldus wrote: 'The forests of Ireland abound with fir which produces resin and incense.'

'Some of the rankings seem odd to our modern eyes... If we look at the lists, it is curious to see pine in the first rank, while elm and alder are relegated to the second.'[14] Pine, it is thought by some authorities, disappeared from Ireland at an early stage.

Today it is thought that all the pines we have in Ireland have been re-introduced from Scotland, after the native stock had become extinct... it is not easy to see that well-grown pine trees could have been common enough at the time the law was codified for them to merit such a high place on the list... elm can only be put down to the second rank because of the extent to which it had been cleared away...[15]

The penalties (*dire*) for trees in the 'noble' group were five *seoit*, or two and one half milch cows or its equivalent (one *set* equals half a milch cow). The compensation to be paid for cutting branches was a year-old heifer, for cutting a fork a two-year-old heifer and for base-cutting a milch cow. In the case of oaks there was an additional fine of a two-year-old heifer for cutting young oak-trees (or perhaps oaks which had been coppiced).

Ash was uncommon until

after neolithic agricultural practices had already been in operation in the country... can it be that before human interference with the woodlands took place, ash could not maintain itself in the high forest, and was confined to the shallow limestone soils? Ash may have invaded the unstable secondary woodland that followed the activities of all farmers.[16]

Alluding to this Kelly says 'hilly arable land, being more liable to partial or sporadic cultivation than the superior *etham remi-bi ethamnaib* (best arable land), could have provided habitat favourable to the spread of ash'.[17]

In the early Irish law tract *Fodla Tire, Divisions of Land*, three types of land are distinguished: *Etham remi-bi ethamnaib* – best arable land; *etham taulchach* – hilly arable land; *etham frichnama* – arable land requiring clearance and manuring. '*Etham taulchach* is described thus: "there is water there and there are ash trees in every second piece of ground. It is good for every plant and for every crop".'[18]

Scots pine is possibly the most controversial of the trees in this group. Some authorities hold that it was extinct well before the tree list was compiled (*c.* seventh century, according to Kelly). On the other hand Mitchell suggests that while the available evidence is that scrub trees were still growing on midland raised bogs in AD 300, the pine may have survived in scattered and isolated growths until the twelfth or thirteenth centuries. It is a controversial point and some authorities claim that the species never entirely died out. That it had a significance beyond the ordinary may be gauged from the fact that, in addition to the words *ochtach* and, later, *gius*, by which the pine was generally known, it was also given the name *ailm*, which has an uncommon and literary meaning. In the ninth century *King and Hermit Dialogue* many species are mentioned to Guaire by Marban and the hermit specifically says: '*Caine ailmi ardom-peitet* – beautiful are the pines which make music for me.'[19] The word *ailm* is also the word used to describe the letter 'A' in the ogham alphabet, which is made up of four groups of five (with the perhaps later addition of a fifth group of diphthongs held to represent the mystic and universal aspect, or centre without dimension, already referred to). *Ailm* is the fifth letter of the fourth group.[20]

Of the 'commoner' trees the alder was plentiful and useful. An inhabitant of marshy ground, riversides and lake shores, its probable use was for hedging and providing material for wattles. Whether or not its relationship to the birch was understood is not known. There is some difference of opinion concerning *sce* (whitethorn or hawthorn). It has been suggested by Binchy that the translation does not refer to the whitethorn, which should be down-graded to the 'slave' category, *fodla fedo*, with the blackthorn, but that *sce*, correctly, refers to the aspen, which should be upgraded from 'slave' to 'commoner', since it is the timber from which some domestic articles were manufactured – bowls, spindles, etc. Whether this theory – presumably based on the inferior quality of hawthorn wood to that of aspen – has substance or not, the fruit and leaves of the whitethorn, even today (throughout Munster certainly), is sought after by children as 'bread and butter', possibly a residual folk-tradition commemorating a useful food-source.

The elm apparently suffered what has been described by Mitchell as a catastrophic decline about the year AD 500, and by the seventh century had virtually disappeared from the woods of Ireland. He goes on to say: 'botanists agree that today the natural habitats of *ulnus glabra*, (the only indigenous species) are confined to cliffs and rough ground (mostly limestone) in upland glens, chiefly in the north'. The limestones of Clare provide the refuge of the last native Irish elms, and even in Clare it was almost extinct 1000 years ago. However, 'elm leaves, bark and wood were all valued by early peoples, and if it had been still common in the countryside, it would have had a higher place in the list'.[21]

The penalties and compensations for any trees in the 'commoner' group were a milch cow, and for branch-cutting a year-old heifer. Alternatively, according to another version of the law, the compensation for fork-cutting was a year-old heifer and for branch-cutting, a sheep. For completely digging out a tree, more serious and fatal than base-cutting, the penalty was the same as for base-cutting a 'noble', that is two-and-a-half milch cows, or the equivalent.

Of the lower orders of tree, the blackthorn was useful for fencing and for its fruit and, no doubt, for its cudgels. The so-called 'sweet-blackthorn' (*draigen cumra*) may have referred to the wild plum (*prunus domestica*), or some similar species. The elder was plentiful and, though despised, was also useful as it is today, as a source of drink (from the flowers and the berries) and, perhaps, dye. A ninth century author comments: 'three signs of a cursed (abandoned) place: elder, corncrake, nettles'. The whitebeam (*findcholl*) was, apparently, categorized because of its heavy, tough wood which does not split easily. It was used for cudgels and clubs. The arbutus (*caithne*; *arbutus medo*) must once have been quite prolific as several places are called after it, e.g. Caithneachan near Killarney, where it is still to be found, Ard na Caithne (Smerwick) on the Dingle Peninsula, where the arbutus is extinct, and Doire na Caithne (Derrynacaheny) in Clare, where it is also extinct.

The precise meaning of *Crann fir* is not known, but in a closely reasoned and compelling discourse, Kelly demonstrates the likelihood that it refers to the juniper. The penalties and fines are apportioned as before.

Of the bush or 'slave' trees the usefulness of bracken is obvious. Furze continues in use as byre-roofing, fencing, dyeing, fodder, fuel and animal bedding. Broom (*gilcach*), a term which may also have included reeds, was useful for making besoms and for producing dye, while reeds, clearly, had many uses – thatching, basket-making, weaving etc. Again the penalties were commensurate; for total extirpation one sheep, but for taking a single stem there was no penalty.

When the Normans arrived in 1169 they brought with them the concept referred to by Osmaston, 'ideas of possession':[22] possession of the land and what stood upon it. This concept, which was not only valid in Norman law, but was the foundation on which it rested, was, in the twelfth century, as alien to Irish law and to Irish leadership as it could possibly be.

CHAPTER THREE

The Normans
and Early Medieval Ireland

'Generations pass while some tree stands' – Robert Browning.

THE EXTENT OF FOREST MANAGEMENT before the Norman invasion is unclear. Timber was certainly used in huge quantities, and was also acquired, presumably as an investment, by the Norse; Tomar's Wood is one example. The Danish kingdom of Dublin – Dyfflynarskiri – exported substantial quantities of timber, perhaps as far as treeless Iceland and the Faroes, with both of which there was considerable trading. All the cities of Ireland – Dublin, Cork, Waterford, Limerick – besides the traditional wattle and clay houses, had timbered buildings. Dublin and Waterford, to say nothing of 'Limerick of the Ships' were known for shipbuilding.

In Irish law title was essentially one of land and, so far as we can judge, ownership of standing timber thereon was vested in the people as a resource through title to the land owned by the septs. When land was owned individually, it was usually the gift of the king (on behalf of the sept), but the recipient appears to have had obligations in respect of maintenance, extending to both trees and worked timber. Under Norman law that right was dismissed and the ownership of land and trees was usurped and, sometimes, differentiated and separated.

MacNeill clarifies the position thus: 'The ancient law tracts indicate that in Irish law, the lordship of land belonged to the political rather than to the economic order of things. It imported authority rather than ownership.'[1] Although Irish kings and overlords had certain powers of grant under Brehon Law, neither ownership nor the power of transference under grant was as absolute as under Norman law. In Irish law it was essentially 'the use of' the land that was granted or transferred. It was liable to re-transferral if the status of the custodian altered.

Chieftains held surplus land, stock and equipment which helped to ensure power and authority. Farmers, especially smaller farmers, might rent land (conacre), paying with stock and services; they could also rent by instalments, working equipment or working and breeding stock. But the land thus acquired did not belong to the donating king or chieftain, who simply administered it. Like law-makers, brehons, and professional men, *ollamhs*, the king held lands associated with his office and might, like them, be donated land by the *tuath* independently of his office. But essentially land was held in common by the people of the *tuath* and no individual could acquire land in the possessory sense. The system was subject to both change and abuse and excluded the majority of menial workers, but it did not give rise to the exploitation of resources as later took place under the system introduced by the Normans. Exploitation might have occurred had there been purpose in it and had the population been sufficiently large, but the issue simply did not arise. When an individual who had administered lands died, the rights of administration, not of ownership, were apportioned severally amongst the immediate kin. Accordingly there could be considerable sub-division and also reversion of land to the pool administered by the king, which, in turn, might lead to further re-arrangement when he redistributed it. Neighbouring *tuaths* of the same sept would co-operate in such matters. Trees and woods were an integral part of the value of a portion of land and, apart from their individual status in law, do not – as happened in England – appear to have had a separate value; certainly no separate title. Tenants could not be arbitrarily handed over to a new owner or lord without their agreement. Irish law did not allow them to be evicted and therefore their consent was necessary in a peaceful, legal transfer.

MacNeill points out that some confusion has arisen by the mistranslation into English of the words '*tuath*' (meaning both the population and the area governed by a king) and '*fine*' (a small family group from whom the king might be elected). He makes the point that the exercise of the king's powers of grant of territory under his jurisdiction was subject to 'prudential limitations'.[2]

Brehon Law survived until the seventeenth century, but by the mid-sixteenth century Tudor forest depredations were beginning in earnest and title to land under the Brehon Law was under challenge from three quarters. Firstly from the English planters and *force majeur*. Secondly from the Norman tradition of possessory title by then 300 years old. Thirdly as a result of the decline and corruption of the Gaelic Order from within and in the face of these two powerful forces. (See Appendix 1.)

While the Celtic Irish were in communication with the Roman Empire both as traders and as enemies, Ireland was neither invaded nor conquered by Rome and, until the Christian 'invasion' in the mid fifth century, its influence on Ireland was very limited, which is not to say that the Irish were unaware of the Roman way of life. Far from it; the declining empire brought Stilicho, Vandal, ex-general and ambassador to Persia, back to Britain to protect it against Crimthan, uncle of Niall of the Nine Hostages, who had conquered Britain 'to the Ictian Sea' (St George's Channel). Whatever of the collapsing Roman administration the Irish may have seen, they left outside Ireland. But there was trade and commerce. There were technical innovations, the coulter and the ploughshare for instance; perhaps the hypocaust and primitive chimneys; the ogham alphabet, and new-style building in stone.

In England, while the Romans extended their granaries, continuing the process of extending 'the cultivated land of Britain at the expense of its trees',[3] they did introduce some elementary forest management. While there are no known records of any specific Roman policy towards forestry in Britain, it may be assumed that established Roman principles of forestry, no less than those of agronomy, were applied. They developed iron smelting and forging, for which, like their towns and cities, vast amounts of fuel were required. Analysis of Roman charcoal indicates that they were using uniform sizes of timber, which, in turn suggests extensive coppicing.

Writing about the situation in medieval France, Osmaston comments:

Claims of ownership and delineation of forest property, without which management is hardly possible, began whether the claimant was an individual or a community such as a village, town or religious body. Frequently ownership would be vested in the king or head of a tribe or people as a result of a custom that 'Waste' land, i.e. land unoccupied by an individual, or even all land was owned by or vested in the king or chief. At the same time rights of usage of the product of the forests, as a relic of immemorial custom, without ownership of the land, might continue provided that the owner's requirements from the property were not affected. Thus it might be that the owner reserved to himself the great trees and the hunting (for its excitement but also for its food value as protein), but allowed rights of common for edible fruits, pasture, deadwood fuel, brushwood and perhaps the less valuable tree species...in ancient France two uses of the forest were recognized, that of the chase which included fishing, and that of forest products. Also the *Capitulaire* of Charlemagne (AD 813) recognized two kinds of forest, those owned by the king and those owned by individuals. The French forest administration is still known as that of *Les Eaux et Forêts*.[4]

This outline suggests a sort of half-way house between the 'communal-type' of ownership that appears to have existed in Ireland and the concept of absolute right of possession held by the Normans. Osmaston defines the object of Norman Forest Law as follows: 'The preservation of hunting (primarily for the king) although the monetary income from fines and charges must have had a substantial consequence.'

Forest management developed from the successful application of Forest Law in England. In both Ireland and England deer and other forest wildlife were a substantial source of fresh meat until relatively recent times (in 1251 alone Henry III ordered 200 bucks from various forests for presents, and 100 bucks and 60 does for Westminster at Christmas). The indications are that they were also an important food-source in prehistoric times; cattle tended to be used more for milk and, in Ireland in particular, were rather a basis for exchange than a source of meat. The early function of forests in these islands was threefold; they were important to the local economy, providing timber for building, implements, utensils and fuel; cleared, they provided space for agriculture (with, of course, consequent supplies of charcoal and potash) and they were a source of food from the fauna (and some of the flora) that inhabited them.

William I (the Conqueror) imposed the concept of Royal Forests, and the Forest Law governing them, on the mixture of peoples who, in a war-torn amalgam of communities with an inheritance of degenerated Roman life-style, lived in England at that time. Was it this common factor, involving similar notions of absolute ownership, that made William's action acceptable? His action was extraordinary in that he laid down a principle giving him outright ownership of all such land. Over the next two centuries this entitlement increased until more than one quarter of England was Royal Forest owned by the king and subject to Forest Law. From the outset, the vast amounts of land involved included towns, villages, manor-farms, land-holdings, commonages and woodlands. William was following, and expanding, the practice in Normandy where the ducal forests were administered in a very similar way. In addition to hunting and the provision of winter meat, the development of the French system seems to have had a political motive, namely ownership and control of woodlands and other areas which could provide the disaffected and those who might oppose his rule with safe refuges. 'The king's chief forester... was a man of at least as much importance as the king's other three civil servants – one medieval historian, Sidney Painter, suggests that "a strong argument could be advanced for the thesis that the royal official who wielded most actual power during John's reign was the chief forester, Hugh de Neville".'[5] The chief forester was clearly someone of great power and authority, for which there must have been sufficient reason. Twenty years after William conquered England many of the Royal Forests he proclaimed were recorded in the Domesday Book.

The word 'forest' was a loose term, perhaps deliberately so considering the potential in revenues involved, and was given a variety of definitions, such as in the *Dialogus de Scaccario* of Richard FitzNigel, written around 1179, *A Treatise of the Laws of the Forest* by John Manwood, which first appeared in 1598, and J. G. Turner.[6] FitzNigel's definition is: 'the king's forest is a safe refuge for wild beasts; not every kind of beast, but those that live in woods; not in any kind of place, but in selected spots, suitable for the purpose'. Manwood, writing 581 years later, says:

A forest is a certain territory of woody ground and fruitful pastures, privileged for wild beasts and fowls of the forest, chase and warren, to rest and abide in, in the safe protection of the king, for his princely delight and pleasure, which territory of ground, so privileged is meered and bounded with unremovable marks, meers and boundaries, either known by matter of record, or else prescription: and so replenished with wild beasts of venery or chase, and with great coverts of vert, for the succour of the said wild beasts, to have their abode in: for the preservation and continuance of which said place, together with the vert and venison, there are certain laws, privileges and offices, belonging to the same, meet for that purpose, that are only proper unto a forest and not to any other place... And therefore a forest doth chiefly consist of these four things, that is to say, of vert, venison, particulars and privileges, and of certain meet officers appointed for that purpose, to the end that the same may be the better preserved and kept for a place of recreation and pastime meet for the royal dignity of a prince.

Turner says:

In medieval England a forest was a definite tract of land within which a particular body of law was enforced, having for its object the preservation of certain animals *ferae naturae*. Most of the forests were the property of the Crown, but from time to time the kings alienated some of them to their subjects... But although the king or a subject might be seised of a forest, he was not necessarily seised of all the land which it comprised. Other persons might possess lands within the bounds of a forest, but were not allowed the right of hunting or of cutting trees in them at their own will.

The Norman and Plantagenet kings, with their adroit political wisdom, encouraged the view that the royal forests were established and enlarged in order to provide them with hunting and fresh meat. Poaching the king's deer, boar and other 'venison', was forbidden. However, from the time of Henry II (when any real threat from dissident and dispossessed Saxons and Danes was well past), public policy in relation to forests was felt to have served its purpose. By the time that Henry encouraged Dermot Mac Murrough, king of south Leinster, to seek the aid of restless Welsh border knights to undertake the invasion of Ireland, a milder forest policy prevailed.

Punishments for forest offences were now less often capital (for, not surprisingly considering their initial purpose, death was a common penalty for breaking forest law), and even mutilation and imprisonment were modified. But fines, which amounted to taxes, were substituted when woodland was assarted (cleared for agriculture). Offenders were not compelled to return the land to woodland, provided an agreed annual rent was substituted. In this way fines became the most important source of direct revenues from the forest. Later on systematized methods of raising money from forests – leasing, sales of wood and so on – were developed, and, until the Tudor period, gradually superseded food and hunting as the economic dynamic of forestry in England. When the Normans came to Ireland they brought this persuasive economic concept with them.

It is evident that the woods around a medieval village, irrespective of the legal system, were an essential part of the local economy. Apart from providing pannage (food and browse) for fattening pigs, and all-season fodder for other stock, they provided timber for building, for repairs, for farm work, for implements and, of course, for fuel, facts of life of which any just legal system of the day would have to take account. In Ireland, where villages were infrequent, but other settlements were sometimes as large as, if not larger than, the English medieval village, such requirements might have varied – the amount of timber used for corduroy roadmaking might have been considerably greater, for instance – but the principle remained. The reliance on forest timber would have been comparable to that of similar populations anywhere in Europe, and was underwritten in Irish law.

Early and medieval communities in Ireland practised the elementary encouragement of growth in certain species suitable for building, for weapons, tools and furniture. But that is very different from the possession, designation and management of forest land under Norman law in England, and which they expected to continue in Ireland.

On this established system of local supply and demand the Norman lords sought to impose their own system – not too strenuously, as it involved raising exchequer revenues from the forests which went ultimately to the English king rather than to themselves. Corruption of both systems was the inevitable outcome.

The Norman settlers were accepted by some Irish rulers and 'obtained' land from them. Each side, naturally, assumed that in such transactions the law with which they were familiar would prevail. One result was that, although some of the lands thus 'acquired' by Normans were designated royal forests, the designation was meaningless without a means of enforcing the Forest Law. By and large such means did not exist. Where it did exist it was not generally in the interests of the Norman knights to enforce it.

The conflict of legal and economic systems that followed the Norman settlements in Ireland encouraged political confusion and contributed largely to the decline of the already demoralized Gaelic order. Kings and chiefs countered the castles and baileys of the

Norman knights with their own and, in order to minimize the effects of the Norman heavy cavalry and Flemish bowmen, began to 'plash' and entrench the forests.

Forests were potential gathering places for opposition. Since the Normans who came to Ireland came as 'allies' of an Irish noble and not as conquerors, this was not a major consideration for them. Moreover they were not led in battle here by their own king; they accepted the *Rí Éireann* and the process which was to become known as *Hiberniores ipsis Hibernis* began. Nevertheless they sought – unsuccessfully in the main – to introduce the brutal forest laws of England. They were also concerned with the provision of timber for building and other purposes and, of course, with hunting and the chase.

The formative base for forest law in England, where it was collated and regulated and enforcement and judicial procedures were established, was the Statute of 1184 (Henry II) known as the Assizes of the Forest or, more popularly, as the Assize of Woodstock. It was of great importance to the development of forestry in England. It established courts and co-ordinated and administered laws of the forest independent of, and distinct from, the courts and Common Law of the land. The statute, however, had no force in Ireland; nor did the laws derived of it. These laws also provided the foundation for the continuing policy of forest management and administration (and a consequential sense of public appreciation) that developed towards forests and forestry in England over a period of some nine hundred years. But neither the management nor the resulting public appreciation of forests extended to Ireland. This lack became increasingly apparent in later years.

Disafforestation – removal from the purview of the statute – as a deliberate function of forest management was also a factor in England from very early on. But even when, because of increased industrial activity, the rate of disafforestation increased in England, established forest management continued and some technically 'disafforested' forests remained subject to Forest Law in some respects. Broadly speaking the Assize of Woodstock forbade hunting in the royal forests, and limited what those who possessed land within a forest could do with their own property.

While the writ of the Assize of Woodstock did not run in Ireland, and the Laws of the Forest which proceeded from it were not capable of being enforced, in some controversial instances, such as in the example below (in which the relevant dates are important), attempts were made to introduce and enforce the law.

The chief assize held in England to pass judgments on breaches of Forest Law was called the 'Eyre'. It was summoned and presided over by the chief forester or one of his agents. In 1175 the chief forester of England was Alan de Neville, notorious for his severity and for the rigour with which he both applied the law and exacted (financial) penalties or forfeits. Of him it was written: 'This Alan so long as he lived enriched the royal treasury...'[7] Under the chief forester, to administer and enforce the laws, were the forest courts and law officers. There were two forest justices of England. Each forest had a warden and under him a lieutenant. Below them came verderers, usually knights who acted as judges in the swainmotes, courts of first instance for forest offences. Higher courts were the Eyres.

Beneath them were foresters who did the day-to-day work of management and enforcement, supported and assisted by woodwards, agisters, rangers and regarders each of whom had specific duties and functions. In addition there were offices such as foresters-of-fee, hereditary posts available to gentlemen who had the right of income from their bailiwick. Such functionaries and titles also existed from time to time in Ireland, but, for the most part, in name only.

In 1219, during the incumbency of Alan de Neville, Thomas fitz (or de) Adam was appointed 'custodian' of the king's forest of Ireland. 'The words given in the patent rolls are *totam forestam nostram Hiberniae*, the word "forest" being in the singular.'[8] Forbes argues that this can only mean that 'the forest or unappropriated land of Ireland was referred to *in globo*, wherever it existed and that the forest laws could be applied to it without any special formality, such as the appointment of a royal commission to give it effect, as was the case with the Assize of Woodstock'. But another possibility sprang to controversial prominence that same year, with a dispute between fitzAdam and Archbishop Thomas Luke of Glendalough.

For whatever reason – to do with the payment of rent, perhaps, or the indulgence of the king – Archbishop Luke felt himself to be exempt from certain provisions of English Forest Law which fitzAdam sought to enforce. The archbishop particularly objected to the provision entitling 'the servants of the forest' (foresters, forest wardens and forest workers as they would now be called) to free food and drink from those who lived within the forest boundary, which meant that he was expected to provide for fitzAdam's officers, a not inconsiderable burden given the considerable area involved, which included at least one monastic site and several settlements.

Matters came to a head when fitzAdam arrested a man for deer stealing and the archbishop sent 'a dean and three priests to demand discharge of the prisoner'.[9] FitzAdam refused and was promptly excommunicated by the archbishop who also complained to the king, threatening at the same time to suspend all religious services as long as the situation continued. The king, knowing where the real power lay, appeased the archbishop and sacked fitzAdam. The following year, 1220, he commanded Geoffrey de Marisco, Justiciary of Ireland, to take 'into the king's hand his forest of which Thomas fitzAdam had custody, and cause it to be safely kept in honour of the Church of Dublin by two of the king's men of Dublin, and two of the archbishop's'.[10] Note that the reference is to 'forest', singular, and that it is to be kept in the honour of the Church of Dublin. It might also be noted that the forests of Wicklow were within, or close to, the Pale and have, since Norman times, been established forests within the meaning of the term (and sometimes within the meaning of the Act).

The function of the chief forester was to make the best use he could of forest land – which, at this time, meant raising as much revenue as possible from it for the king. The role of the 'custodian' in Ireland may be assumed to have been similar, however different his problems. No Irish ruler would willingly admit foreign laws which not only required revenues from them, but infringed basic Irish law. It may, therefore, be taken that it was only where Norman rule held that English Forest Law could be contemplated. Even in such cases the king's writ did not always run, and the knights and barons, barricaded in their isolated castles, often had more immediate and important matters to be concerned about.

It is not unreasonable to suppose that fitzAdam, exercising his authority in Ireland under the influence of the notorious de Neville, would have been anxious to emulate him and prove his own mettle. Dermot Mac Murrough had given King Henry II a strip of land – two cantreds – in north Leinster, between Dublin and the sea, in the area around what is now Bray, Greystones and Shankill. The archbishop – or, at least, the archdiocese – of Glendalough exercised jurisdiction over an area roughly the same size as present west Wicklow. Forest land, as we have noted, was often 'held of the king by someone else', and royal forests might be disafforested on payment of an agreed amount.

The Papal Bull, *Laudabiliter*, maintained that the Church in Ireland was in disarray, requiring drastic reform and overhaul. Laurence O'Toole, Archbishop Thomas Luke's (Irish) predecessor, had been as political an animal and as concerned an Irishman as might be. He had stood between the people of the country and the invader. Archbishop Luke, his imposed successor, was a Norman. Whatever about ecclesiastical reform within, the Church in Ireland was deeply divided politically. The Norman archbishop, whose interests were certainly advantaged by *Laudabiliter*, and whose diocesan lands were contiguous with those of the king, must not only have supported the king's laws, but also had a reasonable expectancy of goodwill, if not more, from the king in return for that support. It seems improbable, to say the least, that he would refuse to acknowledge a declaration that Glendalough was part of a royal forest. Add to that that a sum of 300 marks which passed between the archbishop and the king[11] in consideration for which, it seems, Glendalough was disafforested in 1229, and things become clearer still.[12]

Dermot Mac Murrough's 'grants' of land to the Normans in the Dublin region resulted in the expropriation of three septs in or near what became the Pale. These were the Uí Dunchada, the Uí Faelain and the Uí Muiredaig. This may help explain the forest tradition that, in contrast to the rest of the country, has existed to the benefit of woodlands in general in Wicklow to this day.

At least one other Irish forest was designated a royal forest. According to *Sweetman's Calendar* (Vol. 1, ss.155-6) 'the woods of Drumsgranshane and Ballyconerewey' provided, in 1252, a King's gift of sixty oaks. This wood was probably in Clare. Forbes's theory that *all* the forest in Ireland were so designated is further weakened by this differentiation, and there undoubtedly remained an unbridged gap between any such designation and title that may have been given to a forest, and its governance and management. And, like the country in general, forests in Ireland remained much as they had been before the arrival of the Normans, except where Norman overlordship was sufficient to enable local changes to occur. Forest Law was neither statutorily relevant nor generally introduced. The Normans did, however, introduce rabbits and fallow deer, both of which were to become important in relation to forest development in the future. As in England rabbit warrens – *coniceir* – were important preserves and food sources, and gave rise to such placenames as 'the Warren' and 'Warrenpoint'.

In pre-Norman Ireland perhaps the most common use of timber, apart from firing, was for building. Dwelling houses of all classes, as well as early churches, were usually of wood, that material being easily secured and easy to work with. Most commonly used timbers were deal, oak and yew. In fact the custom of building in wood was so general in Ireland that it was considered characteristic and 'after the manner of the Scots'.[13] When Henry II was in Ireland in 1171-2 he had 'a royal palace constructed for himself of plain wood, built with wonderful taste, in which he and the kings and princes of Ireland kept the festival of Christmas'.[14] It was by pulling down one of the wooden houses in Waterford in 1169 that Raymond le Gros effected an entrance into that city. Until the eighteenth century highly ornamented wooden houses were common in Dublin, Drogheda and other towns (though new ones were banned following the Great Fire of London in 1666).

Dwellings were also often of wicker (*teach-fighti*, a wicker-work house, from *figim*, I weave). Stout poles, standing pretty close to one another, were placed in a circle when the house was to be round, which was the commonest design (but larger buildings were often oval or oblong). These poles were fixed in post-holes (often rediscovered by archaeologists today) and the spaces between them were closed with inter-woven rods and twigs, generally of hazel, which was cultivated for the purpose. The poles, usually of oak or pine, were peeled and polished smooth. The surface of the wicker-work was plastered on both sides and then polished and whitewashed with lime.

According to the Brehon Law a precise method was to be employed for the construction of different types of building. Generally, in the building of a dwelling, after the poles had been fixed in the ground, the spaces were filled with wicker to form the *fraig*, or side wall. If prepared panels were not used, a strip of a certain width was woven all round, beginning at the bottom. Another strip was woven above that and so on until the eave was reached, after which a sloping drip-board was fixed all round and at the junction of each adjacent pair of strips. Another was fixed at the eave, overall. In the houses of the higher classes the doorposts and other special parts of the dwelling and furniture were often made of yew, carved and ornamented with gold, silver, bronze and precious or semi-precious stones. The Brehon Laws prescribed fines for scratching or otherwise disfiguring the yew posts or lintels of doors, the headposts of beds or the ornamental parts of other furniture. Small timber compartments which are thought, variously, to be bothies or houses in their own right or, perhaps, sleeping cubicles from larger halls or houses, are found (often in bogs) from time to time. Such sleeping cubicles are constantly referred to in old Irish writings.

Alcoves and pegs for hanging were usually built into the walls of the houses, often painted colourfully on the outside, with abstract designs it is thought rather than formal murals, in addition to the whitewash. Early Christian teachers adapted this decorative tradition by substituting illustrations from the Bible, used as aids in instruction. When a house was to be rectangular the poles were set in two or more parallel rows and the interstices were filled with wicker in the usual way. It is possible that, in some cases, the poles were set according to an established pattern and that the wicker panels were produced to a standard size accommodating to these. Certainly what appear to be standard wicker panels were used for flooring and for outside pathways.

Such wicker-work buildings were also common in Scotland and Celtic Britain, and are frequently referred to in early writings. There can be little doubt that hazel rods were deliberately cultivated for building and other purposes such as shield-making. The Irish word for shield, *sciath*, is still used in the Gaeltacht parts of Munster to describe a concave wicker basket, often used as a turf-scuttle. Spenser testified to the use of wicker shields in sixteenth-century Ulster: 'Their long broad shields, made with wicker roddes, which are commonly used amongst the said northern Irishe'. Shields were also made of yew and alder.

The collection of hazel rods was clearly a major activity. As the name Baile Átha Cliath demonstrates (the word for a hurdle being '*cliath*' and for a ford '*ath*') wattling had several construction uses. The hurdle at the ford of Átha Cliath brought the *Slíghe Cualann*, one of the five great highways from Tara, across the River Liffey and was, presumably, some sort of wattle bridge. That it was more than a simple ford in the generally accepted sense of the term is evident from the depth and span of the river as calculated for that time. In the late tenth century the Norse of Dublin built a wooden bridge spanning the river just east of the ford to link Dublin (Dyfflin) with the new suburb, Ostmantown, beyond which, in north Co. Dublin as far as Balbriggan, lay the granary of the city kingdom, Dyfflynarskiri.

Wattle buildings could be quite substantial; Henry II's palace may well have involved wattling. Whole townships were built of it. Keating claims that the 'banquetting hall' of Tara was 300ft long, 75ft wide and 45ft high, but this is contradicted by Joyce who maintains that the hall was at least 759ft long, 46ft wide and that it was originally both longer and broader.[15] As to whether it was made of wattle or of timbers no one now knows, but we may be reasonably certain that forest products contributed significantly.

Some buildings, presumably of a 'superior' kind, were built of sawn planks instead of wicker. An instance is the dwelling of St Colmcille, retained for his exclusive use, on Iona. However, timber had to be imported to barren Iona, and it was manifestly easier to import planks than wattles, clay, mud, lime etc. The hostelry or inn built at Rahane in AD 749 was of boards, and we are told that it was so unusually large that it took a thousand boards to build it.[16]

Timber obviously had a special function in roofing. If a house was large and circular the conical roof was supported by a tall, strong post standing in the centre of the floor. The roof of a quadrangular house was much like that of ordinary houses up to a few generations ago: there was a row of supporting poles, or two rows if the structure was very large. Reeds, very often specially cultivated for the purpose, were used for thatching, as was straw. But, for some reason – for they were less efficient – shingles were also used on some houses; costs, availability, custom, status may all have played a part, or it may have been a Norse innovation. These shingles were sometimes of tile and sometimes of thin, fan-shaped boards, generally either of oak or yew, overlapping as in modern slate or tile roofs. Furze commonly provided roof supports, as it still does today in small, old-fashioned farm out-houses and byres. The *Annals of Ulster* state that in the year AD 1008, the oratory of Armagh was roofed with lead.

Timber and associated activities may well have been the subject of comprehensive laws about which we know nothing. We do know that wood-workers in ancient Ireland were numerous, and were accorded a respectable status. An obvious important source of employment for carpenters, for instance, was house-building. The Brehon laws define house-building as *alteach* and a carpenter who devoted himself to such work was called an *altiller*. The native yew, formerly abundant, was valued for its pliant and attractive wood and was used for making a great variety of articles, particularly decorative utensils, and for inlaying and decorating other timbers. Working in yew is recorded as an important trade. It required considerable skill and much training and practice. Native yew is very hard and difficult to work; the cutting tools must have been of a very high quality and those employing them highly skilled. Work in yew was known as *ibaracht* from the name of the tree itself. There are numerous references in Irish writings to 'carvings of red yew'. The skilled craftsman in yew (and possibly other timbers) was called an *erscoraigh* (wood-carver).

The contribution to wood-craft and forest products by the Norse – the Danes and Norwegians – who established the principal cities of Ireland and occupied and ravaged much of the country for more than 200 years, was considerable, but was mainly confined to wood-working and exporting.

During the medieval period Irish forests in Norman hands were frequently used simply as a source of supply for the English market – and, of course, as a source of profit for the Normans involved. Thus was introduced a pattern which was to accelerate dramatically during the sixteenth and seventeenth centuries when forests were felled indiscriminately for profit and because they were fastnesses and refuges for bands of Irish soldiers, hence the coinage of the term 'woodkerne'.

If the Normans brought Norman law with them, in time they acclimatized to a considerable degree and adopted, or partly adopted, some Irish customs, including law: conversely the Irish adopted some Norman customs. The inevitable dichotomous situation was corrosive of both systems. When the Tudor conquistadors arrived they admitted no pusillanimity. They neither sympathized with nor attempted to understand what they considered primitive customs, rather than laws, of savages.

CHAPTER FOUR

From the Later Middle Ages to the Tudor Conquest

'Your ghost will walk, you lover of trees' – Robert Browning.

THE PRINCIPLES OF MODERN forest management evolved in France where coppice working had been systematically practised since the ninth century. As elsewhere, the problem was threefold:

(1) The control of supplies to and the practice of rightholders whose grazing cattle entered forests and destroyed coppice, regrowth and seedlings.

(2) The prevention of inevitable abuses wherever there were small scattered coupes or selective fellings in high forests where merchants could easily defraud the owner and corruption among supervisory forest staff was easy and common.

(3) Successful regeneration by seedlings in both coppice and high forest.[1]

To which might be added a fourth – that this be done profitably.

These principles were brought to England by the Normans. It follows that the administrative and legal features common to forests in England and Ireland from about the thirteenth century onwards were of Norman origin.

It is impossible to consider forestry in Ireland without constant reference to England and the progress of forestry there. English forest management, developed from applied forest law, related purposefully to the requirements of the national rather than the local economy. By the sixteenth century, the demand in England for timber began to exceed supply, and conservation measures were introduced. The attention of producers, processors and industrialists, all requiring vast amounts of timber, turned to the alternative source closest to hand. But while they brought to Ireland all the resources and machinery to fell, export and market Irish timber, they brought no accompanying protective management programme. Nor, except for the woodlands of Wicklow, close to the Pale, where a tradition of active forestry still exists, had the ineffective attempts to introduce forest law in Ireland been accompanied by any significant attempts at management.

Citing Cambrensis as his authority, Hore writes: 'the Irish were accustomed to improve the impregnable character of a wood by cutting down trees on both sides of passages through it, casting some in the way, forming breast-work with others, and plashing or interlacing the lower branches of standing trees with the undergrowth'.[2] It was repeatedly declared that 'the Irish could not be tamed while the leaves were on the trees'; implying both that the woods sheltered them from their enemies, and that the foliage and summer's grass supported their horses and cattle; so that 'the best service that could be done', to quote the phrase of the day, was to attack the wild foe in the winter, and cut and burn down their thickets.

Citing Betham's *Feudal and Parliamentary Dignities* C. Litton Falkiner states: '[The] system was to drive the native population from the plains to the woods; with the result that the Irish territories tended to become ever more and more a succession of forest fastnesses.' The work itself cites a clause from a statute of Edward I (1296) as follows:

The Irish enemy, by the density of the woods, and the depths of the adjacent morasses, assume a confident boldness; the king's highways are in places so overgrown with wood, and so thick and difficult, that even a foot passenger can hardly pass. Upon which it is ordained that every lord of a wood, with his tenants, through which the highway was anciently, shall clear a passage where the way ought to be, and remove all standing timber, as well as underwood.

Norman development, meanwhile, was as methodical as circumstances allowed. The underlying economic motives are admirably summarized:

The Anglo-Normans ... cut through the country smelling out the better lands like well-trained truffle-hounds. The Anglo-Normans were prepared to expend capital on the organisation of their manorial farms and were only interested in land from which they could hope to draw a dividend on their investment. Fertile and well-drained soils were what attracted them... This was the first occasion on which financial considerations directly impinged on land use in Ireland.[3]

To this consolidation and expansion of Norman power there was no effective Irish answer. The basic Anglo-Norman unit was the manor, extending to perhaps 3000 acres which would include considerable amounts of woodland. But gradually, between AD 1200 and AD 1300 'the colonists prospered ... leaving the hill country, the woods and the bogs to the Irish ... who could offer (little) resistance to the heavily armed and well-drilled invaders'.4

The following throws some light on the situation:

In 1305 a legal enquiry was held at Castledermot, at the request of the earl (of Ulster), respecting his title to a certain territory, containing two baronies and a half, in outer Connaught, when it was found, by verdict of the Jury, that '*if* those parts were *cleared of Irish* their value would be 250 marks yearly; but that this expulsion could not be affected without a great power (*magno posse*) of the king's men, and incalculable expenses, exceeding the value of the said land, and principally because the said O Conughur (the legitimate owner) is one of the five chieftains of the Irish, ... So long as the Irish kept to their woods and fastnesses, they were safe enough; for it was only when they risked battle in the field that they were overcome by their opponents'.5

However, in the fourteenth century Irish resistance strengthened and – perhaps more importantly – centralized: this was the so-called Gaelic Revival which, however faintly, helped to stimulate resistance to the Tudors later on. During the Bruce wars the warring armies, Irish and Anglo-Norman, 'between them left neither wood nor lea nor corn nor crop nor stead nor barn nor church, but fired and burned them all'. The Black Death devastated Norman and Irish impartially, halving the population in the second half of the fourteenth century. These events 'all combined to give the Anglo-Normans blows from which they never really recovered ... the Irish resurgence then made life impossible for the open manor and the open village. The lord had to retire into the uncomfortable protection of the tower-house, while the ordinary folk fled back to Britain, leaving their "deserted village" behind them.'6

Forbes states: 'Little reference is made to Irish woods after the thirteenth century until the reign of Elizabeth,' and then he adds characteristically, 'chiefly in connection with the objectionable practice the rebels had of entrenching themselves in the woods and fastnesses of the country... no reliable account of Irish forests as a whole during the medieval period is in existence'.7

Forests were long a source of national income in England, where it was realized that better management meant – in the long term – more revenue. Forest and woodland timber was sold directly, as were licences to fell. Forest grazing, pannage for swine, forest mineral rights, turf, heather and bracken cutting; the rights of passage, tolls – all were marketed, but not at the expense of basic forest management (though this was neglected or subject to corruption from time to time). But from the mid-thirteenth century inefficiency and corruption characterised English forest administration. Forest judges came less and less to hold court; revenues found their way as often into the pockets of intermediaries as to the Exchequer. Eyres were infrequent (none was held during the fifty-four years between 1280 and 1334). They were important affairs attended by all forest functionaries, free tenants of the forest, representatives of all townships within the forest and all who were accused of serious forest offences – if they were still alive. In the instance cited above, involving a hiatus of 34 years, clearly some of the accused had gone to plead their cases before a higher jurisdiction before being summoned to the Eyre. The swainmote usually met two or three times a year. Below the swainmote was the woodmote, a sort of deposition court, which merely noted offences to be tried by a higher authority.

English forests were also afflicted with the problem of outlaws – by no means simply Saxon and Danish dissidents living in and off the forests and on travellers and those who lived in the vicinity. The legendary Robin Hood is simply the best known of these. They also included Edric the Wild and Richard Siward (who was at one time pursued by the sheriffs of ten counties). In Chester Sir John de Davenport, forester of the area, claimed two shillings and a salmon for every master robber taken, and one shilling for each member of his band – a noble recipe for corruption! Deriving, perhaps in part, from this, the transfer of attitudes to Ireland and to the native Irish resisting the Norman invader from their own forests is quite apparent.

In England during the fourteenth century enforcement of forest law became more and more lax, but it was not until the end of the fifteenth c. that a new and somewhat more positive Forest Act was proclaimed when conditions were very different. After the Black Death of 1349 population and prosperity multiplied. Chaotic labour conditions after 1349 had stimulated the wool trade which proved to be of the advantage of the yeoman farmer and the industrialist. Progressively more land came under cultivation and more timber was felled for construction of ships and larger and more luxurious houses were built, largely of wood. Wood and charcoal were the fuels. Forests were rapidly deteriorating and disappearing, a process that was accelerated by the lack of fences and consequent free range of pasture within the woods so that seedlings and coppice regrowth were destroyed.[8]

In 1217, two years after Magna Carta, when the threat from disaffected Saxons and Danes had passed and pressure from the community at large demanded greater personal freedom, Henry III issued a new and concessionary Charter of the Forest, with certain rights retained to the Crown.[9] This Charter became the basis of forest administration for the next hundred years and marked the beginning of the gradual change in policy that came to full term with the Tudors, especially in the Forest Act (1543, Cap. XVII) of Henry VIII. It was prompted by two factors: firstly the national requirement for timber and shipping as England embarked on a challenge for the balance of European power and control of the Atlantic trade routes and, secondly, the shock disclosure of the extent of corruption in forest administration and the increasing shortage of suitable native naval timber.

In spite of controlling legislation, by Tudor times there was a timber shortage in England. Timber became expensive. Henry VIII's Act sought to halt the decline and create new standards of preservation. 'Its provisions included compulsory enclosure of regeneration areas and silvicultural directions on the reservation of seed trees.'[10] This Act was a legislative watershed. It was the third major forest and woodland policy to have been introduced in England in a period of some five hundred years, all reinforced by statutes and law enforcement. Nothing of the kind occurred in Ireland where there was little indication of any attempt at significant recovery in the Gaelic order.

In these circumstances, with the traditional order in neglect and without precedent or encouragement, it is not surprising that a radical – indeed revolutionary – and sophisticated concept such as forest management was overlooked. Vital though this concept may have been to the economic structure of the state, effective central national administration was a *sine qua non* to successful implementation and, of course, in Ireland there was no such effective central national administration. The Gaelic leaders were not equipped to cope with exploitation by foreign interests, even if they had been aware of it.

The prevailing circumstances of political and social confusion and decay made it unlikely that any significant ideas of national identity and progress would advance in Ireland. It is even more unlikely that any idea of a national forest policy would have developed. Even if the idea of such a national forest policy had emerged, whence would it have drawn authority? Both Irish and Norman law were in decline and the country was in a state of more or less continuous chaos. Even if the will were there, Irish law did not provide the means to formulate a national economic policy, and Norman law could not provide an Irish one. Furthermore, in purely commercial – as distinct from any national economic – terms, the optimum market for the forest-owner was often in England, where precisely such a national forest policy created a flourishing market.

The opposition to the Tudor Conquest throughout the sixteenth century enlarged the role of forests as strongholds for Irish. They were dominant features – as Elizabethan directives indicate – for both Irish leaders and English colonists; for the one, secure rallying places, for the other, places of fearful ambush. They were to remain an important strategic factor until the seventeenth century.

The official silence that had descended on Irish forests in the thirteenth century was broken resoundingly by Elizabeth I in the sixteenth when she expressly ordered the destruction of all woods in this country to deprive the Irish of this shelter,[11] and a little later encouraged settlement and colonization and the exploitation of Irish timber to bolster England's resources when, towards the end of her reign, the timber shortage in England became acute. English timber speculators, no less than English statesmen, soldiers and settlers, turned to Ireland. While there was often friction between them about the uses to which Irish timber should be put, two views they held in common; namely profitable exploitation and a means of depriving the Irish of a defence capability.

In 1597, for instance, Sir Henry Wallop informed Robert Cecil, Lord Burghley, Chief Secretary of State for England, that the country 'adjacent to the town of Enniscorthy is full of woods and that he (Wallop) had planted a number of timbermen there, especially cask board makers', all English. Elsewhere in the country surveys of timberland appeared less promising: 'Whereas any woddes doe signify in these plattes ye underwoddes as hazel, holye, alder, elder, thorne, crabtree, and byrche with such lyk, but no great hoke, nayther great buylding tymbr, and the mountayne topp ys barynge save onlye for firres (furze) and small thornes.'[12]

This dour comment, presumably from some English settler, compared with an authoritative description such as McCracken's, suggests that the author was at best disingenuous and at worst dishonest, the evident motive being to discourage others from exploiting timber he wanted for himself.

Everywhere agriculture was greatly dependent on timber – for carts, wagons and implements as well as for fencing and general construction work. Whereas in England coal had, by the middle of the seventeenth century, to a large extent, replaced wood as fuel for manufacture – the absence of major coal deposits in Ireland meant that timber continued to be widely used for industrial purposes, thus increasing enormously the demand on timber supplies as industrial activity, such as tanning, ironworks and charcoal-making, developed. These activities were frequently sited as close as possible to sources of fuel, and to means of transportation. Forests met such requirements admirably, being often close to good roads or navigable rivers. Industries tended to be on rivers or estuaries, especially in the case of ironworks.

While the *Books of Survey and Distribution*, compiled in 1657, and the maps of the Down Survey, give the exact area of every woodland in Ireland,[13] the best account in respect of the Elizabethan period is probably that of Sir George Carew, which mainly concerns Munster.[14]

Payne, the English 'undertaker',[15] stated in 1589 that there was much good timber in many places in this province and that it was so straight and easy to rive, that 'a woodsman with a brake-axe could easily cleave a good oak into boards, which, at 15 foot long and 14 inches broad, by 1 thick, were sold at a rate of 2d. each'. In this year the values of the oak on the forfeited lands in Munster were again pressed on the notice of government; and it was recommended that a high steward should be appointed over the royal manors in Ireland, who should also be a 'wood-ward and chief forester' in this province. But neither this proposition nor the recommendation of the foregoing State paper (the original of which has some notes in Burghley's writing) were attended to when the grants of the forfeited lands (following the Desmond Rebellion) were made. Richard Boyle, afterwards Earl of Cork, who is remarkable for having acquired a vast estate (which was obtained, however, in a manner very different from that described by himself) and who is justly celebrated for the great improvements he effected, bargained with Sir Walter Raleigh, at the time of the attainder of that chivalrous adventurer, to buy his grant of 12,000 acres for the inconsiderable price of £1500 (of which only a third was paid) and immediately began cutting down the woods, in imitation of a notorious English usurer of the day, who inveigled men into selling him their estates, and afterwards sold the timber so profitably that it paid for the land, making, as he said, 'the feathers pay for the goose'. Boyle joined in partnership with one Henry Pyne in purchasing the woods belonging to Lord Condon, an Anglo-Irish owner of a barony named from his family. These partners also bought the timber property of native lords and chieftains, whose simplicity, or ignorance of the market, or perhaps, as in the cases of Raleigh and Condon, impending attainders, led them, as was said of similar sales, to part with what was worth thousands for a song.

The attention of the government in England had been frequently drawn to the public value of the vast quantity of oak then existing in Ireland; and in 1608 one Philip Cottingham was sent over to survey the woods, and report what amount of timber he found suitable to build ships for the royal navy. This surveyor does not seem to have inspected any woods beyond those in the counties of Waterford and Wexford; and, in September, he wrote to the Secretary of State (Burghley), from Mogeely Castle, stating that he had examined the woods belonging to Sir Richard Boyle, and that, although the best and most accessible timber had been cut down for pipe-staves and planks, there still remained much that was valuable for shipbuilding; and he adds that the woods called 'Kilbarrow' and 'Kilcoran', in the County of Waterford, were at that time being cut by Boyle, who had also purchased the forests of Glengarriff and Glenlawrence, in Desmond, with a view to their sale. The destruction that was taking place in the principal forests in Ireland, in which noble oaks, fit to construct ships that would 'carry Britain's thunder o'er the deep', were riven up to make barrels, aroused the attention of Viceroy Chichester; and he drew up a despatch, setting forth in another public and unusual point of view, the ill effects of the purchases made by Boyle and his fellow speculators.

After mentioning that a pretended right was advanced to certain woods in Dowallo (Duhallow), the country of O'Donoghue, by Sir Richard Boyle (who was in such favour that the viceroy merely ventured to hint he was a grasper of lands) and stating that fifty-six tons of Irish timber were about to be sent up the Thames as a specimen, though merchants would not then give 13s.4d. a ton for it, the lord deputy concludes with this significant passage:

> There are forests in this kingdom of many thousand acres, some principal ones ought to be reserved for the use of the Crown, and not wasted, as they now are, by private men, who purchase them for trifles, or assume them upon tricks and devices from the simple Irish, who perhaps have no good title to sell them, or at least know not what they sell. But, finding that private subjects, as mean or meaner than themselves, do for the most part make extraordinary profit of their folly, they oftentimes fall into discontent, and from discontent into rebellion, when the king must be at the charge of its suppression.

As the sixteenth century opened England was on the threshold of that explosion into piracy (at several levels – trade, colonization and statecraft), which gave her such a powerful influence on world events for the next four hundred years, and that later provided the foundation for Lord Palmerston's dictum that 'England has neither friends nor enemies, only interests'. Ireland dominated the western approaches (with the prevailing winds) and access to the Atlantic. An alliance between Ireland and Spain would threaten English ambitions, perhaps decisively. The war with Spain was expensive; if one includes the colonization and planting of Ireland, the cost was enormous. Demands on the English Exchequer had not diminished a century later. Continual warfare placed such financial burdens on the Stuarts that, coupled with their doubtful popularity and inadequate handling of public finances, it led to their downfall and to the rise of Cromwell and the Commonwealth.

Profound political changes in England in the sixteenth and seventeenth centuries were reflected in what happened in their forests. New laws were introduced and major changes in law enacted, resulting directly from the predominant timber requirement mentioned above. Besides the overwhelming commercial and political considerations, analagous to those in relation to oil in our own time, other aspects of timber availability were of importance to western communities. The early and medieval forests of Britain had been administered primarily in the joint interests of the local community and the Crown, but a more sophisticated economy and lifestyle brought changes, and a clearer national forest policy. As these changes occurred, the sophistication that induced them also created demands for timber to provide for the lavish buildings and furniture that were becoming popular and marketable; for smelting; for the hundred and one industrial uses for which timber was essential. By the beginning of the sixteenth century the demand in England for timber was outstripping the supply, even though coal and iron were increasingly contributing to many community requirements hitherto met by timber.[16]

During this period the greatest demand on woodlands was for fuel and not, as might be supposed, for building purposes, although the demand for timber for building – both ships and housing – also began to increase. Iron was obviously of great importance in building, but it took two-and-a-quarter tons of charcoal, or the entire crop of a five-year-old coppice of one acre, to make one ton of iron. As stone and brick gradually replaced timber for building purposes, other expanding domestic requirements – exclusive of shipping – continued to demand more and more timber. The manufacture of furniture and household utensils such as plates, mugs, dishes, bowls, tankards, ladles, pails, forks, knife-handles and so on – called 'treen' – required considerable quantities of wood. Timber was used extensively in agriculture, as it still is today. But it was industrial development, particularly glass-making and iron smelting, that now began to reduce forests significantly. There was scarcely an industry from salt-making to iron-smelting, from pottery to dyeing, in which furnaces, each of which consumed vast quantities of timber, were not constantly active day and night. The clothing industry, too, required large volumes of wood, particularly for dyeing purposes.[17]

The proliferation of iron foundries in England, particularly near London, was creating a scarcity of timber for felling. As a result of this and the other factors already referred to, scarcity prices for timber increased and it became a vital national product. England was by then committed to the struggle for the balance of power in Europe through command of the western approach sea-ways. A significant Act was passed in 1558 curtailing the use of timber in the iron industry. This act forbade the use of any timber of oak, beech or ash, one foot square at the stub or growing within fourteen miles of the sea, on certain quoted rivers which were sea-ways used for the transport of timbers and other commodities, as fuel for iron-making. It was also an offence to use timber for fuel for making iron within a radius of twenty-two miles of the River Thames below Dorchester.

This Act of Elizabeth[18] stated, *inter alia*:

the king our soveriegn lord perceiving and right well knowing the great decay of timber and woods universally within this his Realm of England to be such, that unless speedy remedy in that behalf be provided, there is great and manifest likelihood of scarcity and lack, as well as of timber for building, making, repairing and maintaining of houses and ships, and also for fuel and firewood for the necessary relief of the whole commonalty of this his said Realm.

This Act contained twenty-one sections and was one of the first significant attempts at establishing a modern system of woodland management. It ordered that all woods must be enclosed for four years after coppicing; when a coppice or underwood of twenty-four years' growth or less was felled, twelve standard oaks, known as standills or storrers, should be left per acre. If there were not sufficient oak to make up that count of twelve it could be completed with elm, ash, aspen or beech. Failure to comply with this ordinance was subject to a fine.

In Ireland the twin English policies, one of government and one of her governor-adventurers, were not always *ad idem*. During the fifteenth and sixteenth centuries the Tudor conquerors, mercenary adventurers, and the settlers and exploiters who followed, razed Irish forests for profit and protection. This was done on such an enormous scale that, within little more than one hundred years of its initiation, legislation to curb the devastation was introduced.

It is noteworthy, as McCracken points out,[19] that 'wolves and woodkernes were commonly bracketed together, for they inhabited common ground and represented a common threat to the settlers and there were rewards for the destruction of wolves as well as woodkernes'. Wolves were plentiful. The last wolves died in Ireland about 1770, although they had been extinct in England before 1500. They survived longest in the southwest of Limerick and in Feakle in Clare.

Blennerhasset in 1610 described the wolf and the woodkerne as the most serious dangers to the Ulster colonists and recommended periodic manhunts to track down the human wolves to their lairs. 'No doubt', he observed, 'it will be a pleasant hunt and much prey will fall to the followers'... Tory hunting was practised at the end of the century. Sir William Stewart of Newtownstewart, Co. Tyrone, could write in 1683, 'The gentlemen of the country have been so hearty in that chase that of thirteen in the county where I live, in November last was killed eleven two days before I left home.'

Wild mammals included the native red deer and fallow deer and wild boar (all classed as venison, although the term is now used solely in respect of deer meat), in addition to hare, marten, fox, badger and squirrel. It is thought that the disappearance of the wild boar was due as much to the diminishing supply of acorns, their basic food, as to hunting.

By the mid-seventeenth century two things were apparent in Ireland. These were that timber resources were in danger of being exhausted by continuous exploitation, and that some regeneration was urgently necessary if that was not to happen. In spite of this belated recognition little was done to halt the devastation. In 1602 an English army marching to the siege of Dunboy had made a detour via Kinsale and Timoleague around the woods of Bandon. Thirty years later Boyle could write: 'The place where the Bandon Bridge is situated is upon a great district of the country and was within the last twenty-four years a mere waste bog and wood serving as a retreat and harbour to woodkernes, rebels, thieves and wolves and yet now (God be praised) as civil a plantation as most in England.'[20]

CHAPTER FIVE

Tudor Plantations and After: The Effect on Irish Forests

'The forests, with their myriad tongues shouted of liberty'
– Longfellow.

THE IDEA OF AN HISTORIC IRELAND heavily forested has been authoritatively challenged, especially by Forbes.[1] But while much of the timber covering the country and remarked on by such as Spenser, Giraldus, Fynes Moryson, John Dymmock and others may well have been scrub, it seems unrealistic to argue, like Forbes, that historical Ireland was 'a country no more thickly forested than the Ireland of today'. It is now generally accepted that Forbes, while making a significant and scholarly contribution to the historical background of Irish forestry, from very scanty material, was nonetheless guilty of special pleading; and that he occasionally pursued some of his arguments to conclusions which simply did not accord with the available evidence.

His rejection of the exploitation of Irish forests by the English between 1560 and 1800 may be attributable to a blind-spot. He was, after all, an Englishman in the new Irish Free State, writing at a time when the fledgeling State was displaying a very nationalistic fervour.

While there probably were large areas of scrub woodland, grazing land, bog, brambles, thorns, bracken etc., before 1600, it is not a case of either/or. The existence of such waste-lands does not disprove the coexistence of extensive mature woodlands, of which there is ample evidence. Forbes himself fudges his case by not making clear if his thesis is merely directed against extreme imaginative wishful thinking, of which there was an abundance at the time he wrote in respect of forestry (but examples of which he does not cite), or against mature woods in general. For instance, he states the following and claims that his conclusions are based on deduction:

While no evidence exists for assuming a thickly timbered Ireland down to the last two or three hundred years, it is highly probable that large and numerous areas of scrub-covered ground existed in many parts of the country during Elizabethan times. Some of these might have developed again into forests if given an opportunity. But the forests conjured up by many patriotic and imaginative persons, consisting of thousand-year oaks and yews and mammoth ashes, with pine-covered mountains, 'thick as leaves in Vallombrosa', (correctly – 'Thick as autumnal leaves that strew the brooks in Vallombrosa' – Milton, *Paradise Lost*) may have existed at one time, but not in that dealt with by historians'.

Although his conclusion does not agree with most recent research and he is guilty of special pleading, it is worth giving Forbes's arguments in some detail.

He summarizes his points:

(1) That no reliable account of Irish forests as a whole during the medieval period is in existence. The occupation of the country by the Normans and Tudors had no material effect upon the wooded condition of the country, although considerable quantities of timber were cut for pipe staves, and for conversion into charcoal during the latter period.

(2) The forest laws of the Normans were applicable to Ireland in the same way as in England, but there is no evidence that they were enforced except in the case of Glendalough, and possibly in Wexford in the time of Strongbow. The only deforestation charters recorded are in connection with these two areas. The various references to 'forest' in the State Papers pertained to the unenclosed or unappropriated land of the country, which was regarded as the property of the Crown. This forest had no definite connection with woods or timber.

(3) The Annals, Legends, and other documents relating to the prehistoric period suggest that the condition of Ireland for an unknown period was that of a huge grazing ranch, over which trees were scattered more or less generally. More thickly wooded districts doubtless existed, but no clear indication of their extent can be discovered from the translations made in modern times.

(4) The Irish forest flora does not appear to have changed during the last thousand years. While the pine (*Pinus sylvestris*) may have survived until comparatively recent times, all the reliable evidence on this point is entirely negative in character.

(5) The native forests were gradually destroyed by grazing and fire, continuing over a very long period, and not by any large scale or deliberate destruction.

In this summary the second part of point one is demonstrably at odds with the facts; the first statement of point two is questionable while the remainder seems generally correct; point three contains a very large one-sided assumption; the final clause of point four is arguable and point five is an assertion which is simply not sustainable.

Forbes draws skilfully on a number of sources, in the process revealing and synthesizing much valuable information. But this is done in support of his doubtful main thesis. The general thrust of what he says may be further summarized thus:

(1) At one time there may have been substantial woodlands in Ireland.
(2) Now there are not.
(3) From the medieval period to the present – but especially during the Tudor and post-Tudor period – there is no reliable account of forest reduction.
(4) Therefore there was no significant reduction during this period.

In Tudor England the alignment of forest policy with a national one was part of the process which impelled England unexpectedly to the forefront of the world stage. Across the civilized world (and parts of it not so civilized) world-shaking religious and political events were shaping the destinies of nations. Almost by accident, England became allied to the north European Protestant cause rather than with the south European Catholic one at a time when religion and colonization dominated European events. While any policy, however much it may also have existed in its own right, depended from these superior issues like gew-gaws on a Christmas tree, forestry was recognized as a vital economic factor.

The extent to which the political and economic fortunes of nations, perhaps of the entire civilized world, depended on the continuing supply of quality timber between 1600 and 1862, is hardly realized today. It was the era of preponderant international maritime rivalry and of great wooden ships. 'The relationship of ship timber to sea-power gave it (timber) an importance far above ordinary articles of commerce...In all the maritime nations of that period, the preservation of ship timber was the chief aim of a forest policy.'[2]

There was, accordingly, a precise relationship between sea-power, timber and economic and commercial dominance. Nowhere were the problems involved more acute (though, oddly, sometimes not so well recognized as elsewhere) than in the island power, Great Britain 'where timber supply was inseparably connected with sea-power'.[3] Ireland was also considerably affected by this shipping requirement, but not to the degree commonly thought. While Ireland was colonized primarily for strategic reasons, it was also a prototype for the basic

English policy of colonization in Canada and part of New England; to secure the timber supply.

To the east and south were powerful hostile nations. The old (almost domestic) enemy, France, opposed in politics and in religion, had been superseded by a newer and more threatening one, Spain, grown mighty on the wealth of the Americas, and even more religiously fervid and hostile than her neighbour. The Spanish dominated the Low Countries, and the power of the Spanish Empire threatened England's life-lines to the sea, in particular to the Atlantic ocean. A struggle for control of the north Atlantic and of all that flowed eastwards across it began, between the giant, if rather disorganized, Spain and the more compact, if less powerful, England.

At a time when sail moved the goods of empires on the oceans of the world, Ireland occupied a dominant geographical position. Whoever held Ireland had England open before them. The winds between the two countries were predominantly westerly; as a jumping-off ground for an invasion of England Ireland could not be bettered. It was close, it was self-sufficient, it possessed troops and, above all, it was for long hostile to England both in religion and in politics.

That these things were as evident to the Tudors as they were to the Spanish, history leaves us in no doubt. The English determined to secure Ireland before it became just such a jumping-off place for an attack on themselves. The policy adopted was that of Plantation. For Ireland, three great facts of the Tudor period were, firstly, the strife arising from Irish recusancy in the face of Elizabethan attempts to impose the Reformed religion; secondly the protracted war and its consequences – forfeitures and plantations – arising from the attempted Tudor conquest; and thirdly, the success of that conquest and the final collapse of the Gaelic order.

The 'Plantations', government-assisted settlements by English and Scots on land expropriated from the Irish in Leinster and Munster (later in Ulster and Connacht), affected all three. It is interesting to note that many Scottish settlers were themselves 'dissident' non-conformists, given the option of settlement in Ireland or persecution in Scotland.

The Tudors were businessmen and they decided they couldn't afford to take Ireland by a single massive onslaught. They would open up the country by cutting passes through the woods, bridging the rivers, building roads and keep the roads open by erecting forts and blockhouses along them.[4]

The first Plantation was in Laois/Offaly, in Leinster, when local disturbances gave the English authorities in Dublin an opportunity to intervene. The kingdom was dismembered, the Irish were dislodged and the reigning chieftain was given a contemptible grant of 3000 acres under English law. The rest was parcelled out to settlers from England in lots ranging from 500 to 25 acres, and the area was renamed King's County and Queen's County. It was sufficiently near to the Pale, the defensive area around Dublin already thoroughly settled, to become part of that system and be judged a tactical extension of the Pale itself. The larger grantees were required to pay rent, perform military duties and provide certain services, and all had to undertake to leave their land in tail-male to their eldest son to prevent re-establishment of the Irish system of title, as had already happened in many cases with the Normans.

The next Plantation was a different matter. It was on a large scale, and in a part of the country hitherto extremely hostile to England. Its chief city was the principal mercantile city of the country, and it was an area most susceptible to active intervention from Spain: Munster.

The Desmond Rebellion, led by Sir James FitzMaurice, cousin of the fourteenth Earl of Desmond, was perhaps the first true expression of Irish national self-determination in that age when colonialism, political juggling and religious confrontation in Europe helped to fire the minds of many peoples with aspirations to nationhood. Perhaps the people of Ireland had an older and more legitimate claim than many, even if the reality had degenerated. FitzMaurice revived that crippled spirit by linking the ideas of nationalism and religious liberty in a heady and powerful combination that, once lit, became a beacon for centuries. He also made tactical use of the forest in his war with the English on a grand scale, harried and fought the enemy for years, frequently retreating to a stronghold in the forest of Aherlow.

> All those faire forests about Arlow hid:
> And all that mountain, which doth overlook
> The richest champain that may else be rid.[5]

FitzMaurice's struggle was continued by the fourteenth earl, who was finally forced to take refuge in a forest, whence he was dragged and beheaded. Ironically the Desmonds were of Norman stock, but it was the confiscation of the Desmond lands that, to an important degree, infused Irish nationalist resistance from that time on.

In Munster more than half a million acres were forfeited and made available to colonists, often in blocks of from 4000 to 12000 acres. Blueprints for the development of these lands – 1000 acres for the demesne down to five-and-a-half for cottages – were provided. Edmund Spenser, poet, administrator, advisor and civil servant extraordinary, was given 3000 acres, much of it woodland, where he wrote a great deal of *The Faerie Queene*, including:

> Whylom when Ireland...
> ...that is soveraine Queene profest

Of woods and forests, which therein abound
Sprinkled with wholsom waters more than most on ground.

Raleigh received 20,000 acres (those later bought by Richard Boyle, first Earl of Cork). Thousands of settlers arrived and, without the benefit of local administration (the necessity for which the English authorities had overlooked) immediately began to recover their outlay and establish their fortunes by reducing the woodlands so lauded by Spenser. 'From the woods a continuous stream of timber flowed out – trunks from good, large trees, for ships and houses, branches from these trees and smaller trees for barrel staves, and lop and top and all other woods for charcoal for iron and glass work.'[6]

The Nine Years' War, as it is called, between the Irish, under the Ulster leaders Hugh O'Neill and Hugh Roe O'Donnell, supported by Hugh Maguire, Donal O'Sullivan Beare, a Spanish expeditionary force and others, and the English under several commanders, including the Essexes (father and son) and Mountjoy, was brought to a sudden conclusion with the unexpected defeat of the Irish and Spanish at Kinsale in 1601. An immediate Plantation of a further half million acres in Ulster began.[7]

The Ulster Plantation was to prove the most successful, the most enduring and the most divisive. It differed from previous Plantations in several important respects, principally in the fact that the planters were Scottish, not English, and consequently – though for the most part Lowland Scots – with something of a Gaelic background. They were, however, Dissenters, subject to persecution and dispossession at home, so there was a strong element of 'burning the boats' involved. Moreover, the upper grant limit was 2000 acres, and those landowners who accepted had to bring in their own tenants from Scotland and build defences – a castle and a bawn. A road network was established, as were towns, villages and local industries. The Irish owners were totally dispossessed and killed, driven out or moved to Connacht. 'The Plantation of Ulster was more than the replacement of one group of farmers by another: it was truly an implantation of a different way of life (which) has maintained itself to the present day.'[8]

One explanation for the curious and enduring siege mentality often apparent in Northern Ireland may be found in the following:

When we find a hardy, industrious race transplanted to a country under the sanction and by the authority of the Crown, protected and favoured by the laws and ruling authority of the State, placed under fostering landlords who exhibit a paternal care, and then permitted to hold of their farms under a custom which practically gave them a lease in perpetuity, it would be a strange thing, indeed, if prosperity were not the result.[9]

With the destruction of the short-lived revival of a dynamic Gaelic order, and the Plantation of Ireland, particularly of Ulster, with privileged alien farmers, there began the systematic devastation of Irish forests that led to the substantially forested Ireland of 1600 becoming, by 1750, a treeless wilderness and a net importer of timber.

The planters (and the soldiers who protected them) undoubtedly feared and disliked the Irish woods because of the shelter and refuge they provided Irish troops, as much as they liked them for the profit and land they produced when felled. They thus had three good reasons for felling: security of person, security of pocket, and land.

> The woods and bogs are a great hinderence to us and help the rebels. Much good could be done by Irish churls felling, dressing and burning trees in heaps. This couldst be done whilst leaving sufficient timber for the use of the country, if a tree is left standing every twenty yards. Many people think it would have been well if Ireland had been turned into a seapool than have so charged Her Majesty.[10]

The time was one of intense activity in the forests. McCracken[11] cites an example of wood clearance from about the end of the seventeenth century, when the right of the tenant to cut wood tended to be governed by legislation:

> On the Brownlow estate in Lurgan, Co. Armagh, in the 1670s, to facilitate the clearing of underwood and scrub Arthur Brownlow sold the rights to Edward Hall, who owned an ironworks in Magheralin. By the terms of the agreement Hall paid 4d a cord for a maximum of 700 cords a year. Brownlow's tenants were to be permitted to gather wood for domestic uses and building and Hall was not to take saplings or any wood except that fit for firewood on pain of forfeiture and a fine of six times the wood's value. The cords were not to be converted into charcoal until Brownlow's woodranger had inspected and measured them, and Hall was to work systematically from townland to townland until all the underwood was exhausted.

By 1715 many of the London guilds and companies that had acquired planted lands in Ireland were examining their leases with the object of clarifying the position of timber on their estates. The Irish Society, which had been formed to exploit the newly conquered colony – much as the East India Company was formed a century or so later – decided in 1720 that the timber allowance to tenants did not include even the lop and top of timber felled by the Society. At a time when timber was such a vital ingredient of everyday activity this was a savage restriction (in circumstances brought about by the landlords themselves). By 1781 the situation had become so grave that a series of Parliamentary acts was introduced to try to conserve existing timber and encourage – or enforce – planting.

> In the east-west valleys of Cork and Kerry lay mile after mile of forests that were to enrich the Boyles, the Pettys, the Whites and, for a short period, Sir Walter Raleigh; forests... which would float as the hulls of many of the East India Company's ships, which until the mid-eighteenth century would fill the insatiable furnaces of the ironworks that lined the river valleys, and which would provide the bark for the tanneries of Killarney.[12]

The Boyles' activities ranged across Munster, and were often in defiance of official policy. Vast quantities of timber felled in the province went to the Boyle ironworks, or to make staves. In 1643 Sir Charles Vavasour found the Gap of Barnakill full of oakwoods, ash and birch while, nearby, at Clonegar on the southern banks of the Suir, Arthur Young, over a hundred years later, referred to 'a natural wilderness of tall venerable oaks... the whole wood rises boldly from the bottom, tree upon tree to a vast height, of large oak'.[13] By the end of the eighteenth century both areas were barren of trees.

There were two notable exceptions to the general unsupervised destruction of forests. These were in Carlow/Kildare, where there were woods along the Barrow valley, and at Shillelagh, where a tradition of forestry and foresters is quite lengthy.[14] In each area there were four forest officers, a reeve and four assistants at a salary, respectively, of £100 and £25 a year.[15]

The comprehensive Irish road network played a significant part in the destruction of Irish forests. An extensive, if primitive, road system had existed from prehistoric times, some of it built on the eskers, or raised ridges of sand and gravel deposits (residual arteries of the last glacial age) found throughout the country. Extraction was essential to timber exploitation and this, in turn, required substantial haulage, which led to the question of roads and road surfacing. The existing network was further developed – a process that lasted longer in this country than in England because of the gradual switch there from timber to coal for fuel. The military roads established in this country to help to deal with Irish resistance were also useful for timber haulage. Although up to the seventeenth century, roads in rural Ireland consisted in general of unmetalled tracks which soon broke up under the passage of heavy vehicles, time and weather, the road building and improvement since the Tudor period has resulted in Ireland having a road network said to be more comprehensive, per capita, than that of the United States.

Roads driven into and through forests as access roads and as a means of extracting timber were often called 'passes', deriving, apparently directly, from Brehon Law. Hence, for instance, Tyrrell's Pass in Co. Offaly, called after Captain John Tyrrell who, though an Englishman, was a cavalry officer under Hugh O'Neill; and Deputy's Pass in Co. Wicklow, called after the Lord Deputy who directed the road to be built to facilitate troops protecting workers extracting wood from Shillelagh forest[16] for Westminster Hall.

Some timber was in short supply in Ireland, however, even though it had once flourished here, and had to be imported. Elm was one such, and in 1590 Sir George Carew complained to Burghley[17] that his artillery lay on the ground in Galway and Limerick, unmounted, 'which, for want of elm planks in Ireland, cannot be redressed'. In 1621 there was a request for 130 tons of elm for a similar purpose.[18]

An important and interesting comment demonstrates the intermingling of what, for convenience, may be summarized 'war, gold and timber' in the everyday outlook and attitudes of the average Elizabethan adventurer, much as automobiles, incomes and oil/electricity intermingle in attitudes today:

> The development of heavy cannon in the fifteenth and sixteenth centuries and the consequent consumption of iron, wood, coal and human skills, brought new demands on the mines, the forests and the workshops of Europe. With these came a need for finance, which forced even emperors and kings to resort to the moneylenders... the forester's axe gave not only the wood for gun carriages and ramrods, but also the charcoal to heat the blacksmith's forge and the founder's furnace.[19]

The civil war in England was a cause of great damage to trees and woodlands in both countries, because of the straightforward destruction caused and also because of the impetus given to such industries as iron smelting. According to Dr Lyons, MP, in 'Re-Afforesting of Ireland', *Journal of Forestry* (February 1883):

> The firm hand of Cromwell spared nothing in Ireland that stood in his way; but with the practical sagacity and business tone which essentially decided his action in all things, he turned his attention when his military purposes were served, during the short period in which he remained in Ireland, to the value of his conquest, and did not fail to grasp the important commercial question of the value of the still existing Irish forests...Woodreeves were appointed under the Commonwealth, to protect the various interests of the state.

There now existed legislation of one sort or another designed to protect the strategic and economic interests of England in Irish forests. These instruments, whatever the intention behind them, were ineffective, taking little, if any, account of Irish interests and in practice having the effect of protecting not the woods, but their despoilers. But the questions of woodland management and sustained yield could not, even in colonized Ireland, go unrecognized *ad infinitum*. By the end of the seventeenth century, conquered and despoiled though the country may have been, the need for woodland management became self-evident if the entire country was not to be a waste-land – as, indeed, some contemporary commentators described it. 'They had not left wood enough to make a tooth pick in many places...'[20]

The Plantation system in Ireland led to simple clarification of titles and rights (by the process of dispossession) and was a godsend to an English Crown seeking cheap timber and additional sources of revenue.

This sample of Tudor piracy was, however, sabotaged by the very people the Crown relied on to carry out its policies – freebooters and planters who used the same system to make their fortunes.

Since it was part of the two-pronged Crown policy to reduce the Irish forest refuges available to enemies of the Crown, the planters were able to make their own fortunes at the expense of the second prong of Crown policy (cheap timber for national enterprise), as similar colonists were to do in India, the West Indies and elsewhere. It may be assumed that the policy of clearing the Irish forests was carried out for 'reasons' contained in the comment of the Earl of Cork in 1632, similar to those put forward by John Leyland in his *Itinerary*, made between 1534 and 1543, in which he wrote of Wales: 'Many hills hereabouts hath been well wooded... but now in them is almost no wood. The causes be these... men... destroyed the great woods that they should not harbour thieves...'[21]

The warfare that turned much of Ireland into one vast battlefield almost continuously from the Tudor Conquest throughout the period of the Commonwealth and the Stuart (Williamite) wars, also had a very detrimental effect on forests. 'As the seventeenth century drew to a close, the curtain was finally rung down on the old forests of medieval England. Although it would still be another 200 years before modern forestry started to take shape, from now on the word forest began to convey the idea of timber rather than of hunting and of commerce rather than sport.'[22] Without doubt timber was a vital national economic problem for England. In particular it became a serious naval problem by the mid-seventeenth century. In that period timber supplies in England were in the hands of a cartel which ruthlessly controlled prices and exploited government requirements. Ironically, in England the largest item of expenditure in the Crown forests was always the salaries and wages paid to the forest officers, and the cost of their houses.[23]

In an attempt to bring order and method into a cumbersome and uneconomic forest administration system riddled with corruption, and with little practical relevancy to contemporary requirements, the post of Surveyor General of Woods had been created in the mid-sixteenth century – presumably to administer the new Act of Henry VIII. The experiment was not a success. The new officials did not have the requisite powers; furthermore, since they worked to proposals aimed at improving the supply of timber by controlling felling, rather than by the systematic development and planting of young trees, the measures were at best short-term. Nor did the creation of new officials eliminate the problems of inefficiency and dishonesty.[24]

In England legislation governing forest management resulted from the consideration that strategic supplies of timber were at risk, that they were likely to be increasingly so in the foreseeable future and that timber resources could not, by natural spontaneous methods of regeneration alone, meet demand. The paradox was that meeting existing requirements and conserving against future requirements were, in practice, mutually exclusive. Importation was the obvious answer. But cost-effective imported timber was generally of inferior quality. On the doorstep, however, was the newly conquered and restless colony of Ireland, colonized with ex-soldier and adventurer settlers, where much fine oak flourished.

Initial legislation prompted by this consideration tended to be merely restrictive, such as the English Act of 1621 prohibiting felling of timber within ten miles of the sea or a navigable river. As McCracken points out this Act also applied to Ireland 'but' she adds, 'was never anything more than a dead letter'.[25]

An Act of 1612 that directed that all trees and bushes growing on the highways and passes had to be cleared by their owners is more typical. 'Even by the middle of the seventeenth century in well settled areas the passes had to be kept open to about 60 feet on either side of the road.'[26]

Resulting from this three-level policy the destruction of timber in Ireland was so extensive that from being a net timber exporting country in 1660 Ireland became a net importing country by 1711, when imports of timber totalled 11,000 tons and exports merely 600. Ireland was completely dependent on imports for softwood. Most of this came from Baltic countries (in keeping with Britain's trading policies). So great was the loss of her own hardwoods, these were now being imported also. McCracken states that the increasing demands of the provision trade rather than 'the scarcity of local timber accounted for the virtual cessation of the export of timber after 1700'.

The time was one of intense activity in the forests.

Whether or not the new landlords were able to exploit the timber on their estates commercially, if they wished to develop the estate the wood had to be cleared. There are many examples of seventeenth century leases which require a tenant to clear so much timber annually. An Armagh farmer writing in 1750 declared that he would have a hard bargain of his farm if a memorable storm had not levelled much of an extensive wood on his holdings.[27]

While the seventeenth, eighteenth, and nineteenth centuries saw an explosion of timber-use, England was not to the forefront in European forestry development either in respect of management or legislation. It followed that where England lagged, Ireland received no consideration.

In France the most significant law relating to forests was the law of Charles V known as the *Ordinance de Melun* of 1376. This remained the effective basic law governing forestry in France for nearly three centuries until it was superseded, or rather updated, by the Ordinance of 1669. In pursuance of the principle of sustained yield, it laid down seven practical principles of forest management and codified and collated preceding legislation in terms of forest administration and management to the effect that: 'Masters of the Forest (of whom there were two) investigate and visit all the forests and organize the "coupes" so that the forests can perpetuate and stay in a good state.'

The six main directives of this Ordinance, all of which were incorporated in succeeding legislation until 1669 (when they were expanded and made more precise), were:

(1) Right holders were allowed the 'possibility' or appropriate output of a forest (meaning a regulated, constant yield that did not diminish the stock).

(2) Fellings would be regulated in areas of 10.15 hectares, known as coupes, selected by forest masters, and not by quantities of trees.

(3) Coupes would be delineated, maintained and surveyed.

(4) Coupes would be enclosed after felling.

(5) Standard trees were to be reserved in both coppice and high forest coupes to a minimum number of about eight to the acre.

(6) Trees should be felled with care to facilitate regeneration.

By the middle of the sixteenth century France had achieved, for its broad-leaved Royal Forests at least, an organized forest administration and an ordered silvicultural management designed to prevent abuses by prescribing contiguous coupes, based on clear-fellings with reservation of standards, equal in area and proceeding in a definite direction only returning to the origin after a determined period (not less than a hundred years in high forest or ten years in a coppice).[28]

Two factors led to the Ordinance of 1669, one of them being the universal requirement for naval timber and the second, no less than in England, abuses and corruption which had infiltrated a deteriorating administration over the previous fifty or sixty years. The formal procedures imposed at that time continue to the present.

When corruption was so prevalent in the English forest system, with its own laws and enforcement officers, it is hardly to be wondered at that colonists in Ireland – with cheap Irish timber available to compete with scarce and expensive English timber – were more concerned with their pockets than with uncomfortable English laws. The opportunities in Ireland were clear and Ireland quickly became a timber exploiter's paradise. Here the greatest expense was not the cost of management and wages or housing of staff, but of getting at the timber and getting it out.

Rock Church, King's Surveyor, wrote in 1612: 'How forward every man is in these days to fell down timber and grub up coppices and none endeavours to plant any...'[29] In the same century Moses Cook, author of a distinguished early work on woodlands called *Forest Trees*, wrote: 'There are too many men more inclined to stock up than to plant.'

New industries sprang up overnight, the barrel-stave industry being a major one. It tended to be concentrated in the south – but also along the valley of the River Bann. This was a clear indication of the overseas markets involved – France, Spain and Scotland. The forests of Cork and Kerry 'were to cask nearly all the wine that France and (to a lesser extent) Spain would produce',[30] in the process making fortunes for several of the planter families. In 1611 a London agent called John Rowley wrought havoc in northern woods by trading in pipe-staves, contrary to articles of agreement (with the planters and City of London) which provided that the woods granted for the furtherance of the Plantation were not to be used for commercial purposes.

During the preliminary discussions in January 1650 between the crown and the City of London Companions on the proposed Plantation of Londonderry the question of the ownership of Glenconkeyne and Killetra woods arose. The crown wished to reserve them to itself, but the city declared that only as lords of the soil could they ensure the conservation of the timber, which they stated they did not intend to sell. At the end of January it was agreed that the woods were to be granted to the City in perpetuity to be held of the king in common socage, and that they would be used for building and other necessary purposes in Ireland.[31]

But it later emerged that, in fact, the amount of timber cut and exported for staves would have built the cities of Derry and Coleraine several times over, and amounted to upwards of 200,000 pipestaves a year.

According to Le Fanu[32] the 'trade in pipe-staves... appears... to have been one of the principal causes of the destruction of the woods... the cost of pipe-staves in Ireland was only 40 – 45 shillings a thousand and the trade was... so profitable that it could not be stopped... In 1609 the English Government awoke to the destruction being wrought on this valuable resource and the Lords of the Council gave orders that the timber growing in the king's woods should be reserved for building and repairing the king's ships...'[33]

Characteristically the timber-workers, like the 'developers', were also usually English. Sir Walter Raleigh had two hundred on his Youghal estates,[34] and other owners and surveyors insisted that 'immigrants... and other artificers of timber-work' should be brought from England.[35] McCracken gives an interesting breakdown of one hundred timber-workers on the Bann in 1611 – thirty-two fellers, twenty lath tenders, nine sawyers, eight wainsmen, four timber squarers, four shipwrights, three overseers, fifteen men rafting timber down the Bann and nine men working the cots. Fifty men worked in the woods carting timber to the river using thirty-three oxen and three horses.

The stave industry was enormous. Raleigh complained that a rival of his, Henry Price, made £4000 'privately' i.e. illicitly, from their jointly owned woods. If this – or anything like it – was being siphoned off jointly held areas, it may be imagined what the gross profits were. The industry flourished in spite of futile attempts to control it. 'I find it almost impossible to restrain the making and the working of the timber into pipe-staves without seizing on them when wrought and brought into the port towns which will beget much clamour and offence,' Lord Deputy Chichester complained to the Privy council in 1611.[36] Boyle recorded transactions involving four millions of staves from some of his woodlands between 1616 and 1628, one of which was to pay a Mr Milton of Youghal 6/8d a thousand for the export of 71,000 staves.

Officially the woodlands were earmarked for uses such as shipbuilding, but the profit motive overrode such considerations. In 1612 the East India Company bought, for £7000, part of the woods along the Kinsale River and founded a settlement there for three hundred English, although they were 'Royal Forests' and had been earmarked for the navy only the previous year. The company, which might have been expected to have opposed the stave trade, used the timber from its east Cork woods for both shipbuilding and, illegally, for staves.[37]

McCracken points out that stave manufacture proliferated so much in the Slaney valley that by the 1630s 'most of the wealthy men of Enniscorthy were timber merchants employing over 100 men to raft the wood down the Slaney'.

Although a vast amount of staves was exported from Ireland to France and Spain for the wine trade in the sixteenth and seventeenth centuries, and incalculable as the amount of timber involved was, it cannot have compared with the stave requirements for other purposes. Throughout the entire period the trade, import and export, of agricultural produce increased enormously and most of this was also barrelled.

Besides the strategic and political considerations there was another aspect of Tudor England which played a part in the conquest of Ireland. Tudor England was a land of merchants and businessmen. Unless Ireland, rapidly and unexpectedly, had developed its own dynamic economy (hardly thinkable, much less a possibility, prior to the emergence of Hugh O'Neill), the impact of this nearby mercantile dynamic ensured conquest of one kind or another.

At the beginning of the seventeenth century timber exports went principally to France and Spain. By the end of the century the bulk went to England and Scotland. By 1686 England was taking more than three times the amount of timber it had taken from Ireland twenty years earlier, and was also taking plank. Irish timber was also important to the provision trade with North America, which expanded rapidly during the century when the export of live cattle to England was prohibited in the interests of English farmers, with disastrous effects on the Irish economy. Between 1660 and 1669 the export of live cattle from Ireland to Britain fell from 60,000 a year to less than 1500. As a result Irish exporters turned to salt beef, pork, fish etc., with a natural increase in the requirement for barrels. McCracken[38] shows that while the export of timber decreased during the period the volume of timber used by the export provision trade increased, and the total volume of timber leaving the country remained roughly the same.

The two most important pieces of forestry legislation in respect of Ireland were undoubtedly Charles I, Cap. XXIII of 1635 and William III, Cap. XII, 1698. The first of these was a virtual copy of the definitive English Act of Henry VIII of 1543, which was then (hurriedly) applied to Ireland in an attempt to slow, if not prevent, further decline of woodlands. The second – delayed for five years, so that trees might be planted and have reached four years of age – was the first important forestry legislation for Ireland.

Building methods introduced by the colonists were very different from traditional wattle and clay, and the colonial explosion of building also placed a burden on available supplies of Irish timber. To English developers, prospectors and businessmen, starved and restricted by law in the use of their native woodlands, Irish forests must have seemed like the inexhaustible deposits of the fabled *Iolan bFiodhach* or *Isla des arboles*.

Although none have survived, small towns and villages of half-timbering in the English style were quite common in Ireland in the early seventeenth century. One of the inducements offered to the City of London guilds for participation in the Plantation of Ulster was the availability and abundance of wood for housing. In 1610 each planter was given 200 good oaks to make timber for such buildings as he wished to erect and we have seen that the Charter given to Londonderry in 1613 laid down that the timbering at Glenconkyne and Killetra was to be used only for house construction or other necessary purposes, a condition repeated in the Charter of 1674;[39] both conditions were ignored.

Nevertheless a great deal of building took place in Ulster and local records there show an enormous amount of timber being used for that purpose. After the Great Fire of London in 1666 the demand for Irish oak to help rebuild the city was very great. Yet McCracken[40] points out that the Calendar of State Papers Domestic states that it was permitted to use oak only for roofing, doors, window-frames and cellar floors. 'According to A. G. Henry 150,000 oaks at 10 shillings each, 100,000 ash at 5 shillings each, and 10,000 elm at 3 shillings each were used in the building of Londonderry (the elm must have been imported).

The initial blueprint for the city envisaged 200 houses to start with and a further 300 within six years.'[41] The Calendar of State Papers also records that a Captain Henry Nicall bought a ship for £1000 to ship oak from Ireland.

By 1670 wood was already so scarce that in Dublin its use for building was restricted. A law was promulgated requiring that new houses in Dublin must be built of stone or brick and be slated or tiled. Moreover all thatched roofs were to be replaced within a year and overhanging windows typical of the old half-timbered houses were prohibited, though balconies were permitted. There was very likely more significance to this decree than timber shortage alone. It was probably introduced in Dublin with the fearful evidence in mind of the fire of London, where rebuilding was still in progress, four years earlier, as much as to limit home consumption. During the Cromwellian period timber became so scarce that its use for construction in Ireland was restricted by licence. By the last quarter of the seventeenth century timber was being imported from western Scotland. Even the traditional use of saplings for wattling was forbidden as early as 1705, with marked social consequences. By now timber was so scarce that people were reduced to living in huts made of sods and brushwood, both for themselves and their cattle. 'Timber is so dear and tenants in the melancholy place so uncertain that it is better to sell the ground than to build,' was the cry from Carrickfergus in 1735.[42]

Between 1680 and 1700, or thereabouts, there was a remarkable nationwide decline in hazel growth. Pollen analysis by Professor Frank Mitchell at Littleton Bog near Thurles shows that before 1700 hazel formed 30-40 per cent of all pollen grains but afterwards it formed less than 5 per cent.[43] Mitchell does not indicate if the 30-40 per cent calculation refers to a percentage of the overall *after* the extensive Tudor deforestation or to a fairly constant percentage dating from before that. Another factor to be considered is that hazel pollen is wind-borne and it can be carried very long distances. It is also amongst the earliest of the catkins to come out in spring and, while many species of tree do not flower for several years after they have been coppiced, hazel is an exception, flowering from its second year. Pollen counts could, therefore, exaggerate the amount of hazel in coppiced or cut-back woodlands. And, of course, hazel was certainly managed to some extent because of its value both as a source of food and for its rods for wattling. So much said, there is no doubt that the sudden decline in hazel during these twenty years was phenomenal and must be given consideration. The data can be related to circumstances which may explain another phenomenon – namely the proliferation of the huts of sods and brushwood that were so prominent a feature of the Irish landscape up to the late nineteenth century. To the best of my knowledge these factors have not before been considered in conjunction, but their relationship, especially in the light of the Queen Anne Act of 1705, is too significant to ignore. The following is the relevant portion of that Act (Sec. IX):

And whereas great quantities of young trees are daily destroyed by the making of gadds and withs, and that it will very much conduce to the incouragement of the iron and hempen manufactures, that gadds and withs be no more used in this kingdom, be it therefore enacted ... that from the 1st November no person or persons shall make or use in plowing, drawing of timber, or other work whatsoever, or in wattling the walls of houses, or cabbins, or out-buildings, any kind of gadd or gadds, wyth or wyths, of oak, ash, birch, hazel or other tree whatsoever...

The Act further provided that:

any person by warrant of a justice may search suspected houses and places for any wood, under-wood, poles, trees, clap-boards, barrel staves, poles, rails, stiles, posts, gates or for any gadds, wyths, willows, hedge-wood, bark, rind or coat of any tree, unlawfully barked as aforesaid...

Three apparently unrelated factors may have brought about this perversion of the system; first, the insatiable industrial demand for fuel; second, the new, alien type of house and house construction; and third, the severe penalties for cutting wattles.

Before 1700 hazel was one of the most abundant tree species in Ireland and was one of the few species purposefully cultivated to provide wattles, the essential multi-purpose building material. The disappearance of hazel and hazel-rods must have had a shattering impact on what remained of the social fabric of rural – possibly even urban – Irish society. For the English colonists hazel and wattling had little significance, except, perhaps, that it seemed primitive, and Irish, and was another timber. All the more reason for putting hazel to 'better' use. If in doing so a blow was struck against the 'natives', so much the better. Was it not boasted by the London *Times* that the Irish would soon be as scarce on the banks of the Liffey as the American Indian on the banks of the Delaware? The levelling of the hazel woods, probably the most widespread species in Ireland prior to this, was a significant contribution to this end.

The industries that proliferated with the coming of the planters and the settlers and as a result of the Stuart and Cromwellian wars, demanded one thing more than any other – fuel! In Ireland there was little coal and when the timber began to run out towards the end of the seventeenth century after the tanners, the glassmakers, the shipwrights and, above all, the ironmasters, had been pouring the produce of the forests into their furnaces, sometimes twenty-four hours a day, for more than a hundred years, what remained? Was it then the turn of the hazel woods to feed the furnaces? Was that the real intent behind the legislation prohibiting the use of wattling?

Some of the industries mentioned are worth consideration. Tanning was an ancient Irish industry and recent excavations, particularly at Wood Quay in Dublin, have uncovered what are believed to have been 'tanning pits' dating from the eleventh century. But it was very much older than that. Tanners stripped the bark, usually, but not necessarily, oak, from the living tree. This killed the tree very quickly. Naturally tanners were looked upon with disfavour by other tree users because of the damage they did, and between 1569 and 1767 various attempts were made to control their activity. It became illegal for anyone other than 'a recognized tanner to obtain bark'.[44] In fact, as McCracken[45] points out: 'An attempt was made to control tanners before the question of timber conservation in Ireland was mooted. Setting up of a tannery without a licence was prohibited in 1569 and the tanning of leather in places unauthorised by the Lord Deputy was forbidden.'

Debarking of trees for tanning was forbidden in 1628 and again in 1634. These seem less like attempts at conservation than attempts by pressure groups to reserve the timber for purposes for which de-barking would render it useless. The limitations in respect of siting a tannery may have had to do with the appalling smell involved. An oak-tree of about forty years yielded between nine and twelve pounds of bark per cubic foot, while one of about eighty years would give between ten and sixteen pounds of bark per cubic foot. 'The famous Royal Oak in Conroy Park near Antrim, which girthed 42 feet, yielded 40 pounds of bark (per cubic foot) in 1780 when bark fetched some £6 a ton.'[46]

Prohibition on the export of live cattle from Ireland in the 1660s helped to increase the export of tanned hides which doubled from 106,360 (in 1665) to 217,000, in four years.[47] But by the middle of the eighteenth century the export of tanned hides had declined and this is held due to the cost of bark.[48] Interestingly, there was a considerable market for leather shoes in North America at that time.

The earliest efforts to establish a glassworks in Ireland appear to have been around 1588 when it was suggested to Burghley that the transfer to Ireland of thirteen of the fifteen glassworks then existing in England would help to preserve the English woods and reduce the many woods in Ireland 'than which in time of rebellion her Majesty hath no greater enemy there'. Nine years later Mr George Long (who had made the suggestion) sought a licence for the monopoly of glass-making in Ireland where he had 'kept a glasshouse in the end of Drumfin woods'.[49] While glassworks near Birr supplied Dublin with window panes during the mid-seventeenth century, the majority of glassworks in Ireland appear to have made drinking and ornamental glass – a situation that has not altered today. 'However, while the production of glass was not tied to wood as a fuel, it was an asset to have access to ash trees, as the alkali used in processing glass was obtained from them.'[50] While the amount of timber used for this purpose may have been smaller than the amount consumed by ironworks, it was not inconsiderable. McCracken points out that Richard Boyle used 'twenty-five cords of small cleft wood a week to make eighteen cases of glassware'. We are not told the dimensions of a case, but a cord was a bundle about twelve feet long and four feet in diameter.

So far as ironworks went, it was cheaper and more efficient to send the works after the timber rather than bring the timber to the works. The price of timber was ten or twelve times cheaper than in England, but, since they used English labour, cost of transport was much the same.[51] The iron industry was unquestionably the major industrial predator of Irish woodlands at this time. The number of ironworks was considerable, over 160 of them according to McCracken and other sources. Constant warfare and progress in the manufacture of weapons, especially guns, fuelled the demand for iron. The supply of iron ore and of limestone, which is used as a flux in iron-making, was abundant.[52] More importantly, there was timber to provide the charcoal for smelting. Coal did not become the fuel for iron smelting until the nineteenth century.

Mr James McParlan, MD, in the survey of the County Leitrim (1802)... refers to the highly wooded state of the county in former times. He says: 'Living persons who saw it told me that about a hundred years ago almost the whole county was a continual undivided forest. From Drumshanbo, I used to hear them say, to Drumkerrig, a distance of nine or ten miles, one could travel the whole way from tree to tree by the branches'. The destruction of the forests is attributed to the use of the wood as fuel in iron smelting.[53]

According to McCracken: 'Ideally the best charcoal comes from 25 years old coppice oak.' In England ironmasters practised coppicing to ensure a continuous supply. Generally an acre of coppice gives enough fuel to make a tonne of iron every twenty-five years. But in Ireland, except in Wicklow, no such provision for a continuous supply of fuel was made, 'and the life of an ironworks was limited by the supply of readily available wood. This gave rise to the practice of moving iron works from place to place as supplies of local fuel became exhausted or, alternatively, of using works for a part of each year. McParlan said that the Drumshanbo works in 1717 was ringed with heaps of charcoal as big as three Dublin houses; it took approximately two tons of charcoal to make one tonne of iron bar.'

With the raw material costing so much less in Ireland, Irish iron could be marketed at very competitive prices. The ironworks themselves supported mainly English immigrants. According to McCracken: 'special permission had to be obtained to employ 500 Irish workers at the Mountrath iron works in 1654 until English workers could be obtained, and the Irish had to live within a musket shot of the works'.[54]

In 1610 a Mr Tokefield proposed a rapid and profitable way of removing the woods which were proving refuges for the Irish. It was adopted 'some years later by Sir Richard Moryson, President of Munster, in the south, and Sir T. Philips in the north, when the suppression of the leading rebels permitted...'[55] This was the establishment of ironworks. 'Iron is a brave commodity where wood aboundeth,' warbled Bacon and in the seventeenth century they were, as Falkiner observes, held to be 'the true El Dorado of Irish enterprise'. In his *Natural History of Ireland* (c. 1641) Boate says 'it is incredible what quantity of charcoal is consumed by one iron-works in a year'. Boyle is said to have made £100,000 from his ironworks.

If a fraction of the endeavour that went into forest exploitation for industry had been devoted to forest management, many forests might have survived as important national resources. The last wooded area to be exploited by the ironmasters was the south-west, where timber in the isolated valleys was too difficult of access to have any great commercial value. 'Here the Pettys, Whites and Brewsters opened works in the late 17th and early 18th centuries. By the middle of the 18th century the iron works had become a by-word for the inroads they made into the oak and arbutus woods.'[56]

It is said by some authorities that Irish ironworks failed because foreign iron was cheaper and better. McCracken[57] refutes this: 'contemporary observers are unanimous that they did close because local woods were destroyed'. There is no reason to doubt this. It also seems clear that not only the woods, but the ironworks themselves and the employment they provided could have been preserved if coppicing and management had been practised. It is all the stranger that it was not, since most of the ironmasters were Englishmen who must have been familiar with coppicing. The conclusion must be that there was either a deliberate policy of exploitation, or that the works were conducted with a degree of indifference, perhaps stemming from the abundance and cheapness of the raw materials. Moreover land deforested by iron-working was often good land, more immediately profitable for agriculture. To the problems resulting from absence of compulsory re-planting and preventive legislation was added a rapidly growing population.

In 1698 legislation to grow more trees in Ireland was introduced. After the damage caused to the forests by 'the late rebellion in the kingdom' (the Williamite wars), 'landowners and tenants were obliged for thirty-one years to plant ten plants of four years growth or more of oak, fir, elm, ash, walnut, poplar, abeal, or elder';[58] owners of ironworks were obliged to plant 500 trees a year. Legislation showing a concern for damage being done to trees as a result of the demand for timber (or bark for tanning) was introduced as early as 1635, with further legislation in 1697.

This was the first of seventeen Acts which were a belated and inadequate attempt to halt destruction that had been going on for a hundred years. They were more concerned with England's strategic and economic requirements than with the economic well-being of Ireland or her people, a policy recognized later as being itself inadequate. Many of these statutes actually encouraged corruption. One such Act, of 1765, allowed a tenant to claim the benefit of trees planted provided he lodged a certificate with the Clerk of the Peace of the county – a direct invitation to rack-renting. The last such Act (1791) forbade a tenant to cut down trees except with the consent of the landlord; it also applied to hazel, and firing. It was evidently intended to deprive the people of rights to which they would be entitled in England.

Throughout this period English entrepreneurs were founding or making vast fortunes in the slave trade, more particularly from sugar plantations on the West Indian colonies, whence came the wealth that glittered in the lavish palaces and houses of London, and poured luxury into the incomparable estates and mansions of rural England. Is it surprising that the colonists in Ireland tried to squeeze every possible penny from their Irish estates? Slaves they had in plenty, though white; sugar they did not have, and landowners had to find an alternative.

Awareness of the economic importance to the country of timber prompted sixteenth-century legislation aimed at promoting and renewing tree growth. The following basic principle of forest management summarized the thinking behind it: 'If copses were so divided as that every year there might be some felled, it were a continual and a present profit: seventeen years' growth affords a tolerable fall, supposing the copse of 17 acres, one acre might be yearly felled forever; and so more according to proportion, but the seldom falls yield the more timber.'[59]

Concern about timber supplies in England was not confined to official quarters. 'We doe in all humblenesse complaine unto your majesty of the general destruction and waste of wood, made within this your kingdom more within 20 or 30 last yeares than in any hundred yeares before.'[60]

A number of methods were tried to encourage renewed timber resources, including planned planting. There is a reference[61] to a nursery in 1612, which appears to be one of the earliest on record, but during this century nurseries increased in both number and importance. The development of nurseries encompassed the development and acquisition of knowledge and skills, some of which are still practised. In *The manner of Raising, Ordering and Improving Forest Trees*, for instance, published in 1717, Moses Cooke describes in detail how to store ash keys in layers of sand, a process known today as 'stratification'. Mice, attracted by the seeds and acorns, were considered a major obstacle to the planting and growing of oak by direct sowing.

Conifers did not come into general use until new species were introduced from abroad in the nineteenth century. In the interim, species like Norway spruce, the common silver fir, European larch and Corsican pine had some commercial success, but oak and elm still dominated the timber trade. The one conifer of timber value was held to be Scots pine – long so scarce in Ireland as to be virtually extinct – and even in England it was scarce and unsuitable for most high-grade manufacturing purposes.

Experiments with plantations also began about this time, the earliest about 1550. But they did not become general for another two hundred years.

CHAPTER SIX

Ships and Shipbuilding

'The Gods, that mortal beauty chase, Still in a tree did end their race' – Andrew Marvell.

THE IMPORTANCE OF SHIPS to an island is self-evident. During the Norse period (AD 800 to 1014, approximately) Dublin, Waterford and Limerick were all engaged in shipbuilding while, it is alleged,[1] there was a sizeable slave-market in Cork for 'goods' from Gaul and Spain, all of which meant timber and timber use. But it was not for another 500 years that timber for shipping demonstrated itself as an overall 'national requirement'.

It is interesting that while seafaring in Ireland (especially, but not exclusively, in coastal waters) was quite extensive, the dynamic that had once impelled the hosts of Niall and Crimthann and the missionaries after them to sail the seas of the known world had faded. There had been sporadic bursts of shipbuilding at times, especially just before and just after the arrival of Christianity, and again during the time of the Norse wars and the city kingdoms. But after the Norman invasion there is little evidence of any major Irish fleet – naval or merchant – though there were, of course, ships and shipping on a smaller scale, of which the renowned Ouzel galley was one example.

When England entered the lists against Spain and France in the struggle for power in Europe, it became obvious that ships and shipping were the pivot on which its fortunes would turn.

Of the various fields in which the effects of the timber problem were felt, the navy itself was naturally the most important. The timber supply was inseparably connected with sea power. As long as England could maintain control of the sea, the whole world could be searched for timber.[2]

It is virtually axiomatic that an island which aspires to be a developing power must first become a seafaring, mercantile nation. Lord Coventry, addressing the judges of England in 1635, stated this case succinctly:

The wooden walls are the best walls of this kingdom; and if the riches and wealth of the kingdom be respected, for that cause the dominion of the sea ought to be respected, for else what would become of our wool, lead and the like, the prices whereof would fall to nothing if others should be masters of the sea.

It is hardly going too far to infer that if 'the wooden walls' of England required extension and repair beyond the resources of the managed woodlands of her own territory, it would seem reasonable and proper to such as Lord Coventry that the available and 'cheap' woodlands of Ireland should be employed to fill the gap. Ireland had itself been conquered and planted by the Tudors and the Stuarts in the interests of English foreign policy; what more natural than that her resources should be exploited (or curtailed, as required) in the interests of that policy? Such is the logic of conquest and settlement.

By the end of the sixteenth century the distinction in England between commercial and naval vessels was marked. The practice of adapting merchant vessels for naval purposes in time of war diminished, and specifically commissioned naval vessels were commonplace. While up to then naval fighting, so far as England was concerned, 'was carried on for the most part by private ships gathered for the occasion'[3] these 'privateers' were also built for war at sea – the Elizabethan institutionalized piracy that was the precursor of the later variety. (The Spanish Armada, on the other hand, as the name suggests, largely – but not entirely – comprised a purpose-built or assembled attack force.) By the beginning of the seventeenth century, legislation for the preservation of wood and restrictions in the use of timber had been enacted.

With the development of the East India Company and the expansion of trade with the Far East, followed by the succession of wars with the other naval powers beginning in 1652 with the Dutch, an explosion of shipbuilding was initiated that lasted until the Second World War. For the next two hundred years, until metal ships took over, timber was consumed by the hundred thousands of tons by wooden ships of all classes.

Ships of the British navy consisted of several different classes, from first, second, third and fourth ships of the line, through frigates, schooners, gun-brigs and sloops, to cutters. They were rated according to the number of guns they carried, first-rate ships of the line of 120 guns, down to frigates with 28. Since these ships were very much bigger, and sometimes more numerous, than merchant vessels, naval timber requirements were often greater than those of commercial shipping – as were the rewards for the suppliers.

In building a ship of the line, oak, elm and beech were the main woods used in the construction of the hull. The masts and spars were of pine and spruce. Timber for the masts was usually imported from the Baltic or North America. Timbers, other than mast timbers, employed in building a ship were divided into categories: for instance Great Timber meant enormous pieces required for parts like the stern-post; Compass Timber meant naturally curved pieces required for knees, futtocks, breasthooks and so forth. Plank or planking was used for decks and also for the sides of the ships, both internally and externally, the thickness of the planking decreasing from the keel upwards. Bottom planks were not less than 6" thick and the uppermost were about 3". Bands of heavier timber known as waling ran the length of the ship on and above the waterline; these were formed of the thickest planking. The building of a ship of the line was a very substantial undertaking that usually took about five years. A '74-gun ship... takes 2000 large well grown timber trees; namely, trees of nearly two tons each'.[4] A 74-gun ship would have taken about 4000 tonnes of timber. Bigger ships of the line, 100 guns or more, not only consumed more timber, but required more massive timbers. For lower and more massive masts the proportion was an inch of diameter to a yard of length. Some idea of the requirement involved may be gauged from the fact that a first-rater of 120 guns carried a lower mainmast forty yards long and forty inches in diameter.

Because of the great quantities of timber used, both for the navy and merchant fleets, William Marshall wrote prophetically (though he was not referring to Ireland): 'When we consider the prodigious quantity of timber which is consumed in the construction of a large vessel, we feel a concern for the probable situation of this country at some future period.'[5] In 1756 *Royal George*, a ship of 100 guns, took nearly 7000 tons of timber – that is almost 4000 large trees! And an indication of the woodland requirements involved in the building of a 74-gun ship may be obtained from the following: 'The distance recommended... for planting a tree in the wood... in which underwood is also propagated... is 30 feet upwards. Supposing trees to stand at 33ft, each statute acre would contain 40 trees; consequently the building of a 74-gun ship would clear, of such woodland, the timber of 50 acres.'[6]

It was considered that timber ear-marked for shipbuilding was to be in the proportion of two-thirds for building, one-third for repairs; in contracts for supplying timber to naval dockyards, it was usual to stipulate that knees should be supplied in the ratio of four loads to every hundred loads of oak timber, and in the ratio of six or seven loads to every hundred loads of elm.

British Admiralty findings from 1771 indicate the problem. 'The average annual consumption of the preceding twenty years was shown to be 22,000 loads of oak timber.'[7] By 1800 this annual consumption by the British navy alone had risen to 30,000 loads. (A 'load' was the equivalent of fifty cubic feet. A rough general guide was that the average oak-tree fit for the purpose contained about a 'load' of timber and made nearly a ton of shipping.)

By 1812 it was calculated that the British navy would require the timber of 1000 acres to maintain the navy for one year. Based on that premise it was estimated that if 100,000 acres were planted and managed so that the timber on each 1000 could be felled in successive years, it would, at the then establishment, maintain the naval requirements of Britain indefinitely. The continuous demand for and activity in timber and timber-growing naturally led to the development of important new timber management techniques.

Wooden warships changed very little in design during the two hundred years of their sea dominance. The limitations of timber controlled developments in size, for instance, and, to a great extent, dictated a design which altered little from 1600 onwards. Unlike steel warships of today which are obsolete almost as soon as they have been launched, the effective life of a wooden ship at the beginning of this period was calculated at about thirty years. *Royal William* (90 guns) built in 1719, lasted in service nearly a century without extensive repairs. But by 1800 the life expectancy of a wooden ship had fallen to eight years, and to a naval officer in 1815 is attributed the observation: 'There is no duration.'

But the enemy was not equipped with better ships or guns; the enemy was a far more formidable destroyer of wooden warships than any naval foe, dry rot. 'Different kinds of timber – even different kinds of oak – did not go well together in the same ship. A vessel originally built of English oak, which came in for a small repair and received a dozen knees of Canadian oak, some upper planks of Baltic oak, and a beam of Riga fir, with a few pieces of green inferior English or Irish oak in other parts of her anatomy, was almost bound to decay quickly.'[8] It 'sometimes pulverised whole fleets... It was a baffling enemy, and... dry rot remained unmasted to the end.'[9]

Much timber was also lost by the policy of the English Commissioners of the navy to have three years' supply of timber in the dockyards. In theory this seemed a good idea. In practice it resulted in excessive waste through decay. Timber was generally delivered to the dockyards in the round and was often stored in such a manner as to encourage rapid degeneration.

By the beginning of the nineteenth century 'boiling' or 'steaming' timber so that it might be more easily bent, particularly for ship-building, was common. Transportation was a constant problem and it became customary to convert trees to planking in the woods themselves, where they were sawn immediately after felling and barking, thus producing a considerable saving both of timber and in labour, as the wood was softer when green. It also meant a substantial saving in transport costs as the volume of timber to be hauled to the dockyards and elsewhere was much reduced. The question of furnishing adequate supplies of timber for shipbuilding exercised timber managements concerned about maintaining a sustained yield in the face of increasing demand – not alone because of difficulties of obtaining sufficient quantities, but also (since the size to which ships could be built was limited to the size of the trees providing the timber) of finding the correct shape and size. But the day of the wooden ship was entering its twilight.

An indication of the problems that could arise during the building of huge timber ships, and of the wastage that ensued, may be had from the case of *Queen Charlotte*, a monstrous ship of 110 guns. She was laid down in 1810. By the time she was completed she was so rotten that she had to be entirely rebuilt before being commissioned. While she was being fitted out it was discovered that some of the Canadian oak and pitch pine was in an advanced stage of decay 'apparently due to the fact that in an endeavour to hasten the seasoning of the ship's timbers stoves had been placed in different parts of the hold after it had been finished. Nothing could have been calculated to encourage the growth and spread of fungi more than this and above the waterline she became almost completely rotten.'[10] Originally *Queen Charlotte* was to have cost £88,542, but by 1859 no less than an additional £287,837 had been spent on her in repairs.

Between 1778 and 1800 515 ships with a total tonnage of 23,550 tons, representing a mean of 46 tons, are known to have been built in Ireland. This is insignificant compared with shipbuilding in England where, between 1730 and 1787, the naval dockyards alone used 37,000 tons of timber a year.[11] Merchant shipping consumed even more timber. The East India Company alone increased its tonnage from 45,000 tons to 79,900 tons in the five years from 1771 to 1776.

'The apparently inexhaustible extent of the Irish woods led to the reckless destruction of timber,' states T. P. Le Fanu.[12] But there was, as we have seen, a military policy of deforestation operating as well. Combined with commercial interests, this, to some extent frustrated shipping requirements. There was also constant conflict between those who required Irish timber for the king's shipping and those who wanted it for their own profit.

As an inducement to colonial shipbuilders in the north it was stated that fir poles from Scotland were easily obtainable. In 1608, according to the Calendar of State Papers,[13] Sir Arthur Chichester reported that the woods of Shillelagh were sufficient for the king's ships for twenty years. Cottingham found an abundant supply of timber for shipping in Munster, and, it will be recalled, the same Philip Cottingham of

London, received '£71.3s.4d for hewing and carriage of timber for shipping in the woods of Kilbarro and Kilcorran in Waterford'. The East India Company, as we have noted, established shipbuilding at Dundaniel in Cork, where, in 1613, they built ships of four hundred and five hundred tons burden 'and craved protection against all ill-disposed Irish'.[14] Chichester also commented: 'The Irish build very good ships ... many of our English merchants choose to build here, for foreign trade especially; their oak is very good and they have a very good store of it.'

Of course forest exploitation by colonial interests occurred elsewhere, often on an enormous scale. It was one of the major considerations in the English colonization of North America. But Ireland appears to have been the only significant western European forested country where exploitation was policy. The English historian Arthur Innes wrote: 'There is no evading the fact that the English, who could wax hot enough over the cruelties of the Spaniards in America or in Holland, did without compunction or any sense of inconsistency regard the Irish not even as mere human savages but as wild beasts... English methods, as usual in Ireland, promptly degenerated into massacre and devastation.'[15]

Although Irish timber was a major contributor to the development of English shipping and shipbuilding, shipbuilding in Ireland differed greatly from that in England. Forest exploitation affected it significantly from the seventeenth century onwards and, as with so much else, it was strictly curtailed in order to prevent competition with English produce. There was some commercial shipbuilding, but no naval yards other than some repair yards. There is, of course, an ancient Irish tradition of seafaring, but in historical times, with some rare exceptions, there is little evidence of a seafaring continuum on any large scale, certainly not in the proportions one would expect from an island people.

After the Tudor Conquest in 1601, and the plantations, the bulk of the population were concerned principally with survival. Most of the natural leadership left the country within a century. Development of maritime endeavour was inhibited by the English administration and, increasingly as their normal requirements became greater, the English enlisted – by press gang and otherwise – considerable numbers of men from fishing villages for their navies.[16] Fishing activity, which sometimes involved sizeable fleets, continued, but nothing that even began to compare with what was happening in England.

Down to the end of the seventeenth century the normal craft of Irish inland waters was the dugout canoe, but because of the wood-famine experienced by the ordinary people of the country the dugout canoe became extinct in Ireland at an earlier date than it did in other parts of Europe.[17]

These dug-outs were known as cots and were found all over the country. They were up to fifty feet long, but were generally between twenty and thirty feet. They carried approximately ten men and were purpose built, as a general rule. It is not feasible to imagine one or two men, who must most often have used them, handling cumbersome dug-out craft of thirty feet or more. The larger ones were for transporting goods and livestock and were very common. Cots capable of carrying more than sixty people were used for transporting horses, cattle and timber on the river. Up to twenty tons of timber could be carried in one of these. As timber became scarce the use of these vessels declined because of the absence of the large trees required to make them and they gave way to more conventional planked river boats or, in some cases, canvas covered craft. 'Cot' became a designation in Ireland to denote several kinds of flat-bottomed vessel, from a small, one-man skiff-type to much larger vessels capable of carrying several tons of cargo.

The difficulty in deciding how much shipbuilding was carried on in Ireland is that although it is possible to find records of the tonnage belonging to Ireland or to specific ports these lists give no indication as to the port of origin. Altho᾿ a ship was listed as Irish that does not say that it was built in Ireland.[18]

In the 1770s Dublin shipbuilders (mainly four families, the Kinchs, the Murphys, the Kehoes and the Cardiffs) were faced with circumstances not unknown today. In 1778 their employees demanded a daily wage of three shillings to bring them into line with the wages of English shipyard workers. They threatened to emigrate if their request was refused.[19]

The Connemara hooker (the *bád mór*) is a survivor of a type of sturdy multi-purpose 12-15-ton vessel, which with local variations once flourished along the western and southern coasts as far as Kinsale. The Connemara version, and its smaller cousins the *gleoiteog* (little pet) and the *púcán*, are all that survive, and are now almost entirely pleasure boats and showpieces. They were a product of nineteenth-century trade development, carrying materials from major centres – Cork, Galway, Limerick – to outlying areas too difficult of access, or too costly, by road.

Arklow, like many other centres around the coast, is one location where there has been a tradition of shipping and shipbuilding on a small scale for hundreds of years. The industry waxed and waned from time to time. Wakefield wrote of the vale of Arklow that 'the extent of the woods induced me almost to imagine I was in the midst of one of those immense forests seen only on the continent'.

In 1893, in the early days of the Congested Districts Boards, the O'Donnells turned out two large hookers, *Sophia* and *Star of Murrisk*, for the development of trawling in Clew Bay, Co. Mayo. These sisters measured 40′1″ × 11′4″ × 6′8″

depth and were built for £100 each, a figure that would not pay for a jib sail today'. All of these boats were 'built of massive construction in 1" larch planking, spiked on 6 × 3 inch sawn oak frames, centred 12 inches amidships, and closer forward. An oak or beech keel, about two-thirds overall length of the hull, provided the back-bone to the framework...[20]

A good example of a family business with a name which has become part of a tradition (within the local tradition of shipbuilding) is the John Tyrrell Ltd shipyard in Arklow, founded in 1864. The family had been fishermen and seafarers for generations prior to that and, indeed, the town of Arklow itself developed on sea-trading, with which many of its long established families are associated.

Oak and larch, mostly native, were the principal woods involved. But as in the Claddagh and Renvyle, it was usually brought in by sea. 'Native timber was seldom used exclusively in the boat building industry in Arklow.'[21]

When the firm was first established there was a lot of boat- and shipbuilding in Arklow, mainly small schooners. But long before the nineteenth century Arklow had a substantial fishing fleet. The shipyards developed from cargo-carrying in the early part of the nineteenth century, but shipbuilding on a smaller scale goes back to at least 1715, perhaps earlier. Records, in complete and minute detail, discovered by the late Mr Jack Tyrrell, provide a list, believed to be unique in these islands, of continual shipbuilding from the early eighteenth century. 'There was an identifiable type of fishing vessel peculiar to the Arklow region, as there were identifiable hookers and *púcáns* which were identified with the Connemara region and other types of hooker which were found elsewhere.'

Arklow vessels were generally 40 to 45 feet long, and there was a good deal of coastal trade 'because the roads were bad and the main access around the country was by sea in small ships'. Waterford, Wexford and Arklow had very substantial involvement in the cargo-carrying trade in small ships around the coast and to Britain.

What really developed the shipping industry through Arklow was the export of ore from the copper mines, in particular copper ore residue, sulphur ore. Curiously its worth – and consequently the development of this cargo – apparently was discovered by accident. 'It lay in heaps beside the copper mines and some Arklow captains used it for ballast on the Bristol run. There it was recognized for what it was'. Thus began a trade which lasted for the best part of half a century, stimulating shipping and shipbuilding, which in turn caused a greater demand for timber. The first shipyards arose on the river bank near the closest available timber. Professional timbermen, using horse and wood-carts, extracted and prepared the timber, mainly oak and larch with some fir. Supplies from places such as Ballyarthur, the nearest sizeable wood between Arklow and Avoca, were substantial. The landowner, named Baily, had an established system of replanting, so that stocks were kept up. Ballyarthur wood was mainly renowned for the quality of its larch. There were also, of course, the great oak forests of Shillelagh and Coolattin. Tyrrells bought all the crooked pieces of oak they could, the bigger the better. 'They used to train the trees into a certain shape to provide these bends and crooks.'

The men who went out to the wood and cut down the trees, loaded them onto their carts and brought them in, were called 'timber-men' and were expert at handling the material. They were interesting and exclusive:

Foresters were completely remote; they had very little social life. Their work in forestry was – indeed is – seldom appreciated in the local community. He (the forester) stood out removed from the local people and the land (under forestry or being afforested) was not considered to be in use. If the land being used for forestry is considered to have (agricultural) potential the fact is resented, if it is not considered to have potential the work is considered to be speculative and unprofitable. Invariably they were small, hardy men but what they could do – well! It was a case of give me a fulcrum and I'll shift the world.

A hand spike and a plank were, apparently, their main tools. The timber was handled on an ordinary horse cart with a flat base, but it was a bit higher than normal in that the wheels did not come over the top of the sides. The carts were very strongly made and just behind the horse there was an iron hoop the purpose of which was to keep long logs away from the horse. They brought two, sometimes three – or perhaps only one if it was a 'heavy stick' – angled across the cart. 'If it was oak they were handling they would get up on the tree if possible and lop off the limbs in suitable ways. They knew as much as shipwrights did as to what curves were wanted. They would save the curved boughs and the shapes in them and smaller curves were saved for making smaller knees and that sort of thing.'

Tyrrell cites an example of their work from 1926, resulting from the loss of the main mast of *Invermore*, a family vessel. There is no reason to suppose that the work and effort involved differed materially from what it would have been over the previous couple of centuries. The master of the ship, Tyrrell's uncle, ran her into Wicklow and sent a telegram that he wanted a new mast.

My father and I went out to Carrisfort estate on the south side of the River Glenart. We got hold of Lar McDonagh who was the timber man at that time and his two sons. They found a large tree about 64 feet high and it was 12 inches in diameter at the top, but it was up on Killcarra Hill, half way between Arklow and Woodenbridge, right at the very worst bend in it. There is a little road going up and that is where the larch had to come down. Two of them got it down and into Arklow in the evening of one day. Within the week it was ready and brought by road to Wicklow as a finished mast – a week's work from start to finish.

Tyrrell points out that roots were also used in boat-building and he mentioned one timber-man who habitually supplied them, Miley Byrne. He recalls one vessel, *The Seagull*, built about 1880, the frames of which were all oak roots supplied by Byrne:

> Ballyarthur is all eroded now, of course, and is not being replanted. Avondale, too, supplied a good deal of timber, larch in particular. Parnell had big tracts of timber and he was very much in evidence around this area. My grandfather bought from Parnell and they were good friends. Parnell was a hard-headed businessman when it came to selling. Interestingly he had an invention of his own, a calipers for measuring the diameter of the tree mounted on a long pole so that he could take the measurement half way up to get the correct average.

Parnell sometimes visited the Tyrrell shipyard. On one occasion John Tyrrell asked him to come and 'christen' a new vessel. Parnell enquired if Tyrrell had a name for her and the response came that the ship might be called *United Ireland*. Parnell's reply was 'that's one thing you or I will never see'. Accordingly the ship was named *Irish Leader*. It was a well-known boat for some thirty years on the east coast and was then sold to Galway from where she fished for many years.

Beech is a relative newcomer, having been introduced to Ireland in the eighteenth century, and was used exclusively for keels. 'Normally neither beech nor elm is a long-lasting timber, but they are dense timbers which can take with great tenacity the driving of a number of bolts along their entire length without splitting to hold the frames. Moreover continually submerged in salt water they last forever.'

In one authenticated case a thirty-year-old boat hauled up for repairs required that a portion of the keel be cut out. This proved almost impossible since chisels were 'bouncing off it'. It was the original beech keel.

Local timber which was used would have been, perhaps, a hundred to a hundred and fifty years old. It was bought standing and was left to season naturally in the round. Until the latter part of the nineteenth century Arklow bay extended well up into the valley; Lord Wicklow maintained that if he had dug down a few feet at Shelton he would have found oyster shells. Larch was the only reasonably good timber for masts, but it had the characteristic of twisting.

> We couldn't season masts very well because there was always somebody looking for a new mast and consequently there was seldom time to season them. Often large masts were not seasoned and you could see from a distance whether you had a larch mast or not because it twisted away yearly. The cracks which had been longitudinal, straight up and down, became spiral. It became quite noticeable particularly when a top mast was fitted and eventually it began to move until it went out over the starboard bow.

To build an eighty-foot fishing boat required about sixty tons of oak and forty to fifty tons of larch. Decks were of imported timber.

The oaks required were of the order of fifty cubic feet each and the larches about thirty-five cubic feet. Shipwrights preferred larger oak. It was denser and more economical to use, but scarcer.

A. Dobbs, in his *Observations of the Trade and Improvements of Ireland*, published in 1729, indicates the dearth of softwoods and conifers in Ireland at the time:

Imports (from the Baltic) consist principally of deal boards, timber of all sorts which we cannot do without. As for our importation of wood I am afraid we shall not in a long time, if ever, save in that article, even should we plant, to which we seem generally to have so great a disinclination, for as we increase and improve our demands for it will still be increasing, and when Norway and the Baltic fail, we must look out for another market to buy at a greater expense.

The scarcity Dobbs refers to may have influenced the later decision to plant conifer. The extraordinary consumption of timber by England, and the effects of its demand on this country, may also be noted from the fact that much of the timber that was imported from abroad to Ireland was re-exported to England, in addition to that already being taken from Irish forests for the British market.

By the late 1880s or early 1890s there was a substantial trade in imported timber for shipbuilding. The switch to imported timbers was not only the result of shortage of Irish timber. Imported timber provided greater variety, considered important in boat-building. At that time the principal timbers used for marine building in Ireland were oak, beech and larch. Douglas fir was used to some extent, but it was not a favoured timber since it was not long-lasting.[22]

At that time imported timber, besides being plentiful and varied, was also held to be cheaper than Irish timber. 'But that is no longer the case. It was competitive enough at the end of the last century and well into this century. Now, while Irish timber is not cheap, imported timber is gone out of all proportion,' said Tyrrell.

Larch became scarce for shipping in the early 1960s. An African timber, Iroco, was substituted and was used for the last wooden vessels built by the firm, amongst them the training ship, *Asgard*, with frames of local oak which came from between Arklow and Woodenbridge. The deck beams are of the same oak. 'Teak that we bought for the decks of *Asgard* cost £27.10 a cubic foot and there was over £10,000 worth of teak in her deck. Today that timber would be unobtainable.' Tyrrell continued: 'Even at that it was extremely difficult to get decent trees of about 20-21 feet free of blemish.' This is attributed to lack of regeneration. 'Of course there is no regeneration in Shillelagh or Coolattin, so far as the oaks are concerned, but there was a good deal of timber there which couldn't be got at at that time. The estate hadn't been sold then. Now things are different.'

During the early nineteenth century the import of North American timber, particularly what was called 'pitch pine', developed enormously and it became the standard timber for building commercial vessels for fishing and trading.

I remember going with my grandfather to select logs as a small boy in T & C Martin's before the First World War. We went down to the yard where a Mr O'Kelly was the manager. My grandfather would get the workers to turn over maybe 100 logs before he found the one he wanted. He usually bought about six logs at a time and it would all be perfect timber. The pitch pine was in squared logs and came from Dublin to Arklow by rail. The supply of commercial timber for boat building is sadly diminishing in this country because 90 per cent. of the trees being replanted are fast growing soft-woods and they are not much use for structural work of any kind, particularly for marine craft. But, of course, there isn't the same demand for timber for marine purposes now.

In spite of industrial development, including shipbuilding, by the end of the eighteenth century little that was new had been undertaken in Ireland in terms of forest management and much of the English government's efforts to extract timber from Ireland for shipbuilding purposes were frustrated by local profiteers.

James makes the interesting comment:

There is little evidence that any practical steps were taken in the management of woodland areas and, in retrospect, one can appreciate the reason for this. In the first place there was little knowledge of the subject, largely due to the fact that there had never been any need for it. Second, that as far as the royal forests were concerned, there were so many rights, restrictions and privileges in or over them which had become established over a long period of time, that any attempt at effective management was seriously prejudiced. Third, that since private woodlands were so fragmented, any overall plan of management was faced with major difficulties and last, it is doubtful if those concerned had any clear idea as to what the objects of management should be.[23]

He refers, of course, to the situation in England.

There was a significant difference between France and England in respect of law and management of forests in the national interest. Corruption was nowhere near as widespread in France and management technology proceeded systematically most of the time on a planned, national basis. In England, in spite of forest laws and premonitory management, forestry consisted merely in felling woods and in periodic spasms of feverish planting when supplies of timber ran short (during and after wars especially). The contrast between France and England in this regard is all the more startling in view of England's island dependency on timber and its clear recognition of the fact. This inefficiency reached its nadir in Ireland where no constructive policy existed at all. If England was the European laggard in forest management and development, Ireland was surely the neglected and exploited *terra incognita*.

CHAPTER SEVEN

Estate Forestry in the Eighteenth and Nineteenth Centuries

'The most striking thing on a first sight of the Irish landscape is the total absence of trees of any kind. They are only seen in private parks' – Paschal Grousset, *Ireland's Disease*, 1887.

THE DESCENDANTS OF THE PLANTERS who had effectively deforested Ireland often inherited large estates of confiscated and treeless land. It is often propounded that the country was once adorned with a magnificent cloak of noble high forest despoiled for profit and his ships by the foreign invader. It is an argument that is both simplisitic and inaccurate. It is equally unreasonable to argue, as Forbes, amongst others, appears to do, that well-forested England had beside it a land (more suitable to the natural growth of timber), which, somehow, failed to produce trees in any significant measure except scrub. There is no doubt that there were extensive forests in Ireland before 1600, and there is no doubt that these were gone by 1800. It was an extraordinary decline in a matter of 200 years, and included the startling decrease in hazel from 30 or 40 per cent of the overall timber volume to less than 5 per cent in a matter of forty years.

The facts, for which there is ample evidence, show that timber and woodlands clothed much of the country from before historic times until well into the seventeenth century. These forests clearly went somewhere, and not of their own accord.

By the beginning of the eighteenth century planter landowners were sufficiently secure and confident to want to improve their denuded lands. Some far-seeing individuals appear also to have begun to realize the importance of applying some principles of forest management to their new woodlands. On that haphazard basis they began a programme of regulated timber management. It is appropriate to call this development 'Estate Afforestation', both because it is an accurate description and in order to distinguish it from State Afforestation which, at the beginning of the twentieth century, laid the practical foundations for the major national enterprise of today.

Universally the importance of drainage in woodlands was recognized and the uses and requirements of different species began to be taken into account – for instance, that while beech reproduces from self-sown seeds and requires little attention from the point of view of regeneration, in contrast to most other species it requires to be constantly thinned for the first twenty or thirty years.

Landowners started to spend money on their estates; trees were planted; the need to know more about seeding and the growing and planting of trees was prevalent. Almost thirty books concerned with woodlands and timber-growing were published in England in the eighteenth century alone. In 1745 the Society for the Encouragement of Arts and Manufactures, providing medals and other rewards for the successful establishment of young plantations of exceptional merit, was founded in London, and it attracted the interest of men of the highest influence. A Mr R. Fenwick of Northumberland was awarded a silver medal for planting 102,000 Scots pine in 1764 and two years later, when he put down a further 100,000 two-year trees, he was awarded a gold medal. Interestingly one Jonathan Blackhouse who, in 1812, planted 271,000 larch, oak, spruce and Scots pine, was an advocate of ploughing peat earth before planting 'as this broke up the ground and helped to drain it'.[1] In 1820 1,981,065 trees were planted on 550 acres of wasteland on the Duke of Devonshire's estates in Cumberland; the Duke was given a gold medal.

The Dublin Society (later the Royal Dublin Society) was formed in 1731 to promote and encourage industrial and agricultural initiative and became the Irish equivalent of the arboricultural societies of England and later Scotland, co-operating with them in the promulgation and fostering of forestry in general and reafforestation in particular. Through the Dublin Society and its landowner members who, for all practical purposes, were the only ones other than merchants and entrepreneurs in any way concerned, these ideas and attitudes came to Ireland as well. Between 1766 and 1806, 25,000,000 trees were planted due to the encouragement of the Society, which paid £90,000 in awards and medals.

The premium system to encourage planting was introduced in Ireland by the Rev. Samuel Madden, a founder member of the Dublin Society, in 1739. In 1741 £10 was offered for the greatest number of timber trees raised in nurseries, and in 1742 a plate worth £50 was also offered for the greatest quantity of trees in hedgerows. The money was subscribed by members, the prizes being self-evidently confined to landowners, probably to members.

In 1747 the Society received its first grant of £500, from the Privy Purse, and in 1761 the Irish parliament voted it £2000. Some of this may have gone towards planting, but most went in aid of manufacturing industries which were, of course, great consumers of timber. In 1765 planting premiums and medals were introduced on a more or less regular basis in 1765 for oak, ash, elm, weymouth pine and 'Scotch fir' – £20, £15 and £10 to those who had the greatest number of oaks, not less than 160 per acre, and still in a thriving condition, seven years after the medal had been claimed. A similar scheme was soon provided for smaller landholders, and medals valued at £5 were introduced for renters of land. This was later increased to between £10 and £20. In 1780 £224 was offered in sums of £5 to £20 for propagating forest trees. Under Grattan's parliament, premiums were increased and the system was expanded. In 1783 a premium of £40 was offered

to each person enclosing any quantity of ground, not less than ten acres, with a sufficient fence proof against cattle, and planting the same with a number of oak, ash, beech, elm, maple, sycamore, chestnut, larch, fir or pine, not less than two thousand plants on each acre. The nature of the fence and preparation of the ground must be fully stated in the claim, and security will be required in double the amount of the premium that the said fence and trees shall be preserved, and such trees as shall die, be continually replaced so as to keep up the number of one thousand trees in each acre for ten years. For this premium the sum of £400 will be granted; but if there should be more than ten claimants the sum of £400 will be rateably divided.

Demonstrably this was an exclusive facility which applied only to quite substantial landholders able to afford the considerable security involved.

In 1786 premiums of over £10 were converted to 'the full value of the premium in utensils of agriculture and planting ... made at the Society's manufactory in Poolbeg street...' Ten years later those receiving premiums were required to give security in a bond of £100 that fences 'would be preserved and trees kept at 500 oak and 500 other trees on each acre for ten years'. By 1800 a £100 premium required a written contract with a nurseryman to furnish and plant 'so as to leave at the end of the third year 8000 good growing trees on each acre', of which 3000 were to be of oak and 3000 of larch. A 'sufficient fence' was to be maintained round the plantation for ten years. The shipping – presumably naval – requirement underlying these requirements is clear.

The difficulty of obtaining a sufficient supply of seedlings or young plants was early recognized and in 1765 it was announced that 'the

Society will pay for every person who shall first keep a well-inclosed nursery of forest trees within three miles of ... a county town ... a yearly rent of 20s. per acre for three years'. Trees were to be of two years' growth and the rent had an upper limit of £3 – i.e. nurseries were limited either as to size, three acres, or as to the amount of grant.

Nursery premiums continued for many years. During the forty years that these schemes lasted £12,460.13s.11d was awarded under them. Another £6000 was awarded for planting sallows, willows, and poplars for basket-weaving, involving a total area recorded of 2800 acres, including 108 nurseries.

The premiums were not immune from fraud, and the Society's charter of 1750 empowered the chairman to administer an oath to anyone in order to discover the 'truth or value of anything offered or proposed to the said corporation'. Frauds discovered in 1782 resulted in the requirement that claims be certified by a grand jury foreman. Other conditions followed, not least being that false affidavits were, like perjury in 'other' – presumably civil – cases, punishable. The perpetrators would be debarred from future participation in the Society's schemes and were open to prosecution.

In 1787 an inspector was employed to assess claims at the rate of 'three half-crowns a day whilst on the circuit'. By 1808 the inspector had multiplied to 'proper persons'. But the premiums ended in the same year when it was

> resolved that, in consequence of information having been received that persons who had obtained premiums for planting trees have not preserved the trees according to their engagements together with some suspicious circumstances which throw a doubt on the truth of the certificates and which magistrates are requested to observe and examine before they subscribe, the planting premiums be discontinued until the Society receive reports of the persons to be employed in taking a view of the several plantations for which premiums have been granted.

Apparently the proper persons were not so proper. The Society later noted that 'it looks as if the prevalence of fraud was the chief reason for discontinuing the premiums'.

In 1801 it was stated that farmers complained that the premiums of the Society did not 'extend to smaller spaces than ten acres' as they could not always spare ten acres in one place for planting without encroaching on their best meadows or pasture ground. And, a forecast of enduring attitudes, 'if permission were granted ... to smaller tenants ... it might have the effect of giving the lower class a love for trees and be a greater means of preventing their wanton depredations on them than any penalties'.

But although the Royal Dublin Society was associated with the Society of Arts and had been actively encouraging planting since 1775, its efforts were too little, too late and directed to the wrong quarter. 'Estate Forestry' in Ireland stood on the twin planks, if one may so express it, of dilettantism and cosmetic planting. Exploitation had left devastation too great to be repaired by private means. After the Act of Union, parliament reduced the votes-in-aid to the Society so much that not even the educational and scientific institutions the Society had founded could be maintained, much less planting and manufacturing premiums, 'and all attempts at fostering industries by monetary assistance of necessity ceased'.

Mr Hely Dutton, in a survey of Clare in 1808, remarked, *inter alia*,

giving a premium for oak without limiting, or at least advising the proper soil is so much money thrown away; for some of the plantations I have seen are upon dry, rocky, shallow hills, where larch would have been infinitely more valuable... What a reproach that... whilst the whole county in twenty-five years has had only ninety-six acres planted, an individual in Scotland has in fifteen years planted 3005.5 acres.

The 'planted' north-east appeared to take little interest in forestry. 'The [premium] system seems to have done a great deal indirectly, and by force of example. It did not, however, lead to the creation of a single plantation on a really large scale, and it can scarcely be claimed that it promoted forestry in the proper sense of the term.'[2]

A Board of Agriculture was established in England in 1793. Surveys carried out shortly afterwards produced a report from each county, each report including a chapter on woodlands. At this time forest plantations, in accordance with established usage and requirement, were almost exclusively devoted to the growing of oak and the planting of acorns. But as theory in plantation management developed, together with an awareness of the uses to which other species might be put, ash, elm, beech, sycamore, Scots pine, larch and some other conifers were also promoted. Because of its hardiness and rate of growth, combined with its attractive appearance, larch soon became a popular investment.

From plantation management to nurseries was a natural step and by the early nineteenth century a number of commercial nurseries flourished. 'Tree-farming' had become an industry. Moreover enthusiasm, interest and accumulated knowledge were, by this time, beginning to have an effect on woodland management. Following developments in France and Germany, whence much of the interest, skills and knowledge stemmed, the principle of sustained yield was well-understood, as was the need for maintaining a good growing stock of stools. As a natural corollary to forest activity the role of woodsman as a traditional or hereditary activity had also developed from the mid-seventeenth century onwards, especially after the rise of 'Estate Forestry', and was sometimes associated with the role of gamekeeper.

Even with the best statistics possible, it is difficult to appreciate how all-pervading was wood in the lives of a society lacking plastic, paper bags, wire, tinfoil, reinforced concrete and steel girders, to name but a few modern everyday commodities substituting for timber products. Raffia bags and cane baskets were used for dry commodities, but liquids were for the most part kept in wooden buckets, casks and mugs on a scale impossible to calculate.

The cost of transporting timber overland was exceptionally high and this explains why much inland timber was used for ironworks and, moreover, why it was more convenient for an ironmaster to take his ironworks to the timber rather than the timber to his ironworks. It was reckoned that the cost of transporting oak any distance was equal to the cost of the timber and a haul of over twenty miles raised the price to uneconomic levels.[3]

In both Ireland and England timber continued to be sawn by hand long after it was mechanically sawn on the Continent. Although sawmills operated by wind or by water had been introduced on the Continent during the fifteenth century, an effort to introduce mechanical sawing in London in 1663 'was frustrated by a mob of artisans who feared for their livelihoood and in 1768 a similar mob destroyed a sawmill set up near Limehouse'.[4] In Ireland, as in England, pit-sawing was so prohibitively expensive that it became cheaper to import sawn deals from the Baltic than to cut them at home, even though the cost of transport from the Baltic was greater than the cost of the timber.

Industrial unrest in the timber industry in the eighteenth century was by no means confined to England.

In 1776 some Dublin timber merchants advertised for sawyers and lath splitters from any part of the British Isles (sic) to work in deal yards; and their advertisement was immediately countered by a declaration in the Press by Dublin sawyers that there was no shortage of workers in the city, that most of the employers paid one shilling and 3 pence (1/3d) a dozen for deal cutting and that the merchants who were advertising for new workers only wanted to pay 1/= a dozen and had dismissed employees 'without due cause'. The coopers of Limerick complained in 1769 of a reduction in piece-work rates... demanding a return to the prices of 1765, which had been '2d in each piece of work under the prices of Dublin, Cork and Waterford.[5]

There were, says McCracken, at least 170 timber merchants in Dublin between 1750 and 1800. She presents some interesting statistics on timber merchants, derived from *Wilson's Dublin Directories*. Many of the timber merchants at this time combined the sale of timber with another occupation, usually a related one such as carpentry or coopering, architecture or building. Some timber merchants were women who continued to run the business after their husbands had died and one of them, Bridget Moss, was one of the longest active merchants of all. She took over the firm when her husband died in 1758 and ran the timber yard in Golden Lane until 1786.

Timber measuring was an important trade. One Levi Hodgson of Poolbeg street, a son of Daniel Hodgson who had been appointed City Measurer in 1753, produced a work called *The Compleat Measurer* as a guide for buying and measuring timber. It was so successful that by 1801 it had run to ten editions. It contained elaborate tables showing the costs of different sizes of plank and the various timber items used in house construction, as well as advice on measuring wood and buying and selling. According to Hodgson the most important point to note in buying wood was to see 'that bark, sap and defects such as knots, shakes, etc., were allowed for' except when measuring for freight.

It is self-evident that the planting of trees will not be undertaken until the political situation in a country is reasonably stable, so that he who sows will reap. Furthermore, until it becomes clear that nature's bounty is either failing or inadequate, there is no incentive to plant a crop that cannot be harvested for 100 years.[6]

That statement underlies not only much of the policy and attitudes towards forestry of one hundred years ago, but also today when the life-time of a government is measured in terms of four or five years. It required great courage and commitment to plan and legislate for an undertaking not only bereft of immediate political advantage, but to which there is, even yet, some hostility[7] and which, certainly at the level of state enterprise, demanded a planned commitment to a flexible marketing dynamic at some indeterminate future time.

Most early estate – or any other – planting before the end of the seventeenth century was for shelter belts, orchards and avenues. As McCracken puts it, they tended 'to follow the flag', by which she means they were first introduced in the settled areas around Dublin and Cork. By 1801 there were 132,000 acres of plantations, largely as a result of the encouragement of the Royal Dublin Society. By 1845 this had risen to 345,000 acres and the planting of conifers had begun to outstrip the combined numbers of hardwoods.

Concurrently with the advances in forestry between 1739 and 1806, the Irish parliament developed a policy of fostering and encouraging trade, industry and fisheries, thus providing remunerative employment for an increasing population. However – perhaps ironically – it tended to 'fix the habit of sub-dividing farms, the abundant employment provided preventing it from being unduly noticed'.[8]

The irony was that as the natural resource represented by Irish woodlands and forests declined, timber imports increased. While in some 'Estate Forests' the first tentative steps were being taken towards regulation and sustained yield, such activities came to be seen by the deprived peasantry and smallholders not as a means of reviving a natural, national capital investment, but as yet another means of depriving them of the use of land rightfully theirs. Not unnaturally this helped to entrench an enduring hostility on the part of the smallholder to afforestation. They saw trees being planted on 'their' land. Small wonder they uprooted plantations.

By the beginning of the nineteenth century a timber-growing as well as a timber-marketing industry existed. Conflicts of interest inevitably arose in what was now a commercial enterprise organized on a substantial, if not yet national, scale. For instance while oak could be profitably marketed when it was fifty to sixty years old, it would not be suitable for shipbuilding purposes, particularly for the navy, until it was eighty to a hundred years old. Dr Watson, who was both a woodland owner and Bishop of Llandaf (winner of the first Society for the Encouragement of the Arts planting medal in 1788), observed that while 'large trees sell for more per foot than small ones do, yet the usual increase in price is not a compensation to the proprietor for letting his timber stand to a great age'.[9] Woodland owners were consequently reluctant to hold their oak for sale to the navy because of low prices offered by the shipyards, taking into account the length of time that an owner had to wait until his tree reached the required size – particularly when he could market it profitably without doing so.

Most progress in woodland management in England was the result of private enterprise. At the beginning of the nineteenth century, while woodland management in private hands was reasonably well geared to the requirements of the community, the same was not true of the royal forests and woods belonging to the Crown. The management of these was as inadequate and unproductive as the other was productive and adequate.

The idea of forestry as a managed enterprise emerged in Ireland towards the end of the eighteenth century. Under the Act of Union (1800) the Forest Laws of England could also be said to apply constitutionally to Ireland. But instead of responsible forest management, the English government promptly reduced the grants to the one organization encouraging planting and management in Ireland, namely the RDS.

The Act of Union purported to link a rich country with one that was artificially poor as a result of trade restrictions, exploitation and repression by the former 'protecting' its own interests. 'It shifted the centre of gravity from Dublin, the capital of a poor country, to London, the wealthy capital of a wealthy country.'[10] One of the unforeseen consequences was that London now became the capital; landlords absented themselves from their Irish estates, and tried to ape the style and standards of the wealthy neighbour. 'Thus, not alone could no part of the annual rent be allocated to improvements, but enough of the same rent could not be screwed out of the land to satisfy their illogical scale of living.'

Instead of prospering, the country became steadily more impoverished as capital in the form of rents drained from it. Nothing remained for improvement, and industrial progress never took place. An accumulation of destructive economic measures ensured the degraded impoverishment of the whole country.

In *The Commercial Policy of the Irish Parliament* Binchy demonstrates that the legislation of the English parliament in respect of Ireland between the Restoration and the Union was not only self-protecting and unscrupulous so far as this country was concerned, but actually unconstitutional. The Act of Union, which infringed the rights of the nation, also had social effects, mostly disastrous, some beneficial, all of them far-reaching. The Penal Laws, which had resulted the economic destruction of the nation, were ameliorated and subsequently repealed and the destruction of the smaller industries and stable trades through unconstitutional legislation was gradually curtailed. But the damage had already been done.

One of the major consequences of the decades of persecution and of relentlessly pursued trade restrictions that debarred the people from any trade which might compete with that of England, was that the social and economic dynamics of the major portion of the populace centred upon land to a wholly disproportionate and unwholesome extent, in spite of the fact that tenants, who were by far the majority living on the land, had no security of tenure. Nonetheless land, and their tenuous relationship with it, was often the only security they knew. Their resentment of and hostility towards trees planted on land they rightly coveted is understandable.

The 1793 grant of elective franchise affected land and agriculture and those who worked it in several ways. The number of small holdings and the amount of tillage increased and the number of systematic evictions decreased. But the demands and the oppressions continued. 1793 also marks the beginning of the sometimes violent, but mainly constitutional, movement which was eventually – after about a century – to lead to the people recovering their right to ownership of land.

The dispossessions and plantations, of course, played an enormous part in the formation of the economic structure of the country between 1600 and 1800. Almost without exception the landowners and exploiters of woodlands, the ironmasters, and indeed many of the woodworkers and foundry-workers, were English or of English origin.

In 1810 the Commissioners of Woods and Forests in England, charged with the management of forest and land revenues of the

Crown, took steps to improve the situation there. By 1823 a policy of disafforestation, followed by enclosure and planting with the object of improving the neglected forests, was beginning to achieve results.

In Ireland, in 1827, the land revenues of Ireland were placed under the management of the Commissioners of Woods. Thus the forests of Ireland, at least in theory, now became subject to legislation and administration specifically created with the object of good forest management. The trouble was that there were few forests left to benefit. The total of forest land, including nurseries, scrub and new plantations, was only about 290,000 acres. The differences between the two countries were profound. A series of reports from select committees in 1848-9 disclosed that the situation had been reached in England where almost all the suitable land capable of being used for growing timber had been planted. In Ireland by contrast it had all – or virtually all – been denuded. In any case, just as the frame of mind conducive to realistic forest management became at last apparent in Ireland a change occurred so profound that it altered the whole thinking underlying forest management, administration and economics.

That, in 1812 or thereabouts, the navy commissioners in Britain proposed to ensure supplies of oak for the navy 'indefinitely' indicates the magnitude of the extraordinary change about to take place. It so rapidly and completely ensured the decline of the mighty oak that had dominated forests for thousands of years that by the end of the century oak was scarce – from neglect rather than from usage. About the mid-nineteenth century new species of conifer, which proved to be extraordinarily versatile, were introduced from North America. They could not only substitute for oak in many ways, they were both quicker-growing and cheaper. The so-called 'Heart of Oak' received the first blow. The second blow from America was not long following, and it was mortal. Since the 1840s the question of iron ships had been considered and dismissed in English naval circles, even though oak was scarce, expensive, inferior and despised profiteers monopolized the supply and made fortunes from selling it to naval yards. In January 1862, when the navy of England already outnumbered the combined navies of the rest of the world[11] the fiscal year of 1861 closed on what was: 'Without doubt the maximum [timber purchase by the English navy] on record.'[12]

A mere month later, the altered position was summed up in parliament:

We have learned what if two months ago any man had asserted he would have been scouted as a lunatic; we have learned that the boasted navy of Great Britain, when opposed to iron vessels, is useless as a fighting navy. There is no blinking the question.[13]

What had happened?

In March 1862 a two-day naval battle in the Hampton Roads between Newport and Norfolk, Virginia, near the mouth of the James River, fundamentally altered the whole course of naval architecture and helped to end traditional ship design. It also brought to an end a forest policy nearly four-hundred years old based on the concept of the economic cultivation of oak for ship-building.

Hampton Roads was the decisive naval encounter of the American Civil War. Although involving nine vessels, only two played the vital role mentioned, namely the ironclads *Monitor* and *Merrimack* (formerly *Virginia*), Union and Confederate ships respectively. *Merrimack*, escorted by two gunboats sank the U.S. wooden ships *Congress* and *Cumberland* on 8 March. Three Union frigates went aground. The following day *Merrimac* faced the Union ship *Monitor*, thus precipitating the first ironclad naval battle in history. The outcome was indecisive, both ships suffering damage, but: 'The action showed that iron offered the only adequate resistance to iron.'[14] In Britain it was at once realized that unless the navy was immediately equipped with ironclads, Britain's role as a great sea power was over.

The effect on the forests and those who owned or were associated with them was profound. In Britain, it meant that the 400-year-old principle of nurturing oak for eventual dockyard use came to an abrupt end. 'All the former concern as to the national importance of British woodlands appears to have been completely forgotten; it seems to have passed absolutely and entirely from the recollection of the public and of their representatives in parliament.'[15]

In Ireland the effects were less dramatic, for the obvious reason that there was almost no timber, much less oak suitable for shipbuilding. But, paradoxically, it was followed by a more positive attitude to forestry. Limited reafforestation had already been introduced, but oak was only cultivated on a small scale. The original oak forests were almost all gone and had not been replaced. Most of what had once been woodland was now bare of trees. But interest, even concern, had been stimulated and specific laws applied. The opportunity presented itself in Ireland to encourage silviculture unencumbered by old methods, naval policies in Britain or useless forests of maturing oak. It might also be achieved more or less unhampered by the lust for timber for industry (now nearly exhausted as a result of earlier excesses). Thus a first Irish forest policy coincided with the sudden demise of oak as an economic forest crop. The means and the will to begin serious planned afforestation were there. It would make conifer its dominant species. But questions of land tenure and ownership also became an important issue during the nineteenth century, and this would affect forestry adversely.

By the mid-nineteenth century forestry meant landowners and gentry for the most part and, to the mass of the people, landowners and gentry meant foreigners and grabbers. The common view was that they were usurpers and that anything they did on the land was inimical to the interests of the people – the real owners. This was the view even when what was done was altruistic, as might be argued in respect of forest policy. Unfortunately forestry was a dominant, visible and enduring feature of the alienation of land from the people. It was not – and until recently continued not to be – seen as an activity of benefit to them.

The nineteenth century also saw a powerful resurgence of the spirit of national self-determination. Since 1798 nationalism had been infused with 'dangerous' republican and radical social ideas from America and France. Catholic Emancipation was underway. Literacy, offshoot of Maynooth, was burgeoning (in English, ironically); ideas were spreading. A climate of opinion was being moulded that, ultimately, would lead to action. Then the Famine devastated the land, halved the population, burned into the minds of the people a legacy of hatred and helplessness – and increased the growing awareness of a hunger for rights no less than for food. Prominent among the rights demanded was that to land: Fair Rent, Fixity of Tenure, Freedom of Sale.

As in everything else, religion was a powerful force. 'It was drawn into every matter, whether relevant or irrelevant, and it was used as a reason for withholding leases from tenants who were Catholics.'[16] In the main those of the Protestant faith – in the south mainly Established Church of Ireland; in the north mainly non-Conformists – were the 'haves' and Catholics were the 'have nots'.

The Catholic Relief Act of 1793 (intended to be a liberal step in association with the establishment of Maynooth seminary) was, instead, 'an unmitigated evil'. It purported to extend the vote to Catholics on the same basis as that which their Protestant neighbours in Ireland already enjoyed. They could now own a freehold yielding a minimum yearly value of forty shillings over and above the actual rent. While the intent of the Act was undoubtedly beneficial, the consequences were entirely disastrous. It led to frightful electoral abuses. The accumulated land starvation amongst the peasantry led to the acquisition of forty-shilling titles, corruption and bribery, not least in votes.

The rage for 40/= freeholds was conducive to all manner of offences and vices, to perjury, drunkenness, bribery, rioting and idleness. The tenants had no will, property, judgement or knowledge of their own to guide and govern them. They were the livestock of a great estate and were driven by their landlords to the hustings as a salesman drives his flock into the market.[17]

The vote was, of course, a public one and the electorate was consequently vulnerable to coercion. A tenant voted against his landlord's choice at his peril. The Ballot Act, instituting secret voting, had not then even been thought of.[18]

Landlords had virtual power of life and death over tenants. This ensured that the tenant's vote went against his own opinion and in support of a landlord he regarded as alien both by birth and by religion. By voting in favour of the landlord the tenant voted for continuance of Protestant and alien ascendancy, but if he voted against the landlord he would certainly be cleared off the land by whatever means – and several were open to landlords. The object of the landlord was not to provide tenants with a livelihood, but to acquire political influence through his tenants' votes. When the forty-shilling freehold was disfranchised in 1825 the situation became, if possible, even worse. Leases of any description ended also and were substituted by tenancies-at-will, or yearly tenancies terminable whenever the landlord wished. 'The tenant became a more degraded slave than the most unfortunate negro on the plantations of America...'[19]

It became customary to keep a death sentence, in the shape of a notice to quit which landlords served every year, continually hanging over tenants. Naturally in such circumstances, the land was starved by tenants seeking only immediate gain from exhausted crops. This became accepted practice in spite of the consequences to the land and resentment, fraud, corruption and violence grew.

Eighteenth-century estate papers are full of complaints about tenants, and others, cutting timber. The agent at the Olivers' Limerick estate commented on the theft and mutilation of some ash trees, 'The sight of the same is scandalous to be seen' (G. L. Edlin, 'Review of the Forests of Ireland', *Quarterly Journal of Forestry*, lxi no. 2, 176.) Some landlords felt strongly enough about the depredations made on their woods to offer rewards for the discovery of the offender: Philip Cosby offered 20 guineas – the yearly wage of four woodrangers, or of Lord Gosford's butler – for information after some oaks were cut in his woods at Stradbally in Leix. The Duke of Leinster at Carton, Kildare, offered no reward for information in 1774, but simply announced that he had set man-traps in his plantations. The duke took his planting seriously.[20]

Even where there were leases (usually sub-leases) there were similar disincentives. For instance, a tenant who improved his land or holding was not only denied any remuneration in respect of it, but faced an increased rent because of it. From the mid-seventeenth century 'One finds increasing numbers of leases binding a tenant not only to make a substantial hawthorn hedge around his holding but to plant and maintain trees in it. Nearly always ash was explicitly named and occasionally oak.'[21]

The exception was Ulster, where the landlord and tenant system was on a different basis. There, lands forfeited to the Crown were conferred on English and Scots landowners who brought over their own people as tenants. The colonists, tenants and landlords alike, supplanted the Irish owners and entered into a covenant of mutual support and protection. This became known as 'Tenant Right'; in time, the tenant could acquire the property, and in the meantime, his rights were protected.

When, in the 1830s, landlords refused to recognize the rights of the people of the south to retain a legitimate interest in their land, Thomas Drummond, Under Secretary to the Lord Lieutenant, addressing the magistrates of Tipperary, spoke these memorable words:

Property has its duties as well as its rights; to the neglect of those duties in times past is mainly to be ascribed that diseased state of society in which such [agrarian] crimes take their rise, and it is not in the enactment of statutes of extraordinary severity, but chiefly in a better and more enlightened and humane exercise of those rights that a permanent remedy for such disorders is to be sought.[22]

He was referring to the activities of a restless peasantry goaded beyond endurance to acts of savage violence against landlords and their agents.

In 1790 the population of Ireland was 3,700,000. In 1821 it had increased to 6,800,000. The greater part of this enormous increase was due to the creation of the forty-shilling freeholds for Catholic tenants. The Catholic Emancipation Act of 1829 abolished these, and, as they had been brought onto the land not to cultivate it, but as a kind of human livestock to create political influence for their overlords, the purpose of their creation disappeared and the system of consolidation of small-holdings, which had commenced about 1816, received a further fillip. They (Catholic tenants) were cleared off the land, but not transported. As the land was the only source of wealth of any importance outside north-east Ulster, where high-handed clearances could only be attempted at grave risk and expense to the landlord, the land had still to support them in some form or other.[23]

The Poor Law system introduced in the early part of the nineteenth century further complicated matters, after the Court of Common Pleas in 1840 decided against Lord Westmeath in an action which he brought against one of his tenants for sub-letting conacre. The Court took the view that the relation of landlord and tenant was a substantive one between the conacre owner and the conacre holder. The result, however, as far as the farmers were concerned, was a reluctance to let any land at conacre at all and under the Poor Law system to evict tenants whose valuation did not amount to £4. 'Like the wild animal which is cornered and stands at bay, the cottiers had recourse to intimidation to enforce their rights to live, and in this, much of the agrarian trouble which prevailed in the south of Ireland had its roots.'[24] Another commentator says: 'The whole system was an unmitigated evil. So great a proportion of the population depended on the system [by which the tenant paid his rent in full – in advance – or gave a note in hand for the balance for which he could be sued if it was not forthcoming] for support that it could not be stopped. In 1845, out of a population of 8,295,061, more than one third consisted of labourers and cottiers, both of which classes depended on it.'[25]

Farmers of all persuasions contributed to these dreadful conditions by trading on the necessities of the poor. They were themselves the victims of a harsh system, and they did not hesitate to use the oppressive methods which their own masters inflicted on them.

It was against such a background that the hostility of Irish smallholders and small farmers towards forests and afforestation took root, just when some of the descendants of those who had ravaged and exploited the forests began to take an interest in their reconstitution. The hostility of the smallholders was further aggravated by the Famine of 1845-7. It is, therefore, hardly to be wondered at that even in the late 1980s possessive attitudes towards land and the holding of it were still evident in some rural areas where afforestation remained suspect, and proposals to encourge it were met with evasiveness.

The vexed question of subdivision also affected attitudes to forestry.

[It] is to be distinguished from sub-letting and likewise from the middleman system. It was the outcome of the morbid desire of the average Irish farmer to keep his family permanently around him. The subdivision was a friendly family affair, and utterly distinct from the tyranny of the letter of conacre and of the middleman. The habit was derived in part from the old tribal law and custom that subdivision to supply land for his children was a legitimate and proper proceeding on the part of the tenant.[26]

The work-ethic of the naturally industrious and hardworking farmer was killed by the refusal of landlords to provide security of tenure, homes or capital to equip and work the holding. The landlord maintained that the capital sunk by the tenant on the farm ceased to belong to the tenant and that only while the lease subsisted had he any right to the land. After that the proprietor took to himself the increased value of his property. Naturally this had the effect of killing initiative and causing the land to be overworked. 'Society was in a diseased state. All the elements which go to constitute a normal community – the poor, the middle-class and the rich – were represented. The evil arose from a superabundance of poor and a scarcity of rich people who had the interests of the country at heart.'[27]

This was the backdrop against which the early, fumbling steps in managed forest planning emerged. The struggle was continuous throughout the century and it was not until towards the end of it, when larger and more urgent social and political issues had been compromised in the general direction of order and justice, that the people, through their popular leaders and the activities of the Gaelic Revival, began to discover the potential of forestry. The forests were still firmly in the hands of private ownership, but new ideas were abroad and some of the owners were sympathetic, even nationalistic, the most distinguished being Charles Stewart Parnell.

Between the mid-eighteenth century and the end of the nineteenth century, an active interest in scientific forestry developed. It was a universal interest. Indeed, England was behindhand compared to France and Germany, where the commercial, investment and environmental properties of forestry were concerned. India, too, led Britain in this regard. The British empire had made it possible for England to regionalize its resources – especially natural resources such as forestry, requiring vast acreage – in a manner not readily possible for other continental countries, thus enabling the preservation of the traditional patterns of landed estates and title in Great Britain itself.

The interest in scientific forestry stimulated public activity. The Scottish Arboricultural Society was founded in 1854. Twenty years later the English Arboricultural Society was founded and in 1884, following an International Forestry Exhibition in Edinburgh, a select committee was appointed 'to consider whether by the establishment of a forest school or otherwise our woodlands could be rendered more remunerative'.[28]

The final report of the select committee of 1884 appeared in 1887 and, amongst other things, it recommended that a Board, to be known as the Board of Forestry, should be set up. It was to include representatives from the arboricultural societies of England and Scotland and from the RDS. The intention was that the Board would organize forest schools and courses of instruction. Much of this activity resulted from a working plan prepared by H. C. Hill for the Forest of Dean. Hill had been a Conservator of Forests in India (interestingly much of the advanced thinking on forestry in England at this time was influenced by individuals who, like Hill, had spent time in the Indian service which, in turn, was largely influenced by advanced ideas of forestry from Germany). Hill's plan was 'To introduce a more scientific and systematic system of forest cultivation than has hitherto been adopted.'[29]

It was in Germany and Austria that the correlation between forest management and the community had been recognized and the principles of sustained yield and regulated silviculture were systematized. As early as 1789 Georg L. Hartig had founded a school of forestry at Hessen. And in Saxony, in 1795, Heinrich von Cotta founded another at Zilbach, Thuringia.

In his *Économie Forestière* (de Librarie de l'Academie d'Agriculture, Paris) G. Huffel attributes the concept of volume yield to a forester named Bollem, in the service of the Duchy of Eisenach in the 1740s. Bollem divided the rotation into ten-year periods and allotted to each period 'such an area of forest that the volume yield was kept constant thoughout the rotation'.[30] His work was continued by a Heinrich Hartig, later Head of the Prussian Forest Service. It was improved on by von Cotta and was reinforced by the earlier work of a Thuringian forester named Ottelt. Between them they developed the formulae for volume yield, increment and equated years in forestry on which modern systems of forest management are generally based.

About the same time the Austrian formula for yield regulation, known as The Austrian Cameral Valuation method, was introduced and became standardized. 'By 1825 continental European foresters, at least those in Germany, Austria and France, had grasped the basic principles of forest management ... forests were surveyed and demarcated and divided into felling series, or sustained yield units...'[31] These were principles of great importance to an Ireland where the question of reafforestation was under serious consideration.

The new teaching, demonstrating the use of the uniform shelterwood system of regeneration advocated and practised by Hartig and von Cotta, spread from the forestry schools of Prussia and Saxony to France. The Nancy school of forestry was founded in 1824 with Bernard Lorentz, who had been a pupil of Hartig, as director. Within twenty years the so-called German uniform system, the foundation of modern forest policy, had superseded old systems of forest management.

However, Lorentz anticipated a fall in demand for wood fuel owing to competition from coal. As a result vast areas of coppice with standards were put under conversion to high forest. But, while his anticipation was correct, Lorentz's foresight was somewhat pre-emptive, and the sudden and heavy fall in yield and supplies to industry caused such an outcry that he was forced to resign in 1839.

European foresters also developed the basis for formulae to calculate the economic value of a forest and the income which might be derived from it over an extended period. The best-known of the resulting formulae is the Faustmann Formula, produced in 1849. In 1880, yet another German forester, Karl Gayer, published a seminal and formative work on modern forestry methods, *Der Waldau*.

Gayer emphasized the need to learn from nature and the practical use of common sense in doing so. He recognized that in order to provide sustained yields it is essential for the forest to be maintained continuously stable so that production is not curtailed. He concluded that in order to ensure continuous production all the forces contributing to growth must be maintained in harmony, in an harmonious environment. He therefore advocated mixed and irregular stands of trees and the use of trees climatically suited to the site, which are the basic features of natural forests.

The correlation of scientific management and practical commonsense was probably best developed by Friederich Judeich, who developed the theory he called *Bestandwirtschaft* (individual stand management). His principle of selecting stands for final felling in relation to the objects of management rather than because of age is still basic to any modern system.

There had also been an active policy of modern forest administration in India since 1855. The success of the Indian Forest Service was largely due to the work of three German foresters, Dietrich Brandis, William Schlich and Berthold Ribbentrop, all inspectors general of forests. Retiring members of the Indian Forest Service brought back to England the principles they had learned under the direction of these three men. Schlich left India in 1885 to establish the Forestry Department of the Royal Indian Engineering College in England, where his influence on English forestry at the beginning of the twentieth century was to be profound, his *Manual of Forestry*, together with *Der Forstschutz* by Richard Hess and *die Forstbenutzung* by Karl Gayer, being the works from which modern principles of forestry in England were largely taught. However, the absence in Britain of a 'tradition' of forestry – by which was meant 'tree-farming' on continental lines – limited progress. Paradoxically the almost total absence of forests and of forest tradition in Ireland made it simpler to apply the new principles here with greater expectation of success.

In spite of continuing interest in the subject elsewhere, and among enthusiasts, the development of domestic forest management and silviculture generally was sluggish in nineteenth-century England. The steam engine, the spread of railways, canals and unsurfaced roads did make an impact on a situation which had continued for over two hundred years. But it was not until the first quarter of the twentieth century, when the internal combustion engine became universal and the engine itself reliable, that local self-sufficiency really gave way to an expanded regional or national timber industry.

But Ireland presented a different picture. Neglect and exploitation had so devastated her forests that forestry was, by definition, experimental in any case and new ideas and techniques could be applied with little danger of disrupting any established traditions. Hence the early interest in conifers.

The following, based on Osmaston's summary,[32] are the principal developments that affected forestry in general, and Ireland in particular, during this period.

(1) Methods of control, volume, yield and classification improved and applied.

(2) Scientific formulae governing yield regulation and better management systems introduced.

(3) Continuous appraisal of methods employed in relation to the policies and objectives.

(4) More flexible and irregular silviculture systems favouring natural regeneration.

(5) Changes in methodology reflecting greater awareness of commercial requirements.

(6) Improved thinning techniques and greater understanding of intensity values.

(7) Greater use of native species.

(8) Sustained yield as a primary management objective.

(9) Improvement and extension of roads and new means of transport.

Not all of these applied to both countries equally, of course. But progress in Ireland was slow largely for lack of funds.

From the last quarter of the nineteenth century onwards, it was accepted that coniferous timber would be in greater demand than hardwoods. The prospects for fast-growing, exotic species such as Douglas fir and Sitka spruce, yielding large volumes, were attractive. Ireland was well placed environmentally and geographically, and in general so far as the suitability and availability of land were concerned, to benefit immediately from this change of direction in 'tree-farming'.

Commissioners of His Majesty's Woods, Forests and Land Revenues had been appointed in 1812; in 1848 a Select Committee on Woods, Forests and Land Revenues of the Crown was formed, but it does not seem to have achieved very much. A man called James Morris, for instance, had 'drawn the entire English oak supply into his hands during this period'[33] and there was widespread corruption and profiteering by other timber magnates. Between 1854 and 1885, a critical period spanning the naval revolution and profound social changes and during which one might have expected the Select Committee to be particularly active, it issued no reports whatever. But reports were published for the years 1885, 1886, 1887 and 1890, an indication, perhaps, of that revitalized interest in forestry which was to have important results for Ireland. Progress was limited, but fairly consistent until the Land Act of 1881 brought profound changes. One of the consequences of the transfer of land from landlord to tenant was that estate owners lost interest in their plantations and not only ceased to replenish them, but sold much of the standing timber to travelling sawmillers from Great Britain who proceeded to devastate estates right across the country. The effect of these developments on timber resources already grossly depleted was that by about the early 1920s there were left only 248,500 acres of woodland, roughly two-thirds of the acreage of plantation woodland 80 years before. Just about one-and-a-quarter per cent of the country was under forest.[34]

In 1902 an important departmental committee of the British Board of Agriculture was appointed to enquire into, and report on, British forestry. It endorsed the report of the earlier committees and added recommendations of its own. Among these were that the curricula of agricultural colleges should include forestry as a subject for study and that such colleges should also provide abbreviated courses for young foresters. It recommended that demonstration forests should be established and provision made in them for the education of foresters and woodmen (to be employed as student foresters); and that county councils should arrange lectures on forestry and provide financial assistance for foresters wishing to attend. Emphasis was placed on education and on the involvement of local authorities. In 1904 a school of forestry for woodmen was opened at Park End at the Forest of Dean and it remained open until 1971.

An afforestation conference held in London in 1907 considered a proposal from the Association of Municipal Councils for a national scheme of afforestation to alleviate unemployment. Two years later a royal commission, holding that a national scheme of afforestation would contribute to the solution of the unemployment problem, reported that such a policy of afforestation was desirable and that there were some nine million acres in the United Kingdom which could be planted. This, they argued, would provide permanent employment on the basis of one man per hundred acres planted.[35]

In 1909 development grants for better use of England's natural resources were introduced. In the case of forestry, grants were to be available for schools of forestry, acquisition of land, planting and the creation of experimental forests.[36] In 1911 a forestry committee, consisting of four of the newly appointed development commissioners, was formed. Applications for grants were to be considered under the following:

(1) Expansion of forestry must be founded on effective education and research.

(2) No scheme for large-scale afforestation by the State should be considered until it had first been established where, from an

economic and financial aspect, the most suitable areas for any new forests were situated.

(3) No action should be taken until a sufficient number of trained foresters was available.

Against this general background developed the pattern of modern forestry and forest policy in Ireland and Britain.

The State forestry concept was of special interest to England in view of the strategic timber reserve dilemma brought about by Hampton Roads and its after-effects. It was more convenient to experiment with the idea in Ireland where land and labour[37] were cheaply available and where they might conveniently kill two birds with one stone, i.e. note the results and stimulate Irish forestry development. Irrespective of the outcome of the experiment the results would be useful and at an expenditure more modest than could be contemplated elsewhere in the United Kingdom.

The idea of State forestry in Ireland had been proposed by Dr R. S. Lyons MP in 1884 when he introduced an Afforestation Bill, both far-sighted and far-reaching, although it was unsuccessful. Gladstone and the Earl of Caernarvon, the Irish Lord Lieutenant, sought professional reports on Irish forestry prospects from, respectively, Schlich and Daniel Christian Howitz, a Danish forester (like Schlich living in England). It may be thought ironic, in the Irish context, that the first major 'split' in the modern Irish forestry movement occurred between two foreigners. But it was also these two men who, in their different ways, were responsible for the stimulus that eventually – in spite of many obstacles and set-backs – led to the State forestry programme of today. It was as a result of their advice that the so-called Knockboy Experiment of 1890 to 1898 was undertaken.[38]

From the outset misfortune and controversy dogged the experiment – and, for many years, modern State forestry in Ireland. This first 'and almost the only State experiment in silviculture in modern times'[39] was a disaster. However, it marks the first attempt at State afforestation in the modern sense in these islands (though a co-operative of sorts had been undertaken earlier in Donegal by Dr Lyons).[40]

Particularly unfortunate was the fact that the failure of Knockboy added to the suspicion and dismissiveness with which forestry was viewed in many influential quarters. What might have been a spectacular break-through became instead a shambles. Knockboy began as an untried experiment doomed to inevitable failure if proceeded with along the lines proposed. It could – and should – have been a prototype field-trial based on appropriate scientific assessments for which the relevant data were available and on which a significant State forestry programme might have been based. Undertaken with the best of intentions and supported by enormous political and administrative goodwill, Knockboy failed as much from over-enthusiasm as from any other cause. It was hopelessly sited and the species selected were unsuitable and too many. The political interest, perhaps, but not necessarily, amounted to the 'interference' alleged by Forbes (by which he presumably meant that Chief Secretary Sir Arthur Balfour bought the waste-land and afforested it before any survey by forestry experts was conducted).[41] What is more difficult to understand is why a clearly failing experiment continued, expensively, for ten years.

C. Litton Falkiner's account is probably the clearest and most unbiased.

In 1890 an area of 960 acres near Carna in Connemara was acquired by the Irish Government [by which he means the Irish Administration, E. N.] with a view partly to an experiment in planting and partly to the resulting provision of employment for the people. The property was placed under the Irish Land Commission which spent a sum of nearly £2000 in draining, fencing and planting. On the formation of the Congested Districts Board, Knockboy was transferred to that body, which at once entered zealously on forestry operations on a large scale. In the first report of the Board it observed that 'if the trees grow in this exposed situation close to the shores of the Atlantic it will demonstrate that much of the waste lands of Ireland could be turned to profitable account'. This observation is most just, for a rockier or more wind-swept spot than Knockboy may not be found in all Ireland, and had the trees found root in its sterile sheets of rock, or sustained the unchecked onslaught of Atlantic storms, the difficulties of forestry elsewhere in Ireland must have been small indeed. But no miracle intervened to save the Board from the inevitable failure of an enterprise it had not itself initiated, though obliged to carry it out, and for which it is therefore only partially reponsible. Planting was carried on on a large scale in 1893 and 1894. But in 1895 the report stated that the trees were not thriving, and in the year following it was prudently deemed 'not desirable' to plant any more 'pending further experience'. By 1898 it had been decided not to incur any further avoidable expense, and no mention is made of Knockboy in the subsequent reports of the Board, save that down to 1901 the melancholy figures of expenditure on this experiment are regularly recorded in the Appendix. It appears by the Appendix to the 10th Report that the total outlay of the Board on Knockboy to the year 1900 amounted to £8,703.18s.2d., while the countervailing receipts reached the paltry figure of £24. 4s. 0d. Thus inclusive of the original outlay, the total expenditure on this wholly abortive scheme reached a sum of upwards of £10,500. It is not surprising in view of this unfortunate attempt, that the Congested Districts Board has not been eager to try further experiments, or that its forestry operations have been limited of late to supplying small quantities of trees gratis to small occupiers for purposes of shelter.[42]

Forbes's summary of the beginning of Knockboy is likewise succinct.[43] An enduring sense of frustration was engendered amongst foresters by the failure of Knockboy. Many apparently believed that the failure influenced political will on the subject well into the 1930s. The Congested Districts Board, Howitz, and politicians have all been blamed for the calamity. But it is a fact that the land had been 'looked at', if not surveyed, by Dermot O'C. Donelan of the Irish Land Commission, who reported 'that there could be no doubt whatever that the whole of the plain – 800 acres out of the total of 914 [given as 960 by Litton Falkiner and 1000 by Forbes, E. N.] was well-suited to growing trees' and that 'in seven years there should be good cover in every part of the plantation'.[44]

The tenant of the property, Thomas J. Flannery, is identified[45] as the then parish priest of Carna, whose interest in the welfare of his parishioners apparently encouraged him to unrealistic hyperbole in his efforts to secure them employment. His principal concerns were to obtain a spur to Carna for the Clifden-Galway railway, and to secure cheap loans for fishermen. But forestry was an acceptable substitute and it was he who, apparently on foot of Balfour's statement, offered the land as being highly suitable for forestry. No doubt he hoped it would bring employment in its wake.

At Knockboy they behaved on the principal of the old doctors. They gave you a prescription, which had numerous drugs in it on the chance that you would get in one of them at least what you wanted. At Knockboy they planted thirty or forty species of trees, and without a shelter belt, and, in a word, I understand nothing more foolish was ever done, and they spent £10,000.

But Henry also stated, *inter alia*, 'What I think about bog plantation is that it should be attempted. People in Ireland before they want to do anything must have a dead certainty. They have a dead certainty in the afforestation of waste land.'[46]

From the theoretical point of view the awareness of the need for education in forestry and the establishment of demonstration areas and schools of forestry were welcomed in England. But little seems to have been done in practical terms to rationalize the timber resources of that country. This may have been a result of the reduced demand for timber following the switch to ironclads for the navy, but is also likely to have been because productive English oak forests were largely in private hands. This dilemma arose less because of conflict of interests between strategic reserve requirements and private suppliers (as had been the case when the 'Timber Trust' of the eighteenth century scandalously 'nearly brought disaster to the nation by cutting off the entire supply of English timber at a time when Napoleon seriously threatened England with invasion'),[47] than from the fact that the bottom had fallen out of the market. The private monopolies found themselves with considerable capital tied up in vast and durable crops of unprofitable oak-trees, when the new thinking called for different and more economically produced species. In any event William Schlich, writing in the *Quarterly Journal of Forestry* in 1915, observed: 'very little has yet been done to increase the area under forest. Too much talking and too little action – that is the long and the short of it'. Of course he wrote when unforeseen demands of huge proportions had been imposed on the timber resources of Britain (and Ireland) by the First World War, when it was realized that the fall-off in timber-planting in the preceding half century had reduced resources to a level far below war-time requirements.

Changes in policy included a specific decision about a change in the nature of the raw material, endorsing what had already taken place in Ireland. Prior to the extensive felling which took place during the First World War, established English woodlands consisted largely of broad-leaved trees, particularly oak. The foreign species of conifer introduced during the nineteenth century had, by the outbreak of the war, reached a size and age that demonstrated their value as timber trees and, moreover, their adaptability and ability to grow on poor-quality land. These were Sitka spruce, Douglas fir, the giant silver fir, the western red cedar and Japanese larch, now commonplace. Accordingly when reafforestation in the light of the new policies was considered, attention focused very sharply on these species, already of interest to Irish foresters.

From such considerations, and the case was put forward by Irish forest advocates in the interests of Irish forestry, was conceived the idea of 'state' forestry on the one hand, and the idea of a national programme of forestry for Ireland incorporating it on the other. The distinction between the two is important.

PART TWO

Forestry in Modern Ireland

CHAPTER EIGHT

State Forestry: The New Idea

> 'Next to agriculture, the business of forestry with its related industries is the chief source of prosperity in purely land States' – Theodore Woolsey, *Impressions of French Forestry*.

A MAJOR PROBLEM WITH WHICH FORESTRY has had to contend is a general apathy among landowners based on the view that trees take a long time to grow and, unlike farm crops, give no quick return, leading to the conclusion that forestry does not pay. Thus, apart from the aforementioned 'Estate Forestry' – confined, by definition, to the activities and interests of large landowners, little interest in private planting was shown by the commercial landowner dependent for his livelihood on the produce of his land. Not only was he slow to make an investment of capital, labour and time in such a long-term crop of unproven capability, he was by no means sure of the marketability of either the crop or its by-products. To a considerable extent the medium-sized landowner most concerned felt he was being expected to make an investment in faith rather than in profit, by no means a natural inclination among farmers and commercial landowners. To this hesitancy was added an apathetic approach by existing forest landowners after the foundation of the State when they found that their expectancies (justified or not) under the British Forestry Commission proposals did not materialize, and that markets in Britain were in jeopardy because of the new status of the nation.

If events of history shaped the course of forestry up to the nineteenth century the same, but with emphatic changes, is true from 1900 to the present. The most important of these events was the coming into existence of Saorstát Éireann in 1922. The consequences for forestry were considerable and subtle – to the extent that some may not hitherto have been publicly recognized. Perhaps the most important consequence was a divergence from the programme of national forestry outlined in 1908.[1] Thereafter, until the late 1970s, except for sporadic bursts usually brought about by the dynamism of a few individuals (for instance Seán MacBride, a politician, and Henry Gray, an official, in the periods 1948-52 and 1967-72 respectively), historic events such as the civil war and political considerations such as agrarianism and agrarian-based economics overshadowed the requirement for a comprehensive, sustained forestry programme.

The developments leading to this situation began in the post-Famine period, when the economic situation and prospects were radically different,[2] receiving the vital impetus in the last quarter of the nineteenth century.

The Land Acts were liberally conceived, particularly the Act of 1903, which conferred rights of ownership on tenant farmers and smallholders. They were part of a universal nineteenth-century movement in advanced nations, complicated in Ireland by sectarian and historic alignments and the polarization of landlords and tenants.

For reasons already outlined, attitudes towards woods and forests were clear and divisive. On the one hand were powerful landlords, some of whom had an interest in forestry.[3] But, however progressive, they remained, in the eyes of the people, landlords usurping the land and trees became a visible and resented symbol of their role. Plantations – especially new plantations – were sometimes targets for vandalism. It was of no consequence to smallholders that some landlords foresaw forestry as an economic staple of twentieth-century Ireland, or that in this they had the ear of the administration. That, together, they constituted a powerful forest lobby was of little interest to deprived tenants for whom Hope began to play a thin note. Landowners and administrators were the enemy; trees were one of their symbols, and if they were now, through the Land Acts, doing something just, it was not before their time. And, of course, not all landlords and administrators were part of the forest lobby. Many were indifferent, even hostile.

The 'havoc wreaked on a woodland as a result of the uncertainty caused by the Land Acts at the end of the nineteenth century, when a tremendous amount of timber was sold to travelling sawmills, first by landlords and later by the new owners eager to raise ready money'[4] was a great set-back to such progress in forestry as had been made. For many new smallholders felling woodlands made 'economic sense' and was a 'good thing' in the eyes of their peers; to maintain them was a 'bad thing' and left a tenant open to the accusation of 'getting above himself'. Accordingly, with the opportunity to give way to his hostility *and* be paid for it, his course of action is hardly surprising, especially as the Land Acts conferring right of title seemed to encourage timber clearance. They also inspired similar motives (if for different reasons) in landlords. The enforced selling of lands and estates that had been the entitlement and prerogative of their class for generations was an unwelcome affliction which might be eased, perhaps, by first selling off the standing timber for additional profit.

In England, the predominantly oak forests were in private hands and represented a vast capital loss to the speculators who had been involved in the Timber Trust as well as to others. After the market crashed, many of these forests began to resume the role of wildernesses and game preserves, guarded by gamekeepers and mantraps. Conifer wood was a more economic timber, considered to be more suitable to the revised national strategic reserve (such as it was), and was being imported to Britain in huge quantities, being even cheaper to import than to grow. As the shock-wave receded, commercial domestic forests were again seen to be a national necessity. But the creation of conifer forests on a sufficient scale, involving both conversion of, or substitution for, many of the existing forests and eventual reduction of imports, was not going to be an easy matter in the face of existing broadleaf woodlands and the vested interests of woodland owners and timber merchants.

In 1894 Sir Horace Plunkett, vice president of the newly appointed Congested Districts Board and Conservative MP for Co. Meath, had founded the Irish Agricultural Organization Society (IAOS), perhaps due to a recognition that the Congested Districts Board came under the political control of the Chief Secretary (Balfour) and not under the Civil Service, and was an organization without independent teeth. A Department of Agriculture had been set up in England and in 1895 Plunkett called a conference, which came to be known as the 'Recess Committee' since it was held during the parliamentary recess.

His letter of invitation to all shades of representative public opinion in the country included the following:

The time has come when we unionists, without abating one jot of our unionism, and nationalists without abating one jot of their nationalism, can show our faith in the cause for which we have fought so bitterly and so long by sinking our party differences for our country's good and leaving our respective policies for the justification of time.

Secretary of the Committee was T. P. Gill, former Nationalist MP. The Committee reported in 1896, recommending the establishment of an Irish Agricultural Department. An Act was introduced in 1899 to set up the Department of Agriculture and Technical Instruction for Ireland (DATI) with an initial budget of £170,000.00 a year. The Recess Committee's report incorporated Howitz's recommendations on forestry, and stated at the same time:

There is no reason why the country should not again largely recover its forest area. Of the existing area of so-called waste-land, at least three million acres are calculated by competent authorities to be capable of growing one kind or another of timber; and with this extension of forests would come the very great and valuable variety of industries in connection with the working of wood...[5]

The Report went on to state:

To sum up the chief advantages of reafforestation. A shelter belt along the west coast would protect lands from the violence of Atlantic storms, which carry with them far inland many ingredients injurious to agricultural produce. The planting of mountains would tend to equalize rainfall and temperature, and prevent upland soils being washed away by torrents, and rivers being silted up and lowlands flooded. Forests help the preservation of birds, which prey on insects hurtful to crops. When planted on the banks of rivers they encourage the increase of fish by reason of their shade, the steady supply of water they promote, and the insects and animalculae they bring which trout and salmon and their fry find their best food. They lead to the propagation of forest game (such as pheasants, cock, deer, hares), and become preserves, the letting of which is to the State an important source of revenue. After a certain period, when the trees have grown, and the falling of the leaves has enriched the soil, the grazing of the forests becomes very valuable. There is further the value of the timber itself, and of the bye-products, (bark, charcoal, leaves, grasses, mosses, shrubs, weeds, fallen branches, resin, pitch, tar, turpentine), the intermediate agricultural products (flax, corn, potatoes, roots, fruits, truffles), and the series of wood-working industries (sawing, pole-making, cart and wheel-making, stave-making, handle-making, basket-making, etc). All these have been found to be actual accompaniments of forests in other countries.[6]

This extract, stressing both the social and financial advantages of afforestation, also demonstrates something of the extent to which the prospect of afforestation excited the minds and enthusiasms of many influential people at the time.

But a revealing passage in the Minutes of Evidence of the 1908 Report says:

After some negotiation it was found that this tract [Knockboy] of mountain land could be acquired at a very reasonable price. Then the question arose, was the land suitable for planting, and there was a difference of opinion as to that amongst the experts. They had the opinion of a German expert, and Scotch expert, and I think the opinion of one or two Irish foresters.

Do you know how that opinion ran? – The balance of opinion was not in favour of trying the experiment.

Not in favour of it? – No. Under ordinary circumstances I should say that the project would have been abandoned, only for the fact that no other land was being offered, and Mr Balfour was apparently almost bound to try the experiment...[7]

Dr Wilhelm Schlich submitted his report to Gladstone in 1885. Although it was more cautious than that of Howitz, it also supported the view that waste-land would be more profitable under afforestation than anything else:[8] 'I believe that, as regards the solving of the Irish land question and through it the ultimate restoration of peace and quietness in the country, the afforestation of surplus lands, especially in the Coast districts will be found of considerable importance.'

Howitz elaborated his proposals before the Industries Committee and they were dismissed by Schlich:

> As regards the species of trees to be grown I desire to add a word of warning. I have seen long lists [presumably Howitz's, E. N.] of exotic trees which are recommended for planting in Ireland. If anything of this kind were attempted at the outset, it would be sure to bring discredit on the operation and I feel satisfied that no forester of experience would countenance any such steps... until experience has proved that a species is suited to the climate of Ireland.[9]

Durand states:

> Forbes has dismissed Howitz as a hoaxer and quotes the incident as an example of how ludicrous (were) some of the projects put forward by well-meaning enthusiasts, without practical experience. At best his report is described as 'somewhat optimistic', but Howitz's main mistake was in his naivety of purpose, of redressing Ireland's lack of forests too grandly too quickly.[10]

This is, perhaps, unfair as Howitz undoubtedly influenced many people, some of considerable importance, to see in forestry 'a magnificent solution to poverty in rural areas' – a point of view which, while it has been modified and re-emphasized, has never been disowned.

The evidence of Henry Doran, Chief Land Inspector to the Congested Districts Board, before the 1907 Committee supported Forbes's views. No adequate explanation, except the constraint of the facts indicated, has emerged to show why the Knockboy experiment was continued long after it was seen to be pointless.[11] Nor does there appear to be available evidence to show that adequate protests against this were made by any responsible individual or group.

Knockboy was an important failure, but it had some positive features, the most important being that it focussed political and administrative attention on afforestation in general and State afforestation in particular. But it helped to create a negative climate of opinion in circles where their importance was under consideration. This undoubtedly had lasting consequences, the most obvious being greater caution from then on by politicians and administrators where forestry was concerned.

Two organizations that influenced Irish forestry between 1900 and the foundation of the State came into existence about the same time. When the Department of Agriculture and Technical Instruction (Ireland) was established in 1899, forestry became one of its functions. In 1900 the Irish Forestry Society, of which Dr Cooper was a leading light, was founded 'by a few politicians who had been nursed in Utopian dreams of an undeveloped Ireland, which only required the wand of an economic wizard to make its waste places bloom like the rose'.[12]

Whether this acid comment is justified or not, it is true that many opinion-leaders of the day were riding high on the heady swells of Gaelic revivalism and 'new economics'. Add 'forestry' and the results were surely exhilarating. The Irish Forestry Society was pledged to the view current in England that trees could and should be made to grow anywhere in Ireland. It held meetings throughout the country and, its speakers being people of influence and standing, its ideas received considerable publicity. It enabled the society to bring pressure to bear on the new Department with the not unreasonable view that it should do something 'to create Crown woodlands of a similar nature to those existing in Great Britain...'

The Board of the Department had already (May 1901) unanimously passed a resolution on 'the urgent necessity of the Department giving attention to the reclamation of waste land and reafforesting particularly'.[13] Forbes credits the Irish Forestry Society 'with exerting sufficient influence to have forestry first seriously considered'.[14] But he had reservations about the practicality of their ideas and the enthusiasm with which they propagated them.[15]

A deputation from the Society was received by George Wyndham, now Chief Secretary, to discuss the implications for forestry of the 1903 Wyndham Land Act, and the Council of Agriculture resolved that 'the requirements of land suitable for forestry operations and the administration of plantations on purchased estates, together with arterial drainage schemes and turbary are subjects worthy of special consideration in the coming Land Bills'.[16]

In 1904 Plunkett, while reporting to the Board that forestry operations on any large scale were not practicable, had announced that the Department proposed to acquire lands suitable for forestry operations and establish a school where foresters could be trained. Two years after the 1902 Departmental Committee Report on British Forestry, the purchase of the Avondale estate and the establishment of a forestry school there in 1906 marked the true beginning of State afforestation in Ireland. A forestry programme subscribing to the principles that – 'In the main forest economics unites two distinct concepts – 1) indirect benefits, 2) financial gain'[17] was outlined.

In June 1904 the Department bought Avondale (about 726 acres) in Co. Wicklow together with about 3000 acres of mountain land. The cost was £9,870.14.8. The connection with forestry was notable since Avondale had been built 'and chiefly planted in or about 1779'[18] by Samuel Hayes, possibly the first to publish a book which dealt with aspects of forestry in Ireland, namely *A Practical Treatise on Planting and the Management of Woods and Coppices* (1794). It later became the home of Charles Stewart Parnell who also had a considerable interest in forestry and managed the Avondale forest with care and concern.

Parnell was a collateral descendant of Hayes, and it is surely of interest that Avondale was the 'scene of two distinct pioneer movements in Irish tree planting, one by a private owner, and the other by a Government Department'.[19] It is likely that its association with Hayes and Parnell influenced its purchase by the Department, although Forbes claims that it was 'in many ways not too suitable for its purposes'.

With the recent spectre of Knockboy still close, the Department was understandably initially unenthusiastic about new forestry projects. But by 1906 – following commissioned reports – they were sufficiently encouraged to expend the Departmental Endowment Fund, established by the Act of 1899, on founding the forester training school at Avondale. Some of the most distinguished and active members of the Society were large landowners; two of them, Lords Monteagle and Castletown, were on the Council of Agriculture and were also members of the 1907 Departmental Committee on Forestry, as were the Duke of Abercorn and Viscount de Vesci. With such important landowners promoting it, the Department listened to what they had to say.

Dr Richard Cooper was a much-travelled medical doctor whose enthusiasm for forestry as an economic nostrum for the ailments of the nation infused his crusading spirit. He lectured extensively and as early as 1900 had proposed that a forestry department be immediately set up and that two million acres of poor land be allocated to it and planted.[20] Like many crusaders he tended to extremes. His extremism may have excited more caution than co-operation in the naturally cautious breasts of the administrators he hoped to persuade, but there is no doubt that at the turn of the century influential people of all kinds contemplated forestry in terms of great expectancy. An editorial in the *Irish Times* of 4 June 1901, then the voice of the Anglo-Irish Protestant, urged that forestry should become a branch of the new Agricultural Department, pointing out that in the previous year a mere 629 acres of forest was planted whereas 1451 acres had been cleared of timber stock.

In June 1903 Professor John Nisbet of the West of Scotland Agriculture College, where he was professor of forestry (an ex-official of the Indian Forestry Service and formerly Conservator of Forests for Burma), was commissioned by the DATI to report on the woods, plantations and waste-lands in south-east Ireland. His report was furnished in 1904. He suggested that one-fifth of the area stated in agricultural statistics to be waste might be profitably planted for forestry – about 750,000 acres. On the basis that planting would cost £8 an acre he suggested a ten-year programme costing £5m. Nisbet made an important observation:

The State is the only possible landowner that can be expected to create large compact blocks of woodlands, to be managed on silvicultural principles, with the two-fold object of providing supplies of timber in the future, and of fostering and encouraging rural and wood consuming industries. If this be a duty at all it is the duty of the State and not of the private landowners.[21]

This statement supported the view expressed in a report commissioned from an Irish land Agent, Archibald E. Moeran.[22]

The 1903 Land Act enabled the DATI and the County Councils to acquire land for forestry and for the preservation of existing woodlands. What the Act did not do was indicate how funds for the purpose were to be provided. Consequently, for want of funding, the Department could not offer an alternative to new owners, who saw in the sale of the timber on their properties to sawmillers a source of ready cash, 'so that railway stations and ports were crowded with timber at times, for months on end'[23] benefiting under the Act. Nor, for the same reasons, was the Department in a position to acquire forests or land of its own for afforestation.[24]

In spite of criticisms that there was little point in having a forestry school without forests in which to work (Lord Castletown); that better training was more readily available in England than at Avondale (The Irish Forestry Society); and that 'it is in the power of County Councils to control this Department of Agriculture, whose jobbery and incompetence have become a public scandal and which is now attempting to plunder the country for the benefit of the British Exchequer',[25] the Department supported the view of the Departmental Committee of enquiry into the affairs of the DATI, 'that the Department could not undertake forestry on a comprehensive scale because of a multitude of claims on its resources'.[26] An editorial in the *Freeman's Journal* of 6 March 1903, cited by Durand,[27] commented on the Council of Agriculture as follows: 'the oftener it meets the more manifest becomes its practical impotence. It was never intended to be more than a debating society, as the Act of Parliament that called it into existence took care that no matter within its cognizance should have authority to give a binding direction to the officials whom it was supposed to advise.'

But, following such criticisms, an enquiry made the point that in several cases the views of the Council of Agriculture influenced the actions of the Department in a practical manner. As a deliberative assembly it was the Council's function to discuss only and, if it felt disposed, submit motions to the Board.

To lay out the new forestry school at Avondale the Department selected a man who, to a considerable extent, influenced the course of forestry in Ireland for the next twenty-seven years. This was thirty-seven-year-old A. C. Forbes who, following his appointment as Forestry Expert at Avondale, remained with the Irish forest service until his retirement in 1931. Forbes had been lecturer in forestry at Armstrong College, Newcastle. John Black, who had worked at the Albert College, and Edwin Ellison, agent for Lord Meath's woods, comprised his staff.[28] Forbes became so associated with the subject that his name 'has long been held synonymous with forestry in Ireland'.[29]

Forbes, to use his own words, 'rightly or wrongly, decided to turn it (Avondale) into a forest experimental station on the lines of a Continental forest garden'...[30] 'as a demonstration and experimental area which might prove of service, not only for educational and training purposes, but as one which tree planters throughout Ireland could inspect at any time'.[31] He laid out plots of about an acre each (104 plots on 121 acres) and, between 1906 and 1909, planted various mixtures 'typical of ordinary plantations on different types of soil'[32] of some forty species.

Eight forest apprentices were recruited by advertisement. Courses of study were begun in the autumn of 1904. Rather surprisingly, perhaps, Forbes makes no reference to either students or instructors in his report to the Department.[33]

In 1907 and 1908 two related events occurred which profoundly affected Forbes and Avondale and the future of forestry. In Ireland forestry was the orphan of State services about which, although it excited considerable enthusiasm, no one seemed to know what to do. Funding was inadequate, and not entirely in its own control. Land acquisition for planting was not within its purview. The Estates Commissioners, who administered the Land Acts, were acquiring properties with plantations and woodlands on them, but had no authority to retain these. Willy-nilly the preposterous situation developed where the Estates Commissioners became 'guilty' of disposing of standing timber which the DATI forestry administrators were trying to preserve, but had not the means to acquire. It was a typical bureaucratic jungle in which the 'new boy', in the absence of any enabling legislation, depended for his survival on the leavings or the indulgence of the others, who were quite clear about their own responsibilities. It became obvious that a clear policy on forestry, defined by the Department under which it was administered, was needed.

Thomas Wallace Russell, MP, the new statutory vice president of the DATI in succession to Sir Horace Plunkett, was a forthright and energetic individual who, sharing the concern at the amount of timber-felling in the country, promptly turned his attention to forestry when he assumed office in 1907. He formed a Departmental Committee of Enquiry to enquire into the work of the DATI, including forestry. It urged 'the establishment of adequate machinery and provision of funds for a more active programme (of forestry) but did not feel competent to make detailed recommendations'.34 This committee, under the chairmanship of Mr T. P. Gill, included Lords Monteagle and Castletown, Horace Plunkett, Stephen Brown, Chairman of Kildare County Council, and John Galvin, Chairman of Roscommon County Committee of Agriculture. Arising directly from the views put forward by this committee, the special Departmental Committee on Irish Forestry was quickly set up by Russell in August 1907. Authorities on Irish forestry consider this to be the effective nidus of subsequent events.35 Forbes and Ellison were placed at the disposal of the Committee to enquire into, survey and report on woodlands and plantable land throughout the country.

The Committee was appointed 'to inquire into and report upon the following matters relating to the improvement of forestry in Ireland, *viz.*:

(1) The present provision for State aid to forestry in Ireland.
(2) The means whereby, in connection with the operation of the Land Purchase Acts, existing woods may be preserved, and land suitable for forestry acquired for public purposes.
(3) The financial and other provisions necessary for a comprehensive scheme of afforestation in Ireland.

The Committee consisted of the following:

Thomas Patrick Gill, Esq., Secretary of the Department of Agriculture and Technical Instruction for Ireland (DATI), (Chairman); The Right Hon., Lord Castletown of Upper Ossory, CMG, DL; William Redmond, Esq., MP; Most Rev. Denis Kelly, DD, Lord Bishop of Ross, member of the Agricultural Board; Hugh de Kellenberg Montgomery, Esq., DL, member of the Agtricultural Board; William Frederick Bailey, Esq., CB, one of the Estates Commissioners; William Rogers Fisher, Esq., MA, Delegate for Instruction in Forestry at the University of Oxford; Professor John Ritch Campbell, BSC, Assistant Secretary in respect of Agriculture of the Department of Agriculture and Technical Instruction for Ireland.

(Signed) T. W. Russell, Vice President of the Department of Agriculture and Technical Instruction for Ireland. Dated this 29th day of August, 1907.

The Committee set out clearly 'the main basis on which the enquiry proceeded...' and agreed on the principal objects to be attained by any scheme of afforestation supported by public funds, which were stated thus:

STATE FORESTRY: THE NEW IDEA

The maintenance directly or indirectly of an area of woodland sufficient to produce the supply of timber required by the country for domestic and farming purposes, for the development of industries and commerce essential to its prosperity, and for providing shelter needed for successful agriculture.[36]

In its Report, published in 1908, the Committee proposed that the Department should be given the powers and funds necessary to carry out a general scheme of afforestation, and indicated where the funding might come from. It also advocated such related important matters as inexpensive loans to aid private planting, and new legislation to control felling.[37] At the very outset it emphasized:

In the course of our enquiry we have become so convinced of the urgent need of measures to deal with the the subject of forestry in Ireland, as it is affected by the recent Land Acts and other circumstances, that we have made particular endeavours to quicken our investigation and present an early Report.

The Report contains a memorandum from the Chief of the Mapping Survey Branch of the Irish Land Commission, J. L. Pigot, who stated that the Act of 1903 had led directly to the destruction of woodlands on a large scale and the creation of a situation in Ireland which was 'just what other countries by their legislation or by co-operative societies are striving to avoid'.[38] He cited instances of landowners finding it far more profitable to sell woodlands as agricultural land, having first cut and sold the trees. (It is interesting that here, for the first time in an Irish context so far as can be judged, reference is made to the activities of co-operative forestry, although the reference is to what was happening abroad. Curiously this obvious approach remained submerged – other than for sporadic references during the War of Independence and in the early 1930s and 1940s – until 1985.)

The amount of planting being done was negligible, less than 400 hectares a year – far less than was being felled. The Report found that three-quarters of the timber being cut was being exported, most of it in the round. In view of the dearth of industry, unemployment, and the desirability of creating new enterprises, particularly wood-working industries, the Committee saw this as a serious social and economic loss – a view that received understandable support from manufacturers and sawmill owners.

Considering the difficulties arising from the Land Acts to be very serious for forestry, the Committee felt that the State should assume managerial functions in respect of forests disposed of by a landlord, and that much mountain and waste-land being sold with estates would be suitable for afforestation. They emphasized that if a national scheme of afforestation was undertaken by the State for the future such land would be difficult to acquire again once it passed to tenants. Even though it was of little value to them there would be considerable difficulty in reacquiring it. 'Vendors of estates had, in effect, no alternative, but to sell to timber merchants. They were unable, under law, to retain ownership in trees on land which they had sold.'[39]

But, while Appendix 10 of the Report[40] was specifically decided 'on the facts appearing in the particular case', there is little reason to suppose that decisions in similar cases would have been materially different.

A plantation is, at the present time not infrequently an incubus on the owners' hands. In the altered circumstances arising from the transfer of his tenanted land, he may often be unwilling or unable to protect or improve his plantations... Naturally he endeavours to get rid of it to the best advantage...The tenants will not buy them, although they sometimes ask for them at very low prices in order that they may cut them down...[41]

The Committee referred to the finding of the (1902) Departmental Committee on British Forestry, which suggested that the world was approaching a shortage of coniferous timber.

The unfortunate effects of the Land Acts on forestry were compounded by a ponderous political and bureaucratic machine. Passed with the long overdue purpose of social reform the Land Acts were, *per se*, welcome. Once they were enacted, however, and the adverse consequences that followed their implementation so far as forestry was concerned were seen, there was to hand no statutory means of rectifying this. Accordingly the destruction of timber continued for many years until, in 1909, legislation to control tree-felling was introduced. The 1908 Departmental Committee Report was forthright in its condemnation of the abuses deriving from the Land Acts.

The Report... deplores the disastrous effect upon forestry of the maladministration of the Land Purchase Acts and endorses in emphatic terms the finding of (the) British Committee that the matter of ascertaining the maximum area of land in the country suitable for afforestation is one 'of grave national concern'... Speaking generally, these woodlands have never been managed on commercial and scientific lines. Sport, amenities of residential places, landscape effect, shelter, have been the chief objects for which these woods have been created and kept up. In the course of the minute survey made for this Committee in five of the best wooded counties only in one case did Dr. Nisbet find anything like a definite plan of action forming a basis for the continuance of the work... the investigations which the commitee set on foot, and upon which it has had to rely, cannot take the place of a complete (agronomical) survey of the whole land of the country, which amongst other things would show *what lands have passed into the hands of tenants or tenant purchasers which would be more suitable for forestry than agriculture*. The returns made to the Committee elicit the fact that, exclusive of existing woodlands, from 1,000,000 to 2,000,000 acres – from 5 per cent to 10 per cent of the whole of Ireland... are better adapted for forestry than any other purpose.[42]

The Committee recommended the Report as a basis for immediate action. The report itself is a comprehensive document compiled from submissions and the evidence of forty-eight people, most of them connected with the timber industry or forestry. The witnesses included Augustine Henry, then reader in forestry at Cambridge, who later became an important influence in Irish forestry. The cautious Schlich, directing the attention of the committee to his views that afforestation was important to Ireland for its labour content, observed: 'On the whole I am now more convinced than ever that extended afforestation in Ireland is of great importance to the welfare of the people. The subject has been discussed, during so many years, so often and so thoroughly, that I think what is wanted is "action" rather than further discussion.'[43]

One discovery made by the Committee was that the number of sawmills in the country, many of them owned by English and Scottish timber merchants, had proliferated in recent years because of the availability and low prices of timber in Ireland. The Report stated 'that of 843 saw-mills in Ireland at the present time, 598 have come into existence since 1881; of which 349 started working between 1881 and 1903; and 249 between 1903 and 1907'.[44] The report also stated 'We find that, such as it is, this area (Ireland's 1.5 per cent of woodland) is steadily shrinking. The total area under woods is being reduced; and from the character of the cutting, the quality of the woods left standing is being deteriorated.'

The General Conclusion of the Report is important, clear and concise. Item (2) and the wording of it is of particular importance in relation to the way in which State forestry has developed.

(1) That with the transfer of land under the Purchase Acts to peasant owners, an obligation in respect to existing woods, as well as other obligations connected with landed property, passes from the selling proprietors to the State;
(2) That a comprehensive scheme of forestry, whether undertaken through local authorities, private owners or directly through a forestry department, can only be carried out by or under the direction of the State;
(3) That an exceptional opportunity, which cannot recur, for acquiring land suitable for forestry, and not so suitable for any other use, presents itself now in Ireland, in connection with the Land Purchase Acts; and
(4) That such a scheme, including the presentation and extension of existing woods, and the creation of a new forest area, is a sound investment for the nation, necessary for her agricultural and industrial development, and for the provision of an important capital asset which must otherwise be wasted.

It may be seen that (2) refers to a comprehensive scheme of forestry, not a comprehensive scheme of *State* forestry. It indicates several sources for undertaking it, under the *direction* of the State. This is a very important point and may be taken with (4) which refers to the 'preservation and extension of existing woods... industrial development etc'. Apart from this wording, and the climate of the times of the period, the specific recommendations make clear that absolute, or even majority, State control or ownership was not envisaged. What was envisaged, and what is referred to throughout the recommendations, is a 'National Scheme of Afforestation' under State direction, a very different matter. Perhaps because of the contrary nature of State Forestry development since 1922 this important point seems to have been overlooked, the general and demonstrably erroneous view being that the Report advocated an almost total State enterprise.[45] The view of the committee is given unequivocally in Recommendation IV (1) of its Report: 'The Department of Agriculture and Technical Instruction is endowed with the necessary legal powers for aiding, improving and developing forestry and promoting education connected therewith, and, therefore, for carrying out *the national scheme of afforestation recommended by the Committee.*'

In his memorandum Pigot said:

Once the fee-simple of the land is vested in a tenant-purchaser, the latter has full right to fell and sell timber growing on the land, unless, in so doing, he lowers unduly the value of the land as security for the purchase money advanced. Yet, even if the Land Commission had the right to intervene it lacks the power to prevent. Consequently, trees on purchased holdings are being felled at an increasingly rapid rate, and often in sheer ignorance of the real value of the timber.[46]

In its Report (Par. 55) the Committee made a powerful and compelling case for State action. It may well be a misreading of it, and of paragraphs 54-60, together with the kind of State intervention that in fact developed, which led to the tenor and thrust of the report being overlooked in this respect.

Exceptional Obligation on the State in Ireland.

There are general reasons for State action which we think it well to mention here. But in Ireland there are special and peculiar reasons which put the idea of any other method of dealing with the problem entirely out of consideration. The State has a responsibility which does not obtain elsewhere. At this moment the process of destruction of the woodlands which is going on... is due to the legislation of the State, and, as we have already pointed out, this grievous waste of the woods, with its menace to industries depending on them, must continue unless the effects of this legislation are checked by further State action. But in the past it may be broadly stated that the excessive reduction of the woodland area of this country is due either to what the State has done or to what it has neglected to do... Throughout the 17th and 18th centuries the grantees were allowed to do with timber as they pleased, and what they pleased was in the main to realise with reckless extravagance, with the result that the greater part of the country for which at least shelter might have been preserved, is in its present bare and wind-swept condition. Had provident and intelligent government action been applied to the subject in Ireland, undoubtedly the forest area and the general agricultural wealth of the country would be in a far better position than they now are. It is, moreover, an important factor in the case that the proceeds of the Quit and Crown Rents, which are entirely derivable from these lands and which have yielded a return of upwards of £60,000 a year, have never, since the union of the Irish and the British exchequers, been directly spent in Ireland or applied to Irish purposes, but have been, with the general crown revenues, invested in Great Britain, sometimes even in promoting forestry. There is finally the fact that the State, in abolishing the landlord through the Purchase Acts, is bound to provide and has not provided a machinery to discharge his functions in respect of several matters, including woods, which cannot be left to individual tenant purchasers, and in which the general community, as well as the tenant purchasers, has now a specific interest. Having regard to what is at present occurring in the country, we cannot hesitate to say that, not only does the responsibility lie on the State for taking action, but that if action be not taken at once it will mean a gross neglect comparable with the improvidence of the past and far less excusable.[47]

This very strong statement was made in a context that conceived of State enterprise in the matter involving an eventual forest holding of about 25 per cent to 30 per cent with 70 to 80 per cent remaining in private hands under the direction of some sort of national advisory authority; the British Forestry Commission was to become the solution for that country.

The Summary of the Report also stated that forestry had been deplorably neglected by the government in Ireland with the result that the country, particularly well-suited for tree-growing, had the lowest percentage of any country in Europe; that this was too low for the welfare of the country and that, nevertheless, it was being wastefully diminished chiefly under the influence of the Land Purchase Acts 'with great loss to the country and imminent danger to existing Irish wood-working industries'.[48]

The Concluding Remark of the Report also contains these strong words:

In presenting our Report we desire to express in the strongest way our sense of the obligation which lies upon the State to act immediately in this matter. Grievous mischief, loss, and waste, accompaniments of legislation and other State action, are going on, and ought to be checked without delay. Furthermore, as another accompaniment of legislation a great opportunity for husbanding and developing one of the resources of the country is now available, and if it be allowed to pass without being used, it means not only missing possible gain to the country but producing further actual loss. Legislation is needed in order to carry out the major part of the reform we suggest...[49]

The Committee concluded that, if the land was purchased during the first decade under the Land Act 1903 at an annuity of 3 per cent and planted over a period of forty years, then, provided compound interest during the first eighty years did not exceed 2 per cent, the return thereafter would be at the rate of approximately 4 per cent.

Ireland thus found herself, in the second half of the nineteenth century, in a position to embark on a forest policy employing modern forestry techniques and species. All but one of the essential ingredients existed. That one was, of course, finance: to it might, perhaps, be added Essential Goodwill. That there was goodwill in part is not in doubt, but it was from a quarter that generated massive counter ill-will and massive hostility to the idea of land-use for afforestation.

Factors that produced potentially promising conditions were:

(1) The savage deforestation of the previous two-and-a-half centuries left the country devoid of woodland. Consequently there was no real shortage of land for such a purpose (though plenty of opposition to its use in that respect), nor were there existing traditional hardwood forests demanding maintenance, upkeep and management to diminish a new effort.
(2) The prevailing attitude to forestry in England was a mixture of old-fashioned methods suited to the 'wooden wall' outlook on the one hand, and a new interest in modern continental methods on the other. Little plantable land was available, and there was a strong forest tradition to hamper the introduction of new methods.
(3) According to British law, Ireland was part of the United Kingdom, and, as in Britain, an awareness existed of the economic potential of forestry.
(4) Modern forest thinking and technology were available.

Had the essential goodwill and capital also been available Ireland might, by the mid-twentieth century, have been a splendidly forested country once more, with a secure industrial base founded on a great national resource open to a rising world market. But that did not happen. In those days economic policy was still grounded in the notion of free-trade. Economics was an instrument of politics rather than as today, a century later, the other way round. Political reform, overdue in Ireland as elsewhere in Europe, was the preoccupation of the hour.

Much of value did take place. Even Knockboy was indicative of the interest in forestry that existed. The dynamic 1908 Report was itself the result of sustained, energetic action and pressure from sources as disparate as the Irish Forestry Society, individual landlords like Monteagle and Castletown (both of whom were members of the initiating committee), members of the Gaelic League and Sinn Féin (the voice of Arthur Griffith, journalist and founder of Sinn Féin, was persistent in crying out for action on forestry), all having a common interest in forestry. Had that not been the case then the progress between 1908 and the outbreak of the First World War, when it was significantly interrupted, would hardly have occurred.

The next body to take an active interest in Irish State forest development was the (British) Development Commission set up under the Development and Road Improvements Fund Act, 1909. In their first Annual Report (period ended 31st March, 1911) the Commissioners showed no enthusiasm for financing large-scale afforestation in Britain but took a more favourable view of an application from the Irish Department of Agriculture and Technical Instruction. The Commission put forward recommendations, which were accepted, for advances from the Development Fund for forestry in Ireland on terms providing for no capital repayment or interest charged in the first 30 years, the question of requiring interest and provision for repayment to be further considered at the end of that period.[50]

It appears reasonable to conclude that the Commissioners felt that a new forest undertaking could not bear the dead-weight of compound interest accumulation during the early non-productive years.

The Committee, therefore, recommended that State forestry development should not be made subject to any specific financial objective but that there should instead be a criterion of minimum productive capacity in terms of anticipated timber yield. The minimum standard suggested, expressed in terms of productive capacity of land devoted to coniferous timber, was capacity to produce 40 cubic feet of timber per acre per annum – not by any means a high criterion or one likely to guarantee a good financial return at the lower end of the range of acceptable land qualities. This approach was subsequently adopted in practice in Britain.[51]

Some land was immediately bought to inaugurate this first practical expression of Irish forest policy. This was 1200 acres in Tipperary, soon followed by a further 600 in Co. Wexford. Afforestation was not involved in either case as these lands were already wooded and the Department was simply continuing, as a State forest, what had been in private hands, or 'Estate' forest. Unforested land was also soon acquired with monies from the vote.

A commission was set up in Great Britain with power to grant sums for the advancement of enterprises in practically any form which could be discovered, provided they were not directly remunerative. This was known as 'The Development Commission'. There was little difficulty in qualifying for grants, and forestry, whether in the practical or academic form, was soon brought within the scope of the Commission's activities. After a great deal of discussion on matters of principle the Commission agreed to finance the afforestation of three centres in Ireland in which the Department were negotiating for land, provided that 5000 acres could be secured in each centre, the money, of course, being advanced annually as required. In this particular direction, Ireland was the first part of what was then the United Kingdom to receive any money for afforestation. Both England and Scotland talked round the subject, and had various differences of opinion with the Commission which acted as obstacles to progress.

By the time work had actually started on these areas, however, Europe was thrown into a state of chaos by the outbreak of the Great War. This brought everything in the form of planting to a standstill.[52]

Interestingly the 1908 Committee also noted that, while they were satisfied of the commercial prospects for Irish timber if forestry and afforestation were undertaken on a basis of scientific management and commercial foresight, there were no examples of private woodlands in Ireland being managed primarily for profit, most woods being planted either for cosmetic or shelter reasons, otherwise being used in the main for sport. In other words the situation the Committee found demonstrated not only the results of widespread exploitation and neglect, apathy and hostility on the part of a majority of rural dwellers; they also found that such woodlands as were managed, were being managed for reasons which, in a general sense, had ended in England almost five hundred years before.

The Committee concluded that the amount of land devoted to forestry should be raised from the existing level, estimated at about 1.5 per cent of the total land area, to 5 per cent, a total of about one million acres. They reaffirmed their view as to the respective roles of State and private forestry in the national scheme they proposed.[53]

The overall proposal involved an ambitious programme of afforestation involving an initial rate of planting of up to twenty thousand acres a year *funded* by government loan.[54] It was calculated that receipts would exceed expenditure after fifty years. This proposal specifically raised the question of the proportion of State to private forestry envisaged.[55] The programme went before the Council of Agriculture and immediate action was called for. In putting forward such a programme of national forest development – in its time far beyond anything considered in the United Kingdom – the Committee urged that, apart from any consideration of financial return, regard must be given to 'the wider and less direct results of forestry, to its great influence upon the whole prosperity of rural districts and industries and to its social, economic, climatic and other national bearings... Ireland will not be managing her business as a prudent nation if she does not take every measure open to her at the present time to establish at least such a forest area as to provide a moderate insurance for the agricultural and industrial needs of the country with, perhaps, some residue for export.'[56]

Progress, however, was neither as rapid nor as ambitious as the Committee hoped. Nevertheless the events and influences that coalesced in 1908 bore fruit. The Report went to Parliament and, at the same time, the Treasury sanctioned the purchases in counties Wexford and Tipperary from the Estates Commissioners on the basis that the Treasury was prepared to reimburse 'approved purchases from the Estates Commissioners pending the whole question of afforestation being taken up by the Government'.[57] In 1909 another Land Act incorporated a provision prohibiting the felling of timber on lands purchased under the Land Acts and placed trusteeship in the ownership of woodlands in the DATI and/or County Councils, which were thus empowered to purchase woods by instalments. Arrangements were also concluded between the Estates Commissioners and the Department enabling the Department to be offered suitable woodlands as they came on the market. 'Thus, was developed the practice of the Forestry Authority being offered what was later described as the unwanted lands of other Departments.'[58]

In 1904, the year that Avondale was opened as a forest school, the School of Forestry at Parkend in the Forest of Dean was also opened. Forbes (then Forestry Expert to the DATI) attended the Afforestation Conference in London in 1907 to consider 'that the time has now arrived when the question of afforestation should be seriously considered by the Government, and that it be referred to the Law Committee to take steps for urging upon the Government the necessity for initiating afforestation schemes...'[59] The conference was presided over by Lord Carrington, President of the Board of Agriculture, and considerable importance was attached to it.

In 1909 £6000 was provided to acquire woodlands offered under the Land Purchase Acts. It was a far cry from the £13,000 a year which had been sought, but as Russell, Secretary of the DATI, observed to the Board of Agriculture that year, 'at least a good start on the lines recommended by the 1908 report might be made'.

The Departmental Committee on Irish Forestry cited data on income and expenditure in Europe to demonstrate net forest profitability. 'In this exercise the impact of interest liability in the initial developmental years of a new forest venture was not brought to the fore...'[60] This problem of large initial capital investment with no return for a decade or more was, until recently, still a major one.

Failure to successfully harmonize resource, management, processing and marketing appears to have dogged Irish forestry from the outset. Understandably the thinking of those concerned with forestry in 1909 related to establishing a resource rather than to development of one that, as yet, did not exist. Essentially forestry is a long-term crop planted by one generation and harvested by another. In such circumstances planning must be extended. Management must be both continuous and related to processing and marketing, both of which must be related to consumer needs. It goes without saying that all of these functions must be flexible. What remains for the small 'tree-farmer' is the dilemma adequately summed up as follows: 'The question of how a farmer who plants trees survives for the 10-15 years they provided him with no income... was the reason why there was no interest (in forestry).'[61]

By the outbreak of the First World War there were three separate means of financing Irish State forestry – all of them channelled through DATI. The Endowment Fund serviced the expenses of Avondale and Ballyfad and provided the contribution to County Forest Schemes. The Parliamentary annual vote of £6000 financed purchases, annuities and other charges on properties purchased through the Estates Commissioners, and the Development Fund financed the three existing

State forests.[62] Nevertheless by 1914 only 15,000 acres had been acquired, six years after the Report had called for the immediate acquisition of 50,000 acres.

A 1909 report[63] dealing with forestry and afforestation in Britain included the following:

> It was agreed that a policy of afforestation was desirable and that there were approximately 9,000,000 acres in the United Kingdom which could be planted without substantially encroaching on agricultural lands... a special board of commissioners, having powers of compulsory purchase, should be set up to deal with the work of afforestation.

The figure of 9,000,000 acres included marginal land, mountain land etc., much of it in Scotland. Ireland was held to have some 3,000,000 such acres. A perennial problem was highlighted by Augustine Henry: 'The difficulties of afforestation lie in the cost of acquiring the land suitable for growing forest trees...'[64]

In the course of his Budget speech in 1909 Lloyd George announced that the government intended to introduce 'development grants with a view to making better use of the country's natural resources' including 'schools of forestry, the acquisition of land for planting and the creation of experimental forests'.

The German influence on the new thinking on forestry was everywhere evident and, 'as the first decade of the century came to an end, there were signs that a new appreciation of the technical aspects of forestry was beginning to emerge and this was especially so in the case of forest management and silviculture...'[65]

In 1912 an Advisory Committee on Forestry[66] was appointed, which included Schlich. Its first recommendation was for a detailed forest survey 'with a view to the State acquiring an area of no less than 5,000 acres as an "experimental forest"'.

CHAPTER NINE

The First World War

'Directly or indirectly the forest affects every inhabitant of a country. Its presence means wealth; its absence, the absence of wealth' – Mackey.

THE HISTORY OF IRISH FORESTS from the time of the Norman invasion until the forests were largely eliminated reflects in a curious manner the relationship between the two countries. Not without cause does the word 'fastness', commonly used to describe Irish woodlands, mean stronghold or refuge. While the managed forests of England were being developed into a proud national economic feature, the cradle of the 'Walls of Oak', those of Ireland were retreats and refuges, exploited and eventually obliterated.

The Act of Union in 1800 might have brought relief for the forests – if there had been any left – when, for the first time, Ireland became subject to United Kingdom statutes. However, modern scientific silvicultural methods were also introduced about the same time and some of them were to have important effects on forestry. Accordingly treeless Ireland entered the mainstream of public interest in and concern about forestry in Great Britain, participated in it and, at times, benefited from it and from consequential forest legislation, a situation that continued until the foundation of the State in 1922.

The destabilization of European forestry programmes following naval changes wrought by the American Civil War and imports of new, cheaper timber species which met the general requirements of the timber trade, led to an unprecedented vacuum in European forestry, promptly and efficiently filled in France and Germany by recourse to conifer. But in England imports began to dominate timber and related industries.[1] The effects of this became sharply apparent during the First World War.

In 1909, too, the question of the strategic timber reserve of the United Kingdom, after more than half a century of indifference, was again a matter for concern.

Instead of taking steps to ensure that there were adequate supplies of homegrown timber, successive (English) governments had (between 1860 and 1913) relied on imports from abroad to meet most of the country's requirements and by 1914 the United Kingdom was importing approximately 10 million loads (400 million cubic feet) of unmanufactured timber per annum.[2]

When war was declared, the confidence of the English authorities, who did not expect the war to last, remained unshaken. However, a survey of the country's timber resources was undertaken and it showed that supplies of vital pitwood were sufficient for about two years. Ireland was among the sources surveyed. It was quickly apparent that, like coal and iron, timber was an essential commodity to the economic survival of Britain. Later Lloyd George was to state that Britain had come closer to losing for lack of timber than for lack of food. As the war progressed and the threat to shipping increased (from 885,000 gross tons sunk in 1915 to 3,660,000 gross tons in 1917) urgent action (with, amongst other purposes, the object of saving shipping space) was taken to ensure greater use of native timber on the Western Front and meet the general needs of the Services. It is estimated that during and just after the First World War 450,000 acres of woodland (including some in Ireland) were felled. Yet, according to Thomas Hinde, 'It is a common misconception that the great fellings of the two world wars destroyed the forests of Britain, but it can equally well be argued that these put them to their proper use for the first time in hundreds of years.'[3]

Women were employed in forestry during the First World War. Initially it was on an individual basis and without proper organization or training. Later the Women's Forestry Corps of about 2000 was formed and the employment of women as forest workers was put on an effective basis.

During the war several Orders affecting Ireland in respect of forests and timber were issued. They included The Standing Timber (United Kingdom) Order of 1917, an anti-profiteering measure with the object of preventing the purchase of timber by those who simply intended to hold it until the price rose. Under it sales of timber exceeding £300 were prohibited except under licence.[4]

The First World War retarded progress in Ireland. Avondale closed in 1914, partly as a war measure and partly, it seems, because of a decision urged by Forbes as a result of threatened industrial action by the staff in sympathy with the General Strike of 1913. Severe cut-backs were experienced in funding and 30,000 acres of Irish woodland were felled[5] – although Henry[6] reduces this figure to

10,000 acres. Nonetheless, largely due to the recognized dynamism of Forbes, 'The Department's actual programme until 1918 exceeded anything done by the Government in England or Scotland.'⁷

Today it may not be so well understood that, while the war had been anticipated – at least since 1912 – the situation it brought had not. Britain entered the war on the assumption that it would be over in a few weeks. The prospect of a shipping blockade, such as that imposed by the U-boat campaign, was unforeseen, as was static trench warfare with its horrendous appetite for men, and materials – including timber.

Some idea of the demand on British forests may be gauged from the fact that close to one-seventh of all shipping imports to Britain in 1914-15 consisted of timber, the second largest item after grain. By 1916 prices had increased enormously. Demand for timber from Ireland was also great and planting was curtailed. The neglect of English forestry during the preceding sixty or so years was coming home to roost.

To Victorian England, once shipbuilding, the old *raison d'être* for forests, had vanished, no alternative use except sport presented itself. Minds steeped in the self-righteous philosophy of *laissez-faire* found it preferable to permit the forests to lie fallow, while wood requirements were satisfied with cheap imported timber. 'So intimate is the association in the United Kingdom between sport and forestry that even in an estate that is considered to possess some of the best managed woods in England, the silvicultural details have to be accommodated to the hunting and shooting and trees must be taken down "in different places to make cover for foxes".'⁸

Here is an example of unselfconscious hypocrisy making a virtue of necessity. The forests belong to the landed gentry; the forests are a valuable national resource providing (at a handsome profit) the 'Wooden Walls' of Britain. Suddenly the wooden walls are breached and deemed useless. The resource is no longer national, no longer required, no longer economic. It still belongs to the landed gentry. What to do? Why, what else would a gentleman do but adapt it to suit himself in a gentlemanly way? To sell it he could not, to work it he would not, then let it lie and be a source of amusement.

This was the period when gamekeepers with shotguns, mantraps and other 'tools' of sporting activity were more commonplace in English forests than was a woodsman and his axe or seedling. Woodland management had virtually ceased to exist on many private estates. Old skills died with traditional forestry and both owners and their agents tended to have an antipathy to forests that 'did not pay'. The war changed all that. In economic terms alone, in 1915-16 Britain paid some £37m. more than its pre-war value for imported timber, while felling was also considerable. 'By the middle of 1916 considerable areas were being felled. Private woodlands bore the heaviest burden... owners of (even) low-grade woodlands benefited to a considerable extent since in peace-time such areas would not have realized sufficient money to meet the cost of clearing and replanting.'[9]

Many new species were introduced – especially in Ireland – and had matured as timber trees, sometimes on poorer sites and in a shorter time than hardwoods. By the end of the war, because of the wartime clearances and experience, much of it garnered in Ireland, a change had become possible for forestry in England. Much of the drive behind the change now urged was based on the recognition of the strategic importance of native timber. The required purpose and direction were forthcoming in the Forestry Act of 1919.[10] The final step in the rapid progress of modern State forestry in England occurred in 1923 with the passing of the Forestry (Transfer of Woods) Act, 1923, when the Crown Forests of England, previously administered by the Office of Woods, were transferred to the Forestry Commission.

In 1915 Forbes was given responsibility for Ireland on the newly created Home-Grown Timber Committee, formed as a result of the success of the U-boat campaign. Following the sinking of the *Lusitania* off the Old Head of Kinsale, nearly 900,000 tons of shipping was sunk that year alone. The committee was appointed by the President of the Board of Agriculture (the Earl of Selbourne) in November. Its chairman, F. D. Acland, MP, became a very significant figure in the future of forestry so far as both countries were concerned. The committee, which came to an end in 1917 when its duties and responsibilities were transferred to the new Directorate of Timber Supplies, had the following objects:

(1) To organize supplies of home-grown timber, thus reducing imports (and saving shipping-space if not shipping).
(2) To find adequate supplies of timber for use on the Western Front (which consumed vast quantities in duck-boarding, trench-walls and props alone).
(3) To meet the general requirements of the Services.

The United Kingdom was divided into districts, each under the control of an Advisory Officer; Forbes was responsible for Ireland. Such was the volume of saw-log subsequently generated (even in Ireland) that existing sawmills were unable to cope and the committee had to set up its own.

Another sub-committee, also under the chairmanship of Acland, was formed in July 1916. Its terms of reference were wider and more far-reaching: 'To consider and report upon the best means of conserving and developing the woodland resources of the United Kingdom having regard to the experiences gained during the war'. Both Forbes and Schlich were members. It became known as the Acland Committee and its findings as the Acland Report. Its full title was the Forestry Sub-Committee of the Ministry of Reconstruction.

Acland's 'chief interest in forestry lay in its social aspect since he regarded it as a means of improving the conditions of rural life and of contributing to the well-being of the countryside'.[11] The most important of his committee's proposals were:

(4) The conduct of the war had been prejudiced by the country's dependence on imported timber and in future the United Kingdom must safeguard its supplies of timber...
(8) The first essential was the establishment of a Forest Authority which had the necessary funds and powers to acquire and plant land.

These two points have shaped the course of British forest policy since they were formally adopted in 1919. The Acland Committee visited Ireland in 1916, during the course of its enquiries. The Acland Report was published in May 1917, and its findings dominated the Forestry Act of 1919 which set up the Forestry Commission. Of it has been said:

> Only then (1919) was a stated forest policy combined with the means of its proper execution. It was the first time that that combination had been achieved since medieval times... In the intervening centuries except for Acts... there had been no real national forest policy... Now a positive policy of afforestation and silvicultural improvement over all the forest land to provide a national stock of timber managed to provide continuously a truly useful increment and yield was at least announced and pursued.[12]

This statement, though generally correct, is probably a slight overstatement. The Acts (of 1482, 1543 and 1570) certainly *purported* to introduce and determine an economic forest policy, particularly during the Tudor period.

So far as the war was concerned, while Britain depended on imported timber to an enormous extent, Germany could rely almost entirely on its own forest resources. At the end of the war Germany still had half an acre of woodland per head of population to Britain's one tenth of an acre. In terms of quality and management, there was no comparison.

In 1915 Irish forestry received a blow when the British government, reviewing their financial responsibilities in relation to the proposed Government of Ireland Act (the Home Rule Bill, which, though on the Statute Book, had been deferred for the duration), decided that Land Purchase Loans should begin immediately. The effect of this was that forestry costs and expenses, up to then believed to be interest-free, were now deemed to be interest-bearing loans at 3 per cent *ab initio*. The same rate of compound interest applied to outstanding advances which had not been paid up at the end of a thirty-year advance period. What this meant is that the DATI was suddenly faced with an unanticipated demand for retrospective and future interest. The Department had understood, and had had every reason to, that the money advanced towards afforestation was in the form of interest-free grants.

A further delay was occasioned in mid-war when the meeting of the Council of Agriculture for April 1916 was postponed for a month owing, as T. W. Russell, vice president of the Department, put it, to the 'great calamity which has befallen the country'.[13] The reference was to the Easter Rising of 1916. Russell went on to observe that the farmers had not been connected with the insurrection and claimed that this was an indication of the success of agrarian legislation.

The urgent State concern with forestry evidenced by the Acland Committee was synthesized in the reawakening of strategic interest that suddenly pitched the role and function of forestry in a modern state into focus. The neglect of the following principle in the previous half-century became increasingly evident as the war progressed.

Forestry means the growing of trees in dense masses over extensive tracts of land for the production of timber, which is an indispensable material of civilized life.[14]

The problem was tackled energetically once the issues became clear. The question now became one of determining principles for the future of forestry in the United Kingdom when hostilities ended. Recommendations for a long-term forest policy containing a substantial social element were proposed to substitute for immediate wartime policies containing none. One of the major differences between post-war forest policy in Ireland and in the United Kingdom (to the present day) is the importance given to strategic considerations by the United Kingdom and its surprising absence here, except insofar as self-sufficiency of native timber is an often stated aspect of forest policy. Processing, much less an integrated and comprehensive broad-based timber industry, is another matter. That this should be so is hardly surprising when one considers that, even as recently as the Second World War, when for want of timber imports we were forced to level virtually the last of our mature 'Estate' forests, it was hitherto England's strategic interests that most affected Ireland's forests.

The Acland Committee set out to justify 'social' forestry with the claim that immediate profit was not the most important aspect of national afforestation. They drew a parallel with 'the construction of roads, which are of definite national value, though the capital sunk in them might produce no direct return and could not be recovered'. The committee's final report of 1917 is considered to have laid the foundations of future British forest policy. Reporting on timber resources in the United Kingdom in 1917, it also considered forestry policy in Ireland. A number of its findings were certainly of interest to this country. For instance it was argued that, in the United Kingdom generally, the area of land used for rough grazing but capable of growing first-class conifer timber was not less than three, and was probably more than five, million acres, of which two million acres could be used for growing timber without reducing the home production of meat by more than 7 per cent. It was also argued that if this land was used for timber it would give employment to at least ten times the number of workers previously employed in the area. It was not proposed to plant any arable land, although it was argued that a certain amount of such land should be acquired in order to provide smallholdings for forest workers.

The report recommended that the first essential was the establishment of a forest authority with the necessary funds and powers to acquire and plant the land. The estimated cost, for the first ten years, of acquiring and managing 200,000 acres (of 1,770,000 acres proposed) was £3,425,000. It was argued that it might be necessary to invest £15 millions in the undertaking in the first forty years, after which time it was thought it should be self-supporting. Financial return was found difficult to forecast because this depended on wages, prices, rates of interest and so on. But it was pointed out that if the forests yielded less than the current rate of interest on capital invested, they were, nevertheless, a national necessity. The entire thrust of the report was that 'in the interests of national safety more timber should be grown in the United Kingdom'.[15]

The Acland committee had included, besides Forbes, then Chief Forestry Inspector in Ireland, a number of people influential and knowledgeable about forestry. Among them were Schlich, Lord de Vesci and Lord Lovat (who was to become the first chairman of the British Forestry Commission). None of the Irish interests on the committee could be said, in any way, to reflect the nationalist outlook then not only prominent in Irish affairs, but active in the cause of forestry. The Acland Committee endorsed and adopted many of the proposals and recommendations of the 1908 Committee, Forbes observing laconically: 'The work of this committee was followed by what is usually known as "The Acland Report", so-called after the chairman ... which made recommendations ... following in principle much the same lines as the Irish report of 1908.'[16] Forbes's manner of expression may have had something to do with his lack of sympathy with the idea of 'social' forestry.

While the recommendations of the Acland Committee were generally welcomed, the methods adopted to put them into effect caused considerable controversy, particularly the proposal for a separate forestry authority independent of the British Board of Agriculture. Even before the Acland Report was submitted the British Treasury indicated its opposition to any separate forest authority, principally on the grounds of administrative costs. Both the Scottish Agricultural Society and the Irish Forestry Society supported the Acland recommendations. Nevertheless the importance attached to the Acland Report by the British government may be judged from the fact that, in November 1918, while appropriate legislation was being framed, an Interim Forest Authority was appointed. This body drafted the Forestry Bill of 1919.

In Ireland there was immediate conflict between the adherents of the new proposals and the DATI which, like the British Treasury, but perhaps more in the tradition of the bureaucratic principle 'What you have, you hold', opposed the idea of a forest authority independent of itself. Based on its own experience and results the DATI had, as Henry pointed out, a strong case. Moreover, by 1918 everyone was aware that events were in train that, come what might, must affect the entire future of the country and the Departments of State. In such circumstances the DATI case for retaining control of what, up to then, it had administered efficiently and effectively, was strongly reinforced.

In the event a sensible compromise was arrived at by which the Interim Authority would, in Ireland, operate through the DATI on an interim basis. In spite of this arrangement a clash, involving principle and power and heightened by the political situation, was almost unavoidable; when it came it had an adverse effect on long-term decisions. It also affected the different courses that would be followed with regard to forestry in the two countries over the next half-century; an independent forestry commission was established for the United Kingdom, while in Ireland State forestry, administratively and every other way, remained a subordinate division of one Department of State or another. The irony was that in the United Kingdom the modern concept of State forestry was first undertaken in Ireland.

The Act came into force on 1 September 1919, with the Irish political situation still unresolved. The position was complicated because, while the Act provided that the powers and duties of the DATI 'should, insofar as they applied to forestry, be transferred to the Commissioners',[17] the DATI now resisted this. Eventually, in June 1920, common-sense produced a working arrangement by which the DATI became agents of the Forest Commissioners. Under the previous 'interim' arrangement DATI forest staff had, the previous January, been transferred to the Interim Commission and were regarded as being on secondment to the DATI.[18] (The situation was reversed when Saorstát Éireann came into existence in 1922, but certain forestry funds continued to be paid by the Commission for a limited period.) Forbes became Assistant Commissioner of Forestry for Ireland. Forest land was dealt with in different ways; that bought with funds from either the annual forest grant or with cash from the Development Fund was transferred to the Commission, while Avondale and the lands associated with it, purchased with Endowment Fund monies, were leased by them. The Act provided that a Consultative Commission be set up for Ireland, and this, consisting of twenty-four members – one of them Forbes – was established under the chairmanship of Mr T. B. Ponsonby, who had been a member of the Interim Authority. This commission included representatives of landed interests, of the Congested Districts Board and of the Land Commission.

For England the 1919 Forestry Act marked the beginning of modern State forestry and was one of the last significant pieces of English legislation to affect Irish forestry. 'Stated simply, as a defence measure in 1919, British forestry has now a wider part to play in the economy of Britain.'[19] United Kingdom policy resulted from the acceleration of thought and effort about forestry that the war brought on, and was to have included Ireland. But by the time the Act was ready to be put into effect, Ireland had taken another course. However, many of the principles on which it rested had been piloted in this country and could – and perhaps should – have applied equally here. The Acland Committee Report – from which the Forestry Act developed – followed many of the recommendations of the Irish 1908 Report. 'The (successful) programme already laid down for Ireland was not interfered with, and when the Forestry Bill of 1919 was passed, the Forestry Commission took the forestry work of the Department over as a going concern.'[20]

The Act provided for the appointment of eight commissioners who would develop afforestation and timber production. It also provided that the powers and duties of the various institutional bodies hitherto responsible and relating to forestry (including the DATI) be transferred to these commissioners, who were given extensive powers of acquisition and disposal of both land and timber, as well as powers in respect of education, research, subventions, management and organization.

Forbes, no longer Chief Inspector but Assistant Forestry Commissioner for Ireland, set about creating an organizational structure. He divided the country into two divisions, north and south (coinciding with the Government of Ireland Act about to be introduced), of which the heads were John D. Crozier for the north and Kilmartin for the south. There were five district officers, four of them in the field, one in each of four districts. W. E. Ager was in headquarters and David Stewart, Alaister McRae (both, like so many other executives of Irish State forestry, Scots), Matthew J. O'Byrne and Tim O'Donovan (the latter both Avondale-trained) were in the field. Ager was also British.[21]

Meanwhile important political events took place. On an abstentionist policy, Sinn Féin had won an overwhelming majority of the seats in the general election of 1918 and had set up the first Dáil, which met on 21 January 1919. This alternative 'national assembly' gradually acquired some of the reins of office, if only on a limited scale at first. From the outset considerable attention was given to forestry and a forestry committee was appointed.

With the passing of the Government of Ireland Act, the situation became rather obscure. It seems likely that, although the Forestry Commission Act was technically in force, implementation of the policies for which it was enacted were delayed by the war of independence, the subsequent setting up of the Irish Free State and the civil war. In any event all matters concerning forestry in Ireland, in accordance with the provisions of the Treaty signed in December 1921, were handed over either to the Provisional government, or to the government of Northern Ireland, on 1 April 1922.

In spite of the war of independence, forestry work continued in Ireland between 1919 and 1922. Land was acquired, 'about 8000 acres',[22] and nurseries were expanded. But because of wartime felling the overall woodland acreage was much reduced compared with 1914. In spite of the achievements of DATI during the war years, the war had more serious consequences in the Irish context than anywhere else in the (then) United Kingdom, because of the ratio of trees to population; the small amount of woodland in the country; and the relative significance of felling woodlands in the respective countries. Felling was, in any case, a necessary and progressive step in England, where future forestry requirements demanded a change to conifers; the opposite was the case in Ireland.

By 1919 no one but the politically obtuse could doubt that political structures in Ireland were going to change. It was a question of how and when, not if. A reminder of the heady expectancies from afforestation anticipated at the time might not be amiss. Forestry was seen by some as a remedy for many of the country's economic ills. Those who knew better were those with practical knowledge and experience of forestry, mostly with a background opposed to everything the movement for Irish national self-determination stood for. Not unnaturally they were reinforced in their views when Sinn Féin speakers (again not unnaturally, given that they were both rebels and politicians besides being often ill-advised or over-enthusiastic about the subject) made statements about forestry that were unsupported by fact.

There was a tendency for proponents of the major issue, Ireland *v.* England, to become confrontational on minor ones, such as forestry. The general – and erroneous – view encouraged by generations of alien bureaucratic administrators, that Ireland possessed no mineral wealth or natural resources, must have strongly fuelled the expectation of those who supported forestry.

There were not two, but three, authorities in conflict as to the administration of an Irish programme of national forestry: the United Kingdom Forestry Commission; the DATI; and the Forestry Committee of the new Dáil. While the first two were agreed as to general policy, they remained in conflict as to the means of implementing it. And even the most apolitical individual cannot but have been aware, in such a tumultuous political crucible as Ireland then was, that to devise a long-term forestry programme with any certainty of its being implemented, much less being sustained, was an extremely pious expectation. Yet, improbable though it was in the circumstances, responsibility for forestry policy began now to devolve on the Dáil.

The administrative instruments of both governments proceeded, each in its own way, to formulate such policy as they could. Inescapably and in spite of the 1919 Act, the policy of the Dublin Castle administration tended to be short-term. Robert Barton, a Wicklow landlord, who was Dáil Director of Agriculture, took a particular interest in forestry. The Dáil announced a National Arbor Day in June 1919, and some 25,000 to 30,000 trees were planted. When Barton was arrested his work was taken over by Art O'Connor, who was to become a significant figure in London before and during the Treaty negotiations. T. P. Gill, who was, and remained throughout its entire existence, secretary of the DATI, wisely adopted an equivocal role at this time.

Under the agreement with the Forestry Commissioners he was obliged to ensure that the DATI acted as agents for the Commission. Although the Home Rule Bill was shelved, some other form of national self-determination – popularly republican – clearly was not. Moreover, the 'illegal' Republican government, the Dáil, however unlawful it might be in the eyes of English administrators, could not be ignored. Accordingly, whether he entertained any nationalist feelings or not and, perhaps as part of a policy of enlightened self-interest so far as DATI was concerned, Gill 'is also known to have had regular contact with Agricultural Officials of the illicit Republican Regime'.[23]

Following the Truce and the Treaty in 1921, Saorstát Éireann was established and administration, including forestry services, was transferred to the Provisional Government on 1 April 1922. Forestry was assigned to the Department of Lands and Agriculture. At the same time Forbes, the Assistant Commissioner, was retransferred to the Department. 'The total area of land acquired for State afforestation in the Free State up to 30th September, 1922, was 18,741 acres.' But expenditure had also risen – more than doubled, between 1918 and 1922, from £10,000 to £22,000 – the latter figure, however, including the cost of administration.[24]

The thrust of the 1908 Committee Report had been for a national afforestation programme of about 1,000,000 acres, most of which would be in private hands, and up to the foundation of the State this thinking remained central to the whole question of forestry in Ireland. But the War of Independence and the creation of Saorstát Éireann interrupted this continuum. In some mysterious manner thereafter the question of State forestry in Ireland was metamorphosed; from being one of a million acres mostly in private hands, it became one of a million or two million acres, most of it State planted and owned. It is not clear how or when this change came about. It was certainly a principal reason for differences of opinion between the new men of power and the private landowners. It caused tensions between the old and the new regimes, and between Forbes and Mackey. There is little doubt that Forbes, in the face of powerful criticism, and in spite of any impression he might have given to the contrary, clung to the idea of private afforestation as a major part of any State programme, nor is there much doubt that the idea of an all-encompassing State programme and little else was the generally adopted official view from about 1925 onwards. This may – and probably was – partly dictated by the cathartic events of the break with England followed by the civil war.

The proponents of State *qua* State forestry won the day – though perhaps at the expense of much of value being lost. It seems that the government was faced with a dilemma. On the one hand, at a time when to carry the label 'British' was often to have a matter condemned out of hand without a hearing, especially by the rural population, the option was to follow a plan advocated by Forbes and others, but laid down under British rule (and, indeed, being implemented in Britain), which would, to a very considerable extent, require the co-operation and encouragement of large landowners who had openly supported Britain and decried the new State. On the other hand, it could approve, in principle, global State forestry, but find reasons for delaying implementation which would accommodate both budgetary and constituency considerations. Even if the former option was the more practical and economic from the investment point of view, the latter was the more politically desirable. To some extent such considerations affected what happened between 1925 and 1937, shortly after which the Second World War reduced the prospect of privatization on any large scale for many years.

When Saorstát Éireann came into existence the forests consisted of something less than 250,000 acres, mostly in private hands.[25] With very few exceptions, landowners were unsympathetic to the new regime. Their timber exports to Britain and incomes from timber were now at hazard. By a decision of the Irish government they would not benefit from the enlightened approach (and possible financial encouragement) of the Forestry Commission they had helped to bring about. The forest outlook of the new Irish government smacked to them of a ludicrous mixture of ignorance, prejudice, parish-pump politics and utopianism. Nettled by this and by the break with England they, by and large, proceeded to adopt the course followed by their fellows in England after 1862: they neglected their woods. They refused to co-operate with the government, which responded by ignoring them. The effect was to deprive the country of the use of private forests and the accumulated knowledge and skill of the owners. They also suffered from a genuine – if largely unrealistic – fear that if they caused trouble they might be assassinated or burnt out. The consequences were unfortunate for everyone in the forestry industry in Ireland.

CHAPTER TEN

Saorstát Éireann: The First Decade (1922-32)

'The story of the progress of the forest and its efforts down the centuries to serve man glitters with romance' – Mackey.

BETWEEN 1900 AND THE OUTBREAK of the First World War, more seems to have been accomplished in practical terms than in the previous hundred years. But Mackey adds these cautionary words: 'National forests are expensive things to acquire, costly in brains, labour, time, money. Every additional growing year, while the nation's capital is locked up in them, adds to this cost...'[1]

Seán MacBride, who was very active in the movement for national self-determination and whose interest in forestry then and later was to affect its course profoundly, said: 'Interest in forestry waxed and waned from about the beginning of the 1900s. There was tremendous enthusiasm for it up to about the outbreak of the First World War and again Sinn Féin promoted it. Thereafter it waned again until people like Bulmer Hobson and myself and others reactivated the idea...'[2]

When Saorstát Éireann (the Irish Free State) came into being on 6 December 1922 it inherited, along with the other offices and services of national administration, a forest service with plans and requirements of its own. These were, to a large extent, the result of the work of one man – Forbes, an Englishman, who then found himself in a curious situation. He was head of forest service in a country which had just won its freedom from Great Britain. Moreover, his ideas about forestry did not always coincide with those of his new masters. That his feelings were somewhat mixed – even contradictory – may be supposed. His prospects as a British forestry official, albeit in Ireland (and he had every reason to have expected high honours), were diminished. His choices were to return to England and accept what might, or might not, be available in the forest service there, or to remain in Ireland. He chose to remain and deserves all due credit for the choice. He had been awarded the OBE in 1919; he might, with justification, have expected additional honours from a British government. He was to find in Ireland both obstacles and competitors with which to contend.

When the Forestry Commission ceased to finance forestry operations in the new State at the end of March 1922, the burden of financing forestry operations fell, firstly, on the Provisional government and later on the Free State government. At that time planting on a large scale for timber production by modern forestry methods could normally be undertaken, as was pointed out in the 1908 report, only by the State (though forestry is now so profitable that multi-national corporations and financial institutions find the prospect encouraging). But in 1922 it was the State or nothing, and, in spite of the difficulties of the period, by the end of 1923 the State had increased the area of forest land by about 2000 acres.

Modern forestry is a business and the business is tree-farming, the production – and marketing – of trees by the quickest and most cost-effective means. It is also, of course, of national, environmental, social and economic importance. But except for more recent sporadic bursts, none of these criteria appear to have been dominant since 1922.

Given that any sizeable domestic industry in Ireland is export-oriented and that freight and travel costs are a concomitant expense, how well has the forest industry in Ireland progressed since 1922? The answer is equivocal. It has, as its proponents point out, achieved the minimum 1,000,000 acres of forest recommended by the 1908 Committee, and within the eighty years then allotted. But, say its critics, it could – and should – have done a great deal more much earlier. The planning and organization procedures adopted are also criticized on the grounds that they may have been appropriate to a bureaucratic planting competence, but were inhibiting of a marketing dynamic, and attributable to a lack of clear objectives and sustained policy. A combination of land starvation and planting as an end in itself seemed to dominate such policy as existed, resulting in wastage and uneconomic development; a situation now past.

The report of the Department of Agriculture and Technical Instruction for Ireland, 1923-6, dealing with forestry for the period 1 October 1923 to 31 March 1925, includes the following:

> The general position of forestry and woodland management in Saorstát Éireann during the period under review cannot be regarded as satisfactory. The extensive felling of timber and wood which took place during the period 1914-20 as a result of the conditions due to the European War, came to an end, but the replanting of

the cleared land has only been undertaken in a very few cases. A large number of woods, from which the best timber has been removed, are still in a more or less useless condition, and it is to be feared that many of these will be allowed by the owners to revert to rough grazing through the neglect of fences and other forms of protection.[3]

A view of the future put forward by Howitz in 1884 stated:

Of the 20 million acres of Ireland about one quarter is well suited for forest cultivation, a percentage not at all too great. All the ranges of bogs, all the barren and desolate coastlands, and a great many of the very poor grasslands, are natural forests, and should be made such... the island clime, the proximity of the great ocean current, the Gulf Stream, and the abundance of mold or humus to plant in, as well as the geological formations, are all so favourable to forests that it would be unwise to delay the work till some of these advantageous features shall have disappeared...[4]

With very few exceptions the principal forests of northern Europe today owe their origin to the techniques of forestry developed in the seventeenth and eighteenth centuries and have little or nothing to do with natural hardwood forests of oak, hazel and hornbeam. There was a change in dynamic from one of military (mainly naval) strategy to one of economic strategy; from a social requirement dependent on the first, to one integrated with the second; this gave rise to new methodology which in turn produced experimentation with new species.[5]

Mackey attributes the following comment to the French forestry expert G. Huffel:

The early writers often predicted a wood famine and had not coal been discovered their predictions would have come true, because to supply the equivalent of the present coal consumption of France (1931) more than ten times its present total forest area would be necessary... The demand today is for swift rotation – for wood-pulp, turpentine, resin, rayon, alcohol, and for good quality boards and ties.[6]

As a result of the First World War and of industrial development a shortage of timber in developed countries was anticipated. Misgivings about a future timber shortage were expressed:

Apart from fuel, usually obtained from hardwoods, an overwhelming proportion of the timber now required for industrial purposes is produced by coniferous species, which, with few exceptions, are limited to the temperate regions of the northern hemispheres, North America, Northern Europe and Siberia. The position in Europe is unfavourable. Output cannot be increased from existing forests without seriously diminishing resources...[7]

More than one voice made the point: 'The USA is draining not alone her soft but her hardwood forests. The annual consumption of all kinds of timber is assessed at six times her increment.'[8] Mackey also emphasized that:

Practically the whole of the softwood supply of Great Britain has to be imported, only 3 per cent. being produced at home. The value of these imports, including one and a quarter million tons of woodpulp, exceeds £50,000,00 (1934). The pulpwood import figure for Ireland for 1930 was 355 tons, production – nil'.[9] [He goes on:] (spruce) cultivation on a nation-wide scale can lift our exports from an unsound on to a sound basis. Is the necessary land available to justify the adventure? Will our soil and climate successfully grow the spruce tree? Would a wise people be warranted in undertaking the heavy cost of so ambitious an enterprise?'[10]

An encouraging reply – still valid today – emerged from a respected and authoritative source: 'The world's paper consumption has been increasing with unbelievable rapidity... A thriving timber-growing industry is as basic as agriculture. Upon timber crops are founded permanent local industries.'[11]

In 1922 the new government had to come to grips with the realities towards which their political idealism had impelled them. Like successful revolutionary politicians the world over, they were suddenly face to face with those issues which compel even the most high-minded to expediency. The simple truth, the bottom line that every politician in a representative democracy must face, is that no matter how high-minded your principles, if the voters don't like them you're out.

One does not have to be a political expert to appreciate that long-term programmes, the first returns of which will not materialize for a decade at least (and then only on a basis incapable of servicing the capital), are unlikely to be popular when the statutory term of office of a government is five years. The relevant equation is:

$$\frac{\text{Tree rotation approx. 40 years}}{\text{Govt. rotation approx. 5 years}} \times \frac{\text{Degree of interest}}{\text{Funding}} = \text{No effective national forest policy.}$$

Hence the tightrope, not always successfully negotiated, between balancing the budget and holding on to a majority of votes. The risk of making electioneering promises which exceed the capacity of revenues to meet them is an enticing danger. In long-term and controversial forestry there are few votes; yet it is of great national importance. Therein lies a major dilemma and, by and large, that is the case against forestry as it then was: it was long-term; it was expensive; there was little money adequately to fund it; the farming community were opposed to it, and there were few votes in it.

The case *for* forestry, on the other hand, was succinctly stated.

(1) National forestry means securing to the whole common people the greatest benefit that the modern scientific forest can be made to yield.
(2) This benefit must be secured in the shortest time consistent with the attainment of the national objective. This objective is industrial development, leading to economic balance.

(3) Development and balance in turn imply a steady flow of native raw material from the forest to industry in sufficient volume to keep the requisite number of industrial wheels in continuous motion.[12]

John Mackey's extraordinary book is sometimes dismissed because of its verbosity, forceful opinion, and for the reason that 'Mackey was not a forester'. On the other hand it is sometimes eulogized as a sort of 'bible' of reformation silviculture because of its incisive observation.

Writing in the 1930s, Mackey was an acute observer, committed to the national cause. He seemed to be offended by what he considered inefficiency and 'politicking' and to be informed and concerned about forestry, which he saw as the neglected 'Cinderella of the services'. That he wrote in what has been described as 'an extraordinary mixture of unusually pleasing English prose and of harsh American slang in its most virulent form', should not detract from the importance of what he says.

He repeats the importance to the nation of its own timber over and over, but, unlike the uninformed enthusiast, in a manner that displayed a grasp of the problems involved. These, while not inconsiderable, he considered well worth while:[13] 'To conduct her agriculture and her industries and to maintain the life of her people at a normal level of efficiency and comfort, a nation requires to consume a certain quantity of timber.'[14]

In the *Journal of the DATI*, 1916, Forbes, following the line adopted by the committee, estimated a minimum of a million acres of forest to satisfy our needs in timber alone. In 1924 he wrote in the same *Journal*:

The future of woods and plantations in Saorstát Éireann is a matter that ought to concern all sections of the population. It is to be feared, however, that very few realize its significance, or the results likely to follow indifference to, or neglect of, the subject during the next few years... desperate diseases require desperate remedies; and the time has now come when the Irish Free State must definitely make up its mind whether its woods are to be preserved and extended, or whether it is to become a byeword amongst the nations of Europe as the only country without trees...[15]

As always there was a great deal of (often irrelevant) theory about the subject. The basic facts, propounded by O'Rahilly in 1933,[16] were that (a) forests are a natural resource which, under modern management techniques, are also a source of national wealth; (b) Ireland is a natural forest country which has been deforested; (c) irrespective of argument the country requires a considerable amount of timber and timber products each year, most of which is imported; (d) the country is capable of becoming self-sufficient in this respect more rapidly than is possible elsewhere; (e) the country is also capable of growing a superabundance which could be exported for profit;

(f) development of this natural resource to that end would produce economic and social benefits. These facts, with a modification of the second part of (c), remain much the same today.

Forbes had resumed his former office of Chief Forestry Inspector, under the administrative umbrella of the Department of Lands and Agriculture. But he seems to have become increasingly disillusioned, even bitter. Forestry progress in Ireland during the war surpassed similar efforts elsewhere in the United Kingdom and had helped to lay the foundations for the new, comprehensive policies in the 1919 Act. If Forbes had hoped that the advances which, during the war, put Ireland ahead of the rest of the United Kingdom in forestry achievement, would be maintained, he was to be disappointed. After 1922 Ireland was no longer a participant part of the UK and was outside the scope of the British Forestry Commission. No similar alternative was proposed. The original grand design in which Forbes had been a key participant for Ireland seemed to have dissipated in politics, bureaucracy, civil war – and lack of adequate funding. The opposite appeared to be happening in Britain, in spite of 'the effect of the economic and financial pressure of the post-war period, so that by March (1921) steps were being taken by the (British) government to reduce public expenditure and in this, forestry proved to be no exception'.[17]

The question of private forestry continued to concern Forbes. It was basic to the recommendations of the 1908 Report and encouragement of private forestry in England, where it had always been important, was being promoted at this time.

The report on forestry for the period 1923 to 1925 repeats in more detail the observations made by Forbes in his undated report of 1922-3: 'Woods still in the possession of private owners comprise about 84 per cent of the total woodland area of the country... in which... little planting or replanting has been carried out while the bulk of the valuable timber has been removed from them from time to time during the last 20 or 30 years... a large number will undoubtedly disappear in the course of the next 25 years, and measures to prevent this are urgently required.'[18] In Britain 'A census of woodlands and home-grown timber production had been put in hand in 1924... but it was not until 1928 that the report, in which the results appeared, was published.'[19] This gave a total woodland area in Great Britain of 2,958,672 acres, about 50 per cent of it in private hands. The report said: 'It is unlikely that the area of effectively productive woodland in private ownership will, in the future exceed one million acres.' (In fact, in March 1978, the area of privately owned productive woodland in Great Britain exceeded 2,000,000 acres[20] out of a productive total of 4,218,000.) Without the impetus of the First World War the question of

State forestry in Britain might have remained little more than a talking point for years. In Ireland State forestry was to become virtually the only significant form of forestry, a situation not at all envisaged before the foundation of the State, and in Britain it was to become a partner with slightly more than 50 per cent of the total area under forestry.

Forbes continued to deplore the poor state of private forestry. Writing in 1933 and commenting on the increase of afforestation in Saorstát Éireann to 40,000 acres, an increase of some 22,000 acres since 1922, he said:

> without the private planter the wooded condition of the country as a whole must diminish... The objective aimed at by the Irish Forestry Committee was a woodland area of 1,000,000 acres, equal to 5 per cent of the entire country... on the assumption that the private owner would do his share, amounting to roughly three-fourths of the total.[21]

He was criticized for propounding this policy so assiduously. Mackey summarizes a general feeling current at the time among people, informed and otherwise, who held that those responsible for forestry were not exercising the best use of their authority. It is hard to avoid the conclusion that such well-intentioned critics did not advert to what one might call the 'Founding-proposals', namely the Report of 1908 – or that, if they did, they did so selectively. They may have argued that what was relevant then was not relevant in 1933. Moreover, there was substance in the criticism that politicians had been dragging their feet about forestry. The lack of policy – a situation that, except for administrative changes and the limited 'social' afforestation undertaken under Connolly (to which there was considerable opposition within his department) remained pretty well unchanged until 1949 – was very evident. Mackey wrote:

> I conceive it necessary to review the association with Irish forestry of the late Director of the Forestry Branch, Mr A. C. Forbes, a gentleman trained to English estate forestry, emerging thence into a wider sphere, to whom this departmental committee, in the course of its acknowledgements, admits its indebtedness...
>
> Suddenly, in 1921, at the very moment when Ireland was about to share fully in the great new development of British forestry, he found himself cut off from British expansion, locked up in a separate compartment, in the dark whether the new Government which he was called on to serve had been educated to understand forest values...
>
> Meanwhile the country was being depleted of tree growth... In order intelligently to understand our present forest position it is essential (1) to examine the work of the forestry branch with which he was associated from its inception; (2) the difference in outlook of the late Director, as recorded by himself, according as he viewed (a) the forest in Ireland, (b) the forest outside Ireland.
>
> It is common ground that the first duty of a scientific State department, once launched, is to record the results of its experiments; the second, to educate public opinion to the national value of the task upon which it is engaged...

In the report of the Department of Agriculture for the period 1 October 1923, to 31 March 1926, it states 'the general position of forestry *and woodland management* in Saorstát Éireann during the period under review cannot be regarded as satisfactory'. In the report for 1928-29, running into 198 pages, the forestry review occupies six pages. It refers to nursery work and the general conditions of the plantations. And then you stop. The costs of the Forestry Branch must exist... If any reports have been issued at any time up to 1931 by this branch upon forestry education, research, experiment; if any definite scheme for national afforestation was ever prepared; if a single consultative committee representative of the common people was ever appointed; if any soil classifications have been made; if any comparative tables of growth have been printed; if any data, even a summary, from which an intelligent understanding of the character, economic purpose and present position of the several State forests could be gained, have ever appeared, I am unaware of it.

The Minister responsible stated in An Dáil that it was the business of the forestry branch to grow trees. It exactly described the position of the field service of this branch... but what they are growing them for scarcely seems to have occurred to them.

In the practice of forestry, education and research take primary rank. So when Great Britain, a quarter of a century back, was minded to repair her destruction of our forests her first thought was for education... Mr Forbes, informing us that the united areas of Holland and Denmark do not exceed that of these twenty-six Counties, gives their combined forest areas as 1,490,000 acres... then proceeds, 'There is no reason why the Free State should not be maintaining its woods on about one-sixth of this scale.' Is he serious in putting forward this amazing one-sixth Irish equivalent? Mr Forbes keenly evaluates the economic meanings of the modern forest. He is a writer and prolific reviewer on silvicultural subjects... everywhere he displays eagerness to see nations reap the golden harvest of the forest – but in Ireland it is not particularly necessary, says this English gentleman, that even all *lumber* used by us should be produced here. The general desirability, however, of maintaining a sprinkling of tree growth over the surface of the land appeals to him. This from the technical chief of the forest service of a country with a larger national revenue than half a dozen of the States of Europe, States with forest areas ranging up to 27 per cent., is difficult to understand. Whether it be the United States, England, Europe, New Zealand, anywhere outside Ireland, Mr Forbes's anxiety for forestry progress is constant.[22]

This attack was manifestly unjust. In the early 1930s Forbes was approaching the end of his career with the Departments of State concerned with forestry in Ireland. He was embroiled in a political situation he had little sympathy with, and the progress he had anticipated had been stultified for reasons which must have been very frustrating for him. Pronouncements such as that made in the Dáil on 26 June 1929, by Mr Hogan, were hardly encouraging:

Our policy with regard to forestry is completely unambitious... We think there are about 200,000 acres of plantable land in the country. We have acquired about 30,000 acres of that, and we propose to acquire the balance at the rate of about 5,000 acres per annum ... about 200,000 acres of State-owned forests will produce about two-thirds of the total requirements of the country in soft-woods.

Though it may seem at first glance that the proposition: 'Forestry is vital to the interests of the country' should be enough to generate interest, will and capital, this has not been the general case.

Forbes may not have been temperamentally equipped to sympathize with the delicate problems affecting land and land-use that faced the new national administration. His excellent qualities as a forester may in fact have tended to aggravate the situation in which he found himself. Unwittingly or not he seems to have had some of those characteristics – condescension, a patronising manner, impatience with ways other than his own – associated with British colonial administration. Such characteristics were, of course, deeply resented.

That is not to say that many of his criticisms were unjustified. Under the Irish Free State Agreement Act of 1922 a Provisional government acted on an interim basis until the Free State formally came into existence on 6 December that year. Certainly there was indecision and uncertainty as the new regime took over. And some of it stemmed from the fact that relations were sometimes delicate between the new political masters and the established administration. Added to this were the disruptions of the civil war period and its political consequences.

> The administrative changes may have been beneficial in various directions, but they did not prove particularly favourable to State Forestry in general. Every change-over meant delays, interference and uncertainty regarding executive work of various kinds, and of more significance still was the attitude of the administrative bodies or heads to various matters of policy... there was always the uncertainty when a doubtful decison had to be made about the corns likely to be trodden on and the feet they belonged to. Matters of small importance in the abstract often assumed large dimensions when land was touched on in which some particular person was interested, or some right or privilege demanded which was not in the interest of the executive to grant. Ground was often cut from under the feet of officials in more ways than one, and many decisions were reversed which reacted on work already carried out or in contemplation.[23]

The new Minister for Agriculture was Patrick Hogan. On his first day in the Dáil,[24] 15 November 1922, he spoke about forestry. He stated that it was his intention to continue and expand forestry development as far as funds and suitable lands would allow. He added, significantly, that he had no power to compel private owners to plant.

Most of the powerful landowners, who were the mainspring of forest activity under the DATI and the Forestry Committee, belonged to the old British tradition. The people of the country, still unsettled, were not at all well-disposed towards the preservation of any system that smacked of the alien tradition which had just been shown the door. On the other hand woodland owners, with very few exceptions, felt alienated and aggrieved. Many left the country. Some capitalized their forest (and other) assets as quickly and as effectively as possible. One senses that while Forbes understood the last point, he did not, or would not, understand the first. Nationalists who, prior to the Truce of 1921, had been so enthusiastic concerning the future development of forestry were now engaged in civil war (1922-3).

Other than funding, one theme has until now dominated the question of forest policy. It was enunciated by the Minister for Agriculture in his annual report of 1925-6:

The Department do not desire to acquire for afforestation land fit for agricultural purposes which might be capable of being used to form new holdings or enlarge existing ones. With a view, therefore, to preventing such land being acquired for afforestation they have fixed a maximum price at such a figure as to render its sale to the Department for this purpose an uneconomic transaction.

Conflict of interest over land, where there need be none, prejudice and, perhaps most important of all, lack of sufficient will at appropriate political level, have inhibited forestry programmes down the years. Given that there were few votes in forestry and that a 'brave' politician is often a 'dead' politician, political caution is understandable. As we saw from the equation above, a politician who came out in favour of a vast capital commitment to a programme unpopular with those it was aimed at and unprofitable for upwards of twenty years might certainly be called 'brave'.

Reinforcing the underlying policy, 'Hogan had said that he considered the introduction of other agricultural legislation which he had prepared more urgent than the Forestry Bill.'[25] Given that the country was at that time, and for decades remained, overwhelmingly agricultural and was almost entirely dependent on agricultural produce, it was hardly a surprising statement. But the thinking behind it was to inhibit forestry until the 1980s.

Forest financial policy has often been expressed in such generalities as the following quote from Gray, with its apparently contradictory penultimate and final sentences:

Within the prevailing maximum price at any time it has always been the practice to judge the suitability of land for forestry purposes on the basis of whether there is a reasonable prospect of growing a timber crop on the land, valuation within the maximum price being related to its probable productive capacity. Some overall economic factors have entered into the question. For example, blocks of land in themselves too small or too far removed from existing properties for economic supervision have been excluded and similarly blocks which would have involved abnormally high expenditure on fencing because of a high boundary/acreage ratio. Neither have blocks of predominantly low fertility been received with favour except for purposes of experimental work. It has never, however, been the practice to exclude from forestry development lands on which the prospects of an economic return, on the first rotation, were remote because of high initial expenditure in the clearance of heavy scrub preparatory to planting. Such land is usually of a comparatively high productive quality and its acceptance has been justified by the long-term gain to the national economy of securing in perpetuity fuller utilisation of our natural resources even though the forest crop could not sustain the financial burden of the initial reclamation cost...

The quality of the land devoted to forestry is the most critical controllable factor in determining the ultimate profitability of the undertaking but policy has, with unquestionably good reason, tended to exclude highly productive land on which a forester could show excellent profits and to favour the inclusion of sites of marginal productivity. Within the general policy framework the forest authorities over the decades have never been in a position to be fastidious in the acceptance of land suitability because of the constant difficulty of maintaining and enlarging the intake of land to meet progressively increased planting targets.[26]

The contradiction between the general policy outlined in the first paragraph and the statements in the second is not easily reconcilable unless it is appreciated that it represents the difference between Forest Division policy, on the one hand, and government forestry policy on the other, as Gray understood it to be. Of course Gray was intent on making an attractive case for forestry and may have been attempting to reconcile what was desirable, within the limits of such policy as there was, with the actual facts. The point is illustrated by McGlynn:

In a survey I carried out on behalf of the Department we showed that about 48 per cent of the land that we had planted up to that time could be regarded as uneconomic forestry. We were only given the top of the bogs and remote areas. Our forests today are far too scattered, and in areas that are far too small and inaccessible. A lot of land that should never have been planted was planted. I don't think we should be doing that kind of thing, at least until we have built up a viable and economic sound forest industry with a land base.[27]

McGlynn elaborates a fundamental problem of forestry in Ireland which has hitherto proved the bench-mark for decisive rural attitudes:

The land that was wasted by bad farming was more than suitable for forestry and should have been used for it. Ironically the land that we got was equally suitable for the type of use the farmers were making of it. We never got anything from the Land Commission that they could keep from us. Even if farmers didn't make use of their land they were always land hungry. It meant votes and they were able to apply considerable political pressure. Most of them are hard-headed business-men. If there was a proper policy and a proper education of farmers it could be proved to them that it is far better farming to put land under trees than have a few sheep running wild on it. At least 80 per cent. of the cost of establishing a forest is labour. Rural labour at that and there's plenty of it available'.[28]

These two considerations, political and agricultural, remained paramount. The fact is, however, that the soundly based agricultural economy contributed significantly to unnecessary and ridiculous restrictions so far as forestry, another vital national asset, was concerned, and the results have been held to be inimical to and counter-productive of the national interest.

Forestry was only the Cinderella of other Government departments throughout its whole history. It was under Agriculture, it was under Lands for a long time. Indeed it was ironic that Lands and Forestry were under the one department.

...The whole period we were attached to the Land Commission they gave us the dregs of the land of the country. For a long time any land that was worth more than £4 an acre was considered too good for forestry and we weren't allowed to have it... Even as land became more valuable the price put on land which could be acquired for forestry meant that we never could buy any quantity of decent land. We only got the tail-end of estates or something the Land Commission or farmers just couldn't make use of.[29]

The indications now are that the apathy and lack of interest which, in the past, afflicted opinion-moulders and decision-makers before the question of forestry have also diminished. 'The modern forest is the offspring of liberty. When the Irish people awake from indecisiveness to the meanings of that forest, a new chapter quickening the life of a nation will be opened.'[30]

The first Forestry Act of Saorstát Éireann (Forestry Act, 1928) did not come into force until 1930.[31] It was not as comprehensive or as ambitious as was the original British Forestry Commission proposal for a forestry fund of £300,000 spread over a ten-year period. The Act, however, provided the minister with considerable legislative powers. It was also proposed — although it did not happen until the advent of a new administration in 1933 — that forestry should be transferred from Agriculture to the newly created Department of Lands and Fisheries.

Not alone was the new State obliged to stand upon its own economic feet, under the terms of the Treaty it was also forced to contribute, without any *quid pro quo*, to the British Exchequer (by land annuities amongst other charges). Considerable funds also left the country by way of investment capital. This led to the 'Economic War' of 1933-9 with Britain. Since this, too, was essentially to do with agriculture and the rights of farmers — who backed de Valera to the extent of significantly tightening their belts when England closed her markets to Irish produce — the floating of any other issue already unpopular with the farming community, such as forestry, would hardly have been timely. Yet, according to some experts, it would have been a challenge worth facing. 'At least two more million acres of good forest land could be taken and I regard it as waste because there may be a few sheep running over it and they have to be heavily subsidised to make them pay, whereas forestry could be made pay with a far smaller investment.'[32]

At this remove it is not easy to understand why co-operative tree-farming was not encouraged more. The reasons very likely lie in the attitude of farmers to forestry and the reluctance of governments to challenge the farming community 'unnecessarily' when so much depended on their support during the economic war. 'It would have been far better to have encouraged private forestry more and leave it in the hands of the farmers, but encourage them to plant trees on it. Trees can be grown far cheaper that way than can be grown by an army of foresters. You can't have successful forestry without a land use policy.'[33]

The ideal was thus expressed by Mackey: 'The capital invested in the modern forest yields almost incredible riches. We should, without further deadly indecision, proceed with an initial national scheme to reforest a minimum of 2,000,000 acres... in keeping with this nation's abundant resources... to warrant its claim to pride, wisdom, prudence.'[34]

However, although co-operation was more than once advocated, the means and the will to implement a practical scheme remained out of reach. Political pragmatism demanded that priorities be economically realistic and realizable – though there was not always agreement as to what was practicable. It is reasonable to conclude that forestry was not often seen as a political priority. When it was, as we will see, it was often opposed vehemently by, amongst others, the Department of Finance and by the Department of Lands itself.

I promoted forestry very energetically with de Valera when I was his secretary. And actually I broke with de Valera on forestry. It really was the one thing that determined me that we would have to change the government. De Valera was not willing to take action on forestry. He agreed with it, but then he would say 'My officials say it is too long, or too slow a process. Too long-term an investment.'[35]

The same point was made by McGlynn, but with a somewhat different emphasis, forty years later:

The politicians have little interest because they know very little about forestry and politicians as a rule are not very interested in things that don't have votes. It's as simple as that. Many of our industries which have suffered because of their dependency on the raw material from the forests proved inadequate because of past policy and would not only have survived, but have flourished. And many other industries would have flourished alongside them. How can little industries dependent on inadequate forest supplies in this country withstand competition from multi-nationals?[36]

The important factor remained voting power. It may be argued that when farmers became convinced of the value of tree-farming, and were presented with an economically viable and attractive forestry package, the situation would change and there would be immediate votes in forestry.

No doubt the leaders of the fledgling state felt that projects like forestry, however worthwhile in the long-term, must be subordinate to projects less capital intensive and more immediately visibly relevant; it was hardly expedient to make available the resources necessary for such a long-term project as forestry on the scale required at that time.

Inevitably such forestry policy as existed became focused on planting. And, because of its enduring character, forestry was vulnerable to political rather than administrative attitudes, and this showed.

In spite of such restrictions forestry developed, if conservatively. £31,665 was voted as a grant-in-aid for forestry for the financial year 1922-3, and a total area of 2708 acres was planted or replanted.[37] Corresponding figures for the years 1933 to 1938, to take a random period, are shown in Table 1.

Table 1

	1933/4	1934/5	1935/6	1936/7	1937/8
Outgoings	£74,982	112,310	136,352	156,785	200,748
Incomes	£ 5487	5798	6796	9489	11,357
Acquired lands held in acres	52,604	59,554	77,706	93,248	102,960
Planting in acres	4179	5511	6919	7321	7388

In 1936-7 the Forestry Division exhibited at shows for the first time at a cost of £30.

The formation of a detailed forecast related to processing, market expectations and a view of the overall organization requirements and economic strategy involved, is basic to a modern forestry programme. This provides the basis for decisions about acquisitions, processing, marketing strategy and, above all, capital investment. It is axiomatic that the technical questions of species, planting requirements and research are well understood. The prime requirement is the funding to enable these things to be successfully continued for a full rotation period without significant return on the investment for about twenty years, and no return at all for about ten. Land and land-use are important and emotive ingredients in all of this.

It fell in large part to the professional administrators to outline a programme of forestry and to propose a policy to implement it. The fact that the attitudes of some of these administrators may have been equivocal is simply another aspect of a very complex situation. When the Forestry Act of 1928 was introduced as a bill in the Dáil it was said to be 'bureaucratic'. But by and large, such progress as it was practical to make, was made. It introduced compulsory powers of acquisition for forest lands – powers that, for practical reasons, have never been employed.

This initial forestry legislation was carried virtually without opposition. Although the State now had powers to implement a forest programme, the economy was not in a position to provide the capital required. In only one area did substantial forest development occur, that of land acquisition. By the time the Act was introduced a substantial land-bank (substantial compared to what had existed in the past) had been established. But it was still inadequate and much of it, and of what was later acquired, was unsuitable. 'One can have scrub forests and natural forests of an uncultivated nature which is quite a different thing from modern forestry.'[38]

Because of the prohibition on any land which might possibly be used for agriculture, attention turned to mountain land. In the *Journal of the DATI*,[39] Forbes refers to the difficulties of division of mountain land, often commonage, in small blocks held by numerous owners or occupiers, so that the acquisition of such land by a single ownership became extremely difficult.

The maximum price of land for forestry was £4 to £5 an acre.[40] This was a price-range which Forbes argued would most rapidly give a viable return, i.e. an annual yield of forty cubic feet an acre. Even if this argument was sustainable, and doubts have been expressed about it, the principle fell foul of two obvious defects, the prohibition on the purchase of any potential agricultural land, and the bureaucratic tendency towards delay and to buying as cheaply as possible irrespective of other considerations. Forestry was at that time seen not as a crop, but rather as an undeveloped and undefined potential. Even if funding had been available it is unlikely that forestry would have been allowed to compete with agricultural cash crops irrespective of predicted long-term profitability, or the actual use of the land.

While Forbes's attitude to forestry was essentially commercial, his guiding principles derived from two sources, namely Schlich and the 1908 Committee Report. He was not terribly well-disposed to development that seemed to him less than viable, hence his lack of enthusiasm for 'social' forestry and, in particular, his opinion of acquisition 'for the sole reason that it could be obtained at a cheap rate as a fatal policy to pursue, unless the objects in view were other than financial in character, and any saving on land purchase may prove a very costly business in the long run'.[41] Henry, who contended with Forbes for the title 'Mr Forestry', also emphasized the practical approach.

The administration that introduced the first forestry legislation of the State survived without implementing it for slightly less than two years. However, among the important matters which they did introduce before handing over the reins of office to Fianna Fáil in March 1932 was a planting scheme. This, evidently Forbes's brain-child and equally evidently modelled on a similar English scheme introduced in 1927, was to encourage private planting and proposed a payment of a grant of £4 per acre (minimum five acres), paid in two instalments. It attracted little response, but the principle had been established.

Forbes, who had done so much for forestry in Ireland, retired at the age of sixty-five in 1931. He handed over to John Crozier who, although older than Forbes, had worked under him for a long time. Crozier was acting Director of Forestry for the next three years. On 1 December 1933 the Forestry Division was transferred to the new Department of Lands and Fisheries.[42] The new Minister was Senator Joseph Connolly, who was keenly interested in forestry and who was responsible for the first important policy decision about the use of forestry in the west of Ireland. It was announced that the annual planting target was to be increased forthwith from 3500 to 6000 acres a year. This announcement also stated that the decision was based on a commercial cost-benefit-return basis rather than, as hitherto, for unemployment relief. There appears to have been some political posturing involved since, in view of the small amounts of land acquired, the unemployment (social) relief factor had not been very significant so far as forestry was concerned up to then, and the western planting programme of the new government was unemployment relief under a new name. The really notable aspect was the increase in the size of the planting target to 6000 acres a year.

The transfer of forestry from the Department of Agriculture, where it had been subordinate to agricultural principles, was generally welcomed. It accorded with the views of such diverse opinions as those of the Acland Committee, the Irish Forestry Society, and Forbes, who had already expressed the view that 'agriculture and forestry are not twin brothers with a joint interest, but rival claimants for the same estate'.[43]

In a situation where many of the influential and technical directors of the forest service were English or Scottish, tensions arose. Besides those difficulties which might have arisen because of their background, there also arose personal conflicts. This sometimes gave rise to the bizarre situation of individuals being opposed to one another about the implementation of ideas or systems not accepted by rural Ireland generally, and in a political climate with which the individuals concerned did not sympathize. The case of one such man, Dr Mark Loudon Anderson, is illustrative. Dr Anderson was a graduate in forestry from Edinburgh, appointed, by civil service competition, forestry inspector in Ireland in 1926, having previously worked with the British Forestry Commission. He gave up his post after only two years and returned to Britain (and the Forestry Commission there) in 1928, complaining that his work in Ireland 'was not of a sufficiently definite nature'.[44]

It might not be going too far to suppose that Anderson belonged to that type of individual who works well only in a properly structured and ordered environment, as was certainly not the case in the Forestry Service in Ireland between 1926 and 1928. After four years with the British Forestry Commission Anderson returned again to the Irish Forest Service in 1932 at the invitation of F. J. Meyrick, a fellow Scotsman, then secretary of the Department of Agriculture, 'with the understanding of prospects of almost immediate Chief Inspectorship'[45] and, presumably, the additional understanding that the work would be of a sufficiently definite nature to meet his requirements. The first of these expectancies was not fulfilled for some considerable time, but he became acting director when Crozier retired.

Differences arose between Anderson and his new minister, Connolly, who planned an expansion of the forestry programme. He had ideas about afforesting Gaeltacht areas of the West as a means of revitalizing and providing employment in these depressed areas. His ideas were radical, based on social as well as economic considerations, and accorded, perhaps, more with Forbes's one-time outlook and early confidence, than with caution. Such an approach did not sit too well with Anderson, whose approach was very conservative. Moreover, according to Durand, Anderson's reputation was essentially that of a silviculturalist. He may well have been reluctant to put it at risk by supporting a risky programme with which he did not sympathize.[46]

Before the transfer of forestry to his Department took place, Connolly had asked for an outline of Forestry Service policy. A memorandum (November 1933) was prepared setting this out as follows:

(1) The acquisition by purchase and lease.
(2) The production of plants.
(3) The preparation and execution of planting programmes.
(4) The care and maintenance of established forests.
(5) The control of local expenditure.
(6) The provision of fencing material, tools, etc.
(7) The sale of forest-products, timber, etc.
(8) Utilization of unplanted land by grazing. The letting of shooting rights.
(9) The training of foresters and foremen.
(10) Control of general tree-felling, under the provisions of the Forestry Act, 1928.
(11) Payment of grants for planting.
(12) Advising generally with regard to forestry.

This, a statement of work undertaken, did not amount to a policy – as was being argued by Hobson and others. It accords with the generalization that 'The basic objective of policy over the entire period

[1922-59] was the ultimate creation of a home supply of raw timber sufficient to meet home requirements so far as it is possible to grow at home the types of wood required.'[47]

The memorandum might, instructively, be compared with the Report on Post War Forest Policy produced by the British Forestry Commission in 1943, in which essential principles rather than functions are clearly set out.

> Firstly – The recognition by the government of the importance of growing timber.
>
> Secondly – The need for continuity in the national forest policy including the financial aspect.
>
> Thirdly – The existence of a forest authority for formulating and implementing government policy.
>
> Fourthly – The maintenance of a unified forest service of highly qualified personnel.
>
> Fifthly – The provision of adequate services for research, education and information.

It may be argued that these are not policy either, but principles on which policy is based. The point is that they were seen to be the fundamentals from which policy might be developed. It was not a case of functions masquerading as policy.

Hobson, who, of all the serious-minded and influential people concerned with forestry, might have had some influence on the new government, outlined practical working principles for a forest policy in 1934.

> The forest should be created with the definite aim of making it the basis of manufacturing industry upon the largest scale possible, and within the shortest time. Volume of potential profitable employment should be the test by which any programme is judged. With this definite aim in view we should plant the entire area of available land within the shortest time practicable, having regard only to the ability of the Government to finance the operation.
>
> To carry this policy into effect the following measures would be necessary:
>
>> i. The creation of a forest authority technically equipped for its task, having adequate powers and sufficient means.
>> ii. The laying down of a programme of land acquisition and planting on an adequate scale and for a definite and extended period.
>> iii. A definite policy of forest utility which would govern rigidly the character of the forests to be laid down.
>> iv. A financial policy which would enable the work to proceed as planned and without interruption.[48]

In spite of the reputation attributed to him that he did not delegate easily,[49] Anderson made some useful structural and organizational changes, basically involving decentralization and improved administrative restructuring, giving greater autonomy to district inspectors and inspectors. Connolly did not have Anderson's support in his proposals for afforestation in Gaeltachta, nor, apparently, did Anderson look kindly on the inter-departmental Committee on Public Works which, under the chairmanship of Hugo Flinn, Parliamentary Secretary to the Minister for Finance, Dr Ryan (himself keenly interested in forestry), administered the appropriate funds for forestry as a relief work for unemployment. Largely, one is tempted to think, because of the differences of opinion between the minister and his chief technical advisor and forestry administrator, the question of an overall forestry policy or development plan tended to become fudged, attention becoming focused on the subordinate and relatively minor point of the effects of unemployment relief work on the overall forest programme. The Department made the valid point that planting small areas was silviculturally and economically impractical, although it was to engage in this activity for many years to come. But it failed to put forward any proposals which, advantageously for forestry, would make use of available manpower under the Unemployment Relief Scheme – something which, in view of the minister's ambitious schemes for the West, must surely have been feasible. On this issue the differences between Connolly and Anderson heightened. The already established programme which, they claimed, would be interrupted, if not adversely affected, by the introduction of short-term workers and small or uneconomic plots of land (300 acres were held to be the minimum required to ensure an economic forest centre), cut across the anxiety of the minister and the inter-departmental Committee to develop forestry as far as possible in poorer, but especially Gaeltacht, areas, and provide as much employment relief as possible.

All sections of Irish opinion are anxious to prevent the Gaeltacht from perishing. It is in the Gaeltacht that some of the most extensive areas of wasteland occur, and considerable tracts of this land could be planted and should be dealt with at an early stage. This would mean the provision of employment in the Gaeltacht on a scale which could not be provided on an economic basis in any other way.[50]

This view was challenged by Crozier, acting Director of the Forestry Division, that same year (1931) in a memorandum in which he stated, *inter alia*, that 'Planting on the West coast, except in sheltered positions, is of doubtful economic value and could only be carried out after prolonged experiment.'[51]

Attention focussed on these differences when the public at large were invited to submit lists of areas suitable for forestry. 'Although many areas were suggested they were all regarded by the Forestry Division as unsuitable.'[52]

Into this situation there entered a new factor in the person of Otto Reinhardt, an Oberforstmeister in the Prussian Forest Service. He was also a reserve Hauptmann (captain) in the Wehrmacht, a fact that was later to have certain consequences so far as the Forestry Division was concerned. In 1935 he was appointed Director of the Division on a (yearly) contract basis. Anderson, who until then had been acting director, now finally became Chief Inspector, three years after his return to Ireland.

Up to the outbreak of the Second World War, then, the national forestry programme might be said to have suffered from no policy, no money and no votes.

CHAPTER ELEVEN

The Thirties and the Emergency Years (1933-46)

'Every forest in Western Europe today is entirely regenerated and maintained by the valour, the scientific equipment, the labour of man' – Mackey.

NOT SINCE BEFORE THE FIRST WORLD WAR had forestry been so much in the public eye as in the period just before the outbreak of the Second World War. Both Hobson and Mackey had published forthright works on the subject; the country had settled down considerably from the aftermath of the civil war; Fianna Fáil had a new national programme and was setting about the business of settling some of the still running sores that had attended the break with England. Unemployment, emigration, social and industrial development were its main preoccupations. But forestry was still not getting the backing it needed.[1]

In a lecture to the Forestry Society of University College, Dublin, in 1985, Seán MacBride said:

After the Civil War, re-afforestation also formed an important plank of the old Cumann na nGaedheal party – However, much was said about afforestation, but very little was done... In the 1930s a small group comprising Bulmer Hobson, John Busteed of Cork, Mrs Berthon Walters, Alfred O'Rahilly, Luke Duffy and myself started to urge the implementation of a substantial forestry programme.

This group produced a series of pamphlets called 'Towards a New Ireland'. Two of these were on forestry. We tried to get something done through the Banking Commission and Currency Commission about the whole question of investment in our resources.[2]

But, while forestry was advocated by such people, there was, as yet, no forestry policy; nevertheless some practical advance was made.

Funding, as always, was the most difficult problem. It was all very well to argue that returns would amply justify expenditure –

but that was a long way off. Funding for a forestry programme on the scale being advocated (up to 2,000,000 acres), came down to a question of increasing the National Debt or increasing taxation, and on this hurdle most attempts at interesting government with a view to implementation fell.

The advantages are, in the short term, employment and consequent fillip to the economy; in the long term the development of (associated) industries with additional employment; a sustained contribution to the national economy from the matured forests, and, it is alleged, a more equable climate resulting from extensive afforestation. There are also environmental/ecological benefits. Until such time as a commercial dynamic superseded a planting one, administration on the scale required could only be funded and supervised by the State. But these criteria did not generate sufficient capital to develop both forestry *and* forest economics encompassing the concepts – (1) indirect benefits, (2) financial gain.

Commenting on the Report of the Minister for Agriculture, 1931-2, in which the Minister calculated that to meet our timber needs an area of 500,000 acres would be needed, Mackey said:

> More than double our present acreage under timber of all kinds throughout the country – The truth is that so far from being forest-minded, we have developed a slave mind as regards timber. We are content to continue importing multifarious wood-products, and to go on suffering from a dearth of timber in our houses, and in the amenities of life. Even Dr Schlich said in 1886: 'If the whole of the timber were to be grown locally, it would be necessary to increase the area under forest to about 20% of the total area' – i.e., 3.4 million acres of forest for the Irish Free State.[3]

This aspect was further reinforced: 'A country with only 18.7 per cent of forested land cannot afford to allow further deforestation. Only decadent nations have no considerable area of valuable forest in either public or private hands.'[4] Mackey puts it succinctly: 'Directly or indirectly the forest affects every inhabitant of a country. Its presence means wealth; its absence, the absence of wealth.'[5]

Mackey was by no means the only significant critic of the official approach to forestry at that time. 'The same blighting apathy, from which we are not yet free, has for almost a hundred and forty years delayed the developments necessary to save rural Ireland from decay.'[6]

Anyone who has had the privilege of flying at a moderate (5000 feet) height over that part of France from Normandy south beyond Bordeaux to within a few kilometres of the Pyrenees can not but be overawed at the extent of forestry beneath. Blocked in well-defined squares (*coupes*), it stretches unbroken not for acre upon acre, or even kilometre upon kilometre, but for tens of kilometres in every direction to the arc of the horizon, and marches its curve from span to span on every side. This is forestry such as no one has seen in Ireland for hundreds if not thousands of years. And yet, tellingly, it consists entirely of managed, man-made, relatively 'young' forest. It is also clearly profitable. The following indicates both the importance attached to forestry in the French community at local and central government levels, and the fact that afforestation is a sound modern economic policy given appropriate circumstances and encouragement.

O'Rahilly records[7] that the County Council of the Garonne in this region reported to the Forestry Department of the French Ministry of Agriculture, 1882:

> This is one of the most beautiful pages in the history of civilization and progress – a region which thirty years ago, was one of the poorest and most miserable in France, but which may now be ranked amongst the wealthy and prosperous. Where thirty years ago a few thousand poor and unhealthy shepherds were walking about on stilts to raise themselves from the unwholesome flats... are now villages with sawmills, wood-working factories, charcoal kilns, turpentine distilleries; and for more than 70 miles are seen these vast forests interspersed with fertile agricultural lands, where farmers and foresters by the thousand are finding a healthy and prosperous existence.

O'Rahilly also cites The Recess Committee report:

> shelter belts along the west coast would protect lands from... Atlantic storms, which carry with them far inland many ingredients injurious to agricultural produce... The planting of mountains would tend to equalize rainfall and temperature, and prevent upland soils being washed away... There is the value of the timber itself, and of the bye-products (bark, charcoal, leaves, grasses, mosses, shrubs, weeds, fallen branches, resin, pitch, tar, turpentine), the intermediate agricultural products (flax, corn, potatoes, roots, fruits, truffles), and the series of wood-working industries (sawing, pole-making, cart and wheel-making, stave making, handle making, basket-making, etc)... In this extract stress is laid on the social advantages of afforestation as well as its financial returns.

The consensus in favour of an active afforestation policy continued until the outbreak of the Second World War. An editorial comment in the first issue of *Irish Forestry*, the official publication of the Society of Irish Foresters', had this to say: '...those nations which do not meet their timber requirements from within their own borders, and have to depend to a large extent upon imports, would do well to set their house in order in this respect'.[8]

Reinhardt introduced an energy and dynamism not apparent since Forbes's early days, but he was handicapped by being employed on contract, first annually and later on a three-year contract of which he served only fifteen months before the outbreak of the war.

Reinhardt's appointment was in line with the precedent of seeking continental expertise on which to pattern forestry. As we have seen

German influence on forestry in the UK was considerable. It was said of Schlich (who died in 1925), 'To him, rather than to any other one man, belongs the credit of the spread throughout the Empire of modern ideas on forest policy and silviculture... he was able to pass on much that was based on German forest practice...'[9] In the fourteen years between 1889 and 1904, while some eight German works on forestry were translated into English and published in that country and eleven others used as principal reference material, only one book on French forestry, *Elements of Silviculture* by G. Bagneris, who had been a professor at the Forestry School of Nancy, was translated into English.[10]

When it is noted that 'Schlich... tried to encourage (in England) the idea of systematic forest management which had been taught in Germany... when he looked back on this work he regarded it as the least successful of his career and he attributed this to the frequent changes in the ownership of private estates...',[11] his influence in Ireland may be assessed. Forestry in Ireland was different, lending itself more readily to the adoption of continental methods. There were experiments in State afforestation and, equally important, an official outlook that thought of forestry in those terms. It might even be argued that Schlich's teaching fell on more receptive ground in Ireland than in England, and, no doubt, contributed to the relatively better (State) forestry performance in Ireland up to – and during – the First World War.

Henry, first professor of forestry in University College, Dublin, was more impressed by achievements in France and sent many of his students there to study French methods. But one way or another, both academically and in terms of theory, Ireland did not lag in its professional attitude to forestry. Practice was another matter.

Reinhardt was appointed in November 1935 and, with repostings, remained director until September 1939. The outbreak of the war while he was on leave in Germany prevented his return to Ireland. He had, however, made his mark and, had it not been for the war and its disastrous consequences, it is possible that his plans would have placed forestry in this country on a more advanced level than has been reached even today. He was, according to the then Secretary of the Department of Lands, M. Deegan, 'the ideal director'.[12] During his period of office, expenditure, receipts, acquisitions and planting all increased dramatically, although it is fair to say that the returns for the year preceding his appointment, 1934-5 (with the exception of receipts), also show an increase on previous figures. Progress during his period of tenure may, to some extent, be gauged from Table 2.

Table 2[13]

Year	Expenditure	Receipts	Acquisition	Planting
1933/4	£ 74,982	£ 5487	6984 (acres)	4179 (acres)
1937/8	£200,748	£11,357	13,712 "	7389 "

In 1934, the year Mackey's *Forestry in Ireland* was published, and presumably as a result of Connolly's interest in the subject, the proposal was put that there should be a Minister of Forestry, but this did not happen. Mackey was a strong proponent of the idea that there should be at least two million acres of forest plantation, and that work should begin immediately to that end – a proposal also advocated by MacBride and others. Hobson examined the administrative and financial aspects of forestry and published a statement of policy. In general such proposals, with their marked social content, found a favourable response from Connolly. 'This year we expect to plant over 8,000 acres and preparations have been made to reach a figure of not less than 10,000 acres in 1936-7.'[14]

Bulmer Hobson reinforced this interest in forestry with comparisons from other countries in Europe, such as Czechoslovakia and Finland, which had thriving industries based on forestry.[16]

But by the early 1930s the situation that had looked so promising up to the First World War now seemed bleak. Expectancies under the English 1919 Forestry Act and the Forestry Commission had not materialized and the structure recommended by the 1907 Committee had been somehow 'lost'. Private forestry was in a state of neglect and disarray. There were differences of opinion between the political, administrative and technical heads of the forestry division and powerful critics of State forestry were raising their voices against the system. Funding was scarce, policy more so. For the technical forester the only answer was to muddle along as best he could and this, in general, meant simply to plant what he could where he could. Both Deegan, secretary of the Department, and Crozier, then acting director, tried to bring some order into a difficult situation. In particular they were concerned to temper confrontational situations and also what they considered to be excessive, even 'amateurish' enthusiasm, with pragmatism, on which they claimed departmental policy rested.

Though progress was limited, the question of a forest policy was very much alive during the five years from 1934 to the outbreak of the war, but that of private afforestation became increasingly uncertain. Now that the case for State forestry was no longer in doubt, and with characteristic candour, Forbes saw whither the over-emphasis on it (to which he had been a major contributor) was leading, he devoted much of his time and his public utterances to encouraging private forestry, drawing the wrath of Mackey and others on his head for so doing.

THE THIRTIES AND THE EMERGENCY YEARS (1933-46)

When Connolly became the minister responsible for forestry in 1933 the situation had deteriorated to the point where Forbes could write: 'The serious decline in private planting which has taken place during the last fifty years... shows little sign of abating... it is to be feared that planting on an extensive scale must either be carried out by the State or remain undone. In a country like Ireland this is a misfortune...'[16] All of this is notwithstanding that the 200,000 acres[17] of forests in private hands still comprised by far the largest proportion of the total forest land in the country. Of the 50,000 or so acres of State forest, little more than half was planted in the early 1930s. Effectively, therefore, the only usable or marketable timber in the country up to the outbreak of the war remained in largely indifferent private hands which were permitting this limited resource to diminish.[18] Such was the situation when Reinhardt arrived.

In Britain the Forestry Commission had developed a State afforestation programme along the proposed lines. Private forestry was also active. An organization called the Society of Foresters of Great Britain was formed in 1926 'to advance and spread in Great Britain the knowledge of technical forestry in all its aspects'.[19] Similar group action was undertaken in respect of development. Between 1928 and 1978, in spite of the intervening war, the forest area in Britain increased by almost 2,000,000 acres, from 2,958,672 acres to 4,937,500, of which total – in 1978 – 2,000,000 acres was private forestry.

If the last ten years before the First World War were concerned with forestry and the establishment of a forest authority, then the last ten years before the Second World War were correspondingly concerned with the disposal of the timber that the country was now producing... The Commissioners appointed an Inter-Departmental Home Grown Timber Committee in December 1931... and in September 1933 the Committee issued its interim report.[20]

In England, marketing was quickly recognized as the key to a viable forest industry and in 1932 a number of concerned organizations held a conference to consider that question.

A Home-Grown Timber Marketing Council, formed in 1934, had five branches dealing with marketing information and prices when a National Home-Grown Timber Council was appointed by the Forestry Commission in 1935. The Council – welcomed as 'the most important step in the advancement of British forestry since the establishment of the Forestry Commission'[21] – became a limited company concerned with 'economics, statistics, trade-information and propaganda and a limited amount of research'.[22] In the same year the planting programme was increased 50 per cent to provide work for the unemployed and in December 1936, 'The Royal English Forestry Society appointed a committee to prepare a report on the methods which should be adopted to encourage better forest management on private estates.'[23] It was concluded that 'a real comprehensive improvement depends in the first instance on the arousing of forest consciousness among owners, and in providing them with the means to acquire knowledge'.[24] A booklet was published by the society that year to encourage private owners to take more interest in forestry, and in 1938 and 1939 conferences were held as a result of which an Experimental Advisory Committee was formed to ascertain the demand for advice. But its work was halted by the outbreak of war

In considering the progress of afforestation in Ireland during this period it is, for two reasons, necessary to dwell on events in England. The 1919 Forestry Act affected both countries insofar as future policy was envisaged, but in spite of this common springboard, the courses followed in each country in the intervening period were very different. In England, although the purpose was to initiate and develop State forestry, the emphasis continued to be on private planting and landowners, on co-ordinating their endeavour and on encouraging and making available to them the benefits of modern techniques and expertise, the kind of activity that Forbes unavailingly tried to encourage in Ireland. The emphasis in England was also on marketing, where it was of practical concern for Britain's developed forest infrastructure and forests. In Ireland the problems were more those of planting than marketing, while the task of co-ordinating the unwilling interest and co-operation of most private owners was left cautiously to one side. Unfortunately the key question of marketing was so far over the horizon that it was not given adequate attention in forward planning. Coupled with an uneasy situation, fuelled by lack of policy, this was to prove a serious omission in time to come.

During his tenure of office Forbes continued to press the intent of the 1908 Report, the Acland Report and the 1919 Act, to the irritation, no doubt, of those who did not wish to be reminded of them. Mackey and O'Rahilly, taking account of the status of most forest owners, would – possibly correctly – have considered this approach unrealizable and therefore pointless. However, whatever the genuine clamour for more State action on forestry, in a situation where four-fifths of the country's forest land, and virtually all of the mature standing timber, was privately owned, it simply did not make sense for the State to stand by and see this asset frittered away. As Forbes, in vain, tried to point out, it was inherent in every considered official approach to the question of afforestation that, as in England, private forestry would be the principal element. But, far from being developed as the major partner – or even being sustained as a reservoir of mature timber – in a national afforestation programme, private forestry was neglected following the establishment of Saorstát Éireann, and fell into decline.

Besides offering limited grants for private forestry, the 1928 Forestry Act included measures to extend the State's discretionary powers in relation to private woodlands, the most urgent being control of felling and compulsory replanting. The alternative to proper private forest management, that of the State purchasing private woodlands – later to become a feature of State policy – did not recommend itself to Forbes, presumably because it was self-defeating of a programme envisaging a large measure of private afforestation. It would also, of course, diminish the normally healthy tension between private and State enterprise.

A 'conceptual' national forest programme at this time may, therefore, be said to have had three main approaches. First was Forbes's 'professional' one, with its emphasis on shared forestry, as recommended by the 1907 committee. Secondly, there was the 'expansionist' approach of Mackey, Hobson and others, with the emphasis on a virtual State monopoly and the planting of a massive 2,000,000 acres forthwith. Thirdly, there was the State combined commercial/social approach of Connolly, which was a limited and cautious blend of the other two, the social aspect being limited to Gaeltacht areas.

It was Mackey and other outsiders, and not the government or its agents, who raised the important question, still relevant today, of an agronomic survey of the country. 'An agronomic survey complete, the Forest Authority in a State where compulsory powers of land acquisition are imperative, is in a position to put on paper right away the detail for the whole scheme for the planting of two million acres.'[25] The question of compulsory acquisition was a dilemma for any government. The 1928 Act provided for it, but the powers were never used. It is hard to credit that it was ever the intention to do so except, perhaps, in the case of uncooperative private owners should the necessity arise in the national interest, as it might have during the Second World War. These provisions were perhaps included simply as a tactical precautionary measure, probably at the insistence of Forbes, who would have had other uses in mind. The reality of the situation was that with farmers hostile to forestry the likelihood of any government employing compulsory powers to acquire their land for forestry was so unrealistic as to be remote. The importance of an agronomic survey was also emphasized by McGlynn, who strongly advocated private forestry.[26]

Two ancillary developments at about the time of Reinhardt's appointment in 1935 were (1) a scheme to encourage public awareness of forestry, and, (2) the first intake of professionally trained graduates in forestry since 1930.[27]

In spite of the tenuous nature of his contract – and perhaps because he was a professional forester of ability who got on well with everyone

– Reinhardt seems to have helped to reduce the tension between the political and administrative elements in the department. There may also have been some easing of the tension that had developed between the English or English-trained experts and the Irish permanent staffs as a result of his appointment. Reinhardt introduced continental forestry procedures to offset a situation which might, for want of an appropriate dynamic and the required sustained effort, have begun to stagnate. Tension did not, however, entirely disappear. When Reinhardt proposed that a forester and a trainee might spend some months studying in Germany, Anderson, his number two, and a Scot, offered to arrange a 'more satisfactory and helpful' visit to Scotland. Reinhardt prevailed.[28] After his appointment as professor of Forestry at UCD in 1958 Professor Tom Clear maintained the continental connection, and several parties of students visited Germany to study forestry as had Henry's students in France.

Reinhardt well understood that the purpose and function of a modern forest is to supply timber to the market-place profitably. While there was no comparison with the steps being taken in England, under Reinhardt the sale and marketing of timber in Ireland was taken seriously for the first time. It also signposted new problems ahead, including that of the commercial purpose of modern afforestation. Modern commercial forestry means that trees are a crop like any other. Like any other crop they are planted, husbanded, harvested and marketed. The essential difference lies in the time-scale involved. These fundamental facts do not seem to have been as clearly and firmly fixed as they ought in the minds of some of those responsible for State forestry up to and after the outbreak of the Second World War.[29] State forestry suffered from land and financial starvation. Coupled with an approach of acquisition for acquisition's sake, this led to planting on sites that were completely uneconomical and affected the economic viability of State forests. Under the 1928 Act the minimum area to qualify for a planting grant was five acres. Because of the effects of the war this fell to one acre by 1944 and the planting grant had risen to £10 an acre. In 1958 the minimum acreage was reduced to half an acre in Congested Areas. One result was a sprouting of wholly uneconomic tiny copses in private hands, which in time added fuel to the general rural opposition to forestry as a whole.

Departments of State are not geared to operate on commercial principles. They, and the civil servants who run them, belong to a bureaucratic system charged with furthering the use of public money in the interests of the public, to whom they are publicly accountable. Their activities must accommodate to a bureaucratic, not a mercantile, philosophy. Modern forestry, however, is another matter. It is a resource. It is a crop. It is by its nature commercial

– however long-term – and, perhaps, social. But only the State could undertake such a vast scheme as a national programme of forestry. Nevertheless, if such a programme is to become anything more than an uneconomic end in itself, a change in dynamic from the bureaucratic to the mercantile becomes essential at some median point in the development of a viable cycle.

Moreover, where forestry is a State enterprise, the time factor imposes on successive generations of bureaucrats not one requirement, but two different ones very difficult to meet, especially during the initial phase of the programme. These are the marketing orientation referred to above, and the more complicated problem of sustaining the overview (spanning many political, administrative and technical changes) necessary for the required shift in basic philosophy from a planting to a marketing requirement at the appropriate juncture, while remaining sufficiently flexible to respond and adjust advantageously to changing criteria in the market.

The Minister for Agriculture in the late Government said (Prlt. Debates, Vol. 30, No. 6): 'The Forestry Branch exists for the purpose of growing trees. They are concerned with the economics of forestry as such, but they are not concerned – that is the business of other departments – with the use of the trees afterwards.' I do not know what the words 'the economics of forestry' as used here may be intended to mean, but the whole statement amounts to a definite pronouncement proceeding from a responsible source, and running counter to accepted authority.[30]

The proposal that forestry should have a social aspect was resisted by the bureaucrats (for the reasons already noted) and this seems to have helped thrust such questions as those of marketing and commercial procedures into the background, far down the road. It has been argued that an amalgam of caution, lack of funds, a mixture of anxiety and over-enthusiasm (at both political and administrative levels) to inaugurate and establish *any* forestry procedure, encouraged the planting programme to become an end in itself, and also contributed to insufficient concern being devoted to future marketing.

The Second World War placed a burden on the nation's timber resources in some ways greater than that experienced in Britain. While the general requirement was smaller – because of the population factor – and the volume of timber required much less, the timber resources of the country were also much smaller. The lessons of 1914-18 were forgotten. Were it not for estate woodlands the situation during the Emergency of 1939-46 would have been grave so far as timber was concerned. As it was, the State, thrown back on its natural resources, was compelled to use these woods – the neglected remnants of private estates and virtually the only mature timber available – drastically, albeit under force of circumstance and with enlightened enabling and compulsory replanting powers. At the end of the Emergency, private forests in the country were no more than 16 per cent of the small total. The replanting (much of it done by the State, which took over some private woodlands), while it could not replace the mature trees that were removed, did ensure that the overall woodland acreage did not decline.

When the Second World War broke out, reserves of good-quality timber were tiny. State plantations were still under twenty years old, and most private woodlands were neglected and unproductive. The Emergency helped to produce an effect which, if not galvanic, at least led to an increased awareness in the community at large, particularly amongst the agricultural sector, of the need for reafforestation. Laws were introduced to prevent spoliation and to encourage replanting. The Army, and particularly the wartime labour-force, the Construction Corps, played important roles in respect of forestry at this time.[31]

When imported timber again became available after the war a bottle-neck, which has been a problem since the 1950s, became apparent, namely the absence of an adequate processing and marketing strategy for home-grown timber. This was complicated by other factors; due to inadequate processing resources during the war period, home-grown timber had acquired a poor-quality reputation, which tended to endure long after the cause had disappeared; and more home-grown timber was coming on stream. In Britain the hundreds of years of involvement with trees and 'tree-farming' was an available resource with substantial 'invisible' dividends, and forestry quickly reasserted itself after the war.

Naturally a body that had developed forestry virtually from scratch would not, whether it was itself equipped or not to continue to do so, wish to see the glamour of marketing successfully pass to any other outside body, equipped or otherwise. To what extent such an attitude may have played a part in inhibiting enterprise and the development of a marketing strategy for Irish timber it is hard to say. What does seem to be the case is that full awareness of the necessity to transfer from a bureaucratic acquisition and planting programme to a mercantile one does not seem to have been part of any long-term strategy. This point was put to MacBride.

Nobody but the State could have undertaken a planting programme of the size contemplated and it therefore had to have compulsory acquisition powers and therefore it had to be run by bureaucrats initially. But there comes a stage when the crops are maturing when it becomes a commercial proposition. What must then be decided is how the timber will be used... That's where the governments have fallen through completely. Even on the very modest plantation rate which has resulted from my proposals in 1948, they made absolutely no plans for utilization. It's like planting a field of wheat and having no millers, completely ridiculous. You know to the day, nearly, when timber is going to come on stream and you should have everything waiting for it. Nor did we do anything about marketing.

Forestry is too important a matter economically and commercially to be left to bureaucrats who, through no fault of their own, are not commercially oriented. There should be the equivalent to a Forestry Commission or something like Bord na Móna who have done such a good job of work.[32]

Had nothing else occurred to obscure it this want of adequate commercial advance planning would probably have become evident when the first substantial croppings of the new State plantations took place. But when the Second World War broke out in September 1939, if State control of forestry had not already existed, it would then certainly have been imposed. In the event it became greater than ever just about the time the question of marketing should have received serious consideration under peace-time conditions. The fundamental change of control in wartime was neither possible nor desirable. While the need for wartime constraints disappeared in 1945, the problem did not.

While Ireland was far from being self-sufficient during the Emergency period, resources and ingenuity were energetically tapped to provide acceptable standards so far as most basics were concerned. New departures and innovations in agriculture and horticulture showed what could be achieved. But with forestry matters things were different. War does not allow for long-term planning, especially when coupled with labour/financial shortages. Whatever the long-term prospects and benefits of afforestation it was not a first priority during the Emergency. And the situation, as we have seen, was not promising.

Making his presidential address at the inaugural annual general meeting of the Society of Irish Foresters in February 1943, Anderson touched, if obliquely, on the lack of marketing facilities available to private forestry.[33]

Mackey was moved to a furnace of indignation by what he alleged was an unacceptable level of indifference to forestry by government.

Instead of setting down what the forest means I set down some things that it does not mean, numbering them one to six.

(1) Speaking in An Dáil, June, 1929, the then Minister for Posts and Telegraphs is reported as saying – I would draw attention, I trust usefully, to this amazing interpretation of modern forestry – 'it is generally accepted that the profits from afforestation are very small, but this loss (*sic*) to the State will, however, be compensated by improved climatic conditions, not to mention scenery'. First, then, to start with this Minister, the forest does not mean the establishment of a State Department to provide sylvan reaches along this river, to enhance the distant prospect from that hotel, to grow great beeches with boughs stretching widely out on all sides for the wind to rustle.

(2) It is not the business of a forest authority to provide for the establishment of shelter belts. Grants in money or kind to individual occupiers for the provision of these valuable screens is the business of a Department of Agriculture, obtaining its plants from the national nurseries.

(3) The forest has no remotest association – save through its research laboratories – with the great and nationally important industry of fruit culture.

(4) Forests are not plantations of trees to be laid down quincunxly in a crazy patch like the famous Kentucky quilts of 'Aunt Jane'.

(5) The modern forest ends many dreams, including the dream that all trees are born free and equal. The forest does not breed communists. There is a commonwealth, but no communism in the forest.

(6) The scientific forest does not contemplate the provision of game coverts. The 'gunner' may be a useful member of society, but he must find range for his gun outside the modern forest. Its concern is with production, not destruction; with life, not death; with industry, not idleness.[34]

Mackey also made a useful and practical comparison with what was then being done about forestry in New Zealand – a country with many relevant features in common with Ireland.

When Reinhardt returned to Germany Anderson became director, a post he retained throughout the war, until 1946. The isolation experienced during the Emergency tended to aggravate existing problems. Connolly had been succeeded in turn by Frank Aiken, Gerard Boland and Thomas Derrig. The Emergency ended with Seán Moylan as Minister for Lands. Imports of timber had declined. Inadequately treated or inadequately dried native timber was all that was available for general use. In 1940 the government announced that it intended to secure sufficient planting to satisfy the entire home requirement – namely five hundred thousand acres.

Ironically it was the past neglect and consequent decline of the mature woodlands of Estate Forestry that was the single biggest reason for the impoverished state of Irish forestry in 1939. State forests on the whole were immature and Estate forests, with their owners, seemed to have drifted off into some paradoxical Anglo-Irish Celtic Twilight of their own.

The war reduced private forestry in the country from some 210,000 mature, but largely neglected, acres in 1939, to around 94,871 acres, of which only about 40,000 were 'fully and satisfactorily stocked' in 1950.[35]

There was one notable development as a result of the war and the restrictions imposed by it. Seed for afforestation, most of which had hitherto been imported, was in short supply when war broke out. As a result seed supply from some native sources was expanded rapidly and successfully and continues today. Comparison of Table 3 with the figures given earlier for 1933-4 and 1937-8 demonstrates that while expenditure and receipts were higher than ever before, acquisition and planting declined substantially during the Emergency.

Table 3[36]

Year	Expenditure	Receipts	Acquisitions	Planting
1939/40	£215,382	£17,556	12,722 acres	6815 acres
1940/1	£204,374	£45,102	7987 "	5961 "
1944/5	£230,982	£94,096	3307 "	4230 "

In 1944, four years before his government was defeated, de Valera asked his ministers to submit an Advisory Memorandum on post-war policy. Connolly included a section on afforestation and timber in a comprehensive memorandum in which he stated: 'As to the future of our activities we must, as quickly as possible, try to climb to a programme of 20,000 acres per annum and, if possible, go beyond it.'[37]

CHAPTER TWELVE

A Policy Defined
(1944-58)

'Ireland is basically a forest country' – Mackey.

THE QUESTION OF FORESTRY PROVIDING unemployment relief has been controversial since State forestry was first mooted in these islands. On the face of it it seems to be an undertaking ideally suitable for this purpose, but in practice it has not always worked as well as expected. As a means of providing local, seasonal and permanent employment it has obvious limits. The number of men that can be employed directly in forestry locally varies with the size and stage of development of the forest, before a levelling-off is reached. It can provide winter work for farm employees, extending their productivity, usefulness and incomes. As a means of alleviating large-scale urban unemployment it has not been successful except as a very temporary relief measure.

On the other hand, the employment that could be generated by forestry is very extensive if one takes into account consequential 'down-stream' or indirect industrial and service development – saw-milling, pulp-milling, transport, shipping, processing, joinery and wood-goods, building, paper-mills and other subsidiary industries, besides transportation and logging. This is where the concealed employment value of afforestation lies, and it requires to be taken into account in an assessment of forestry employment prospects. Direct employment benefits are mainly related to permanent employment in large-scale State and commercial forestry and employment resulting from 'tree-farming' by farmers and smallholders.

In Britain, at times of serious nationwide unemployment, considerable importance was attached to forestry as a means of relief employment throughout the country. The London Afforestation

Conference of 1907 was held specifically to consider 'what steps could be taken to plant new woodland areas and so provide work for the unemployed'.[1] But little in this direction 'seems to have been achieved and Lord Carrington (President of the Committee and President of the Board of Agriculture) is reported to have said "that the Conference has knocked on the head the delusion that afforestation was a great cure for want of employment"'.[2]

Two years later, in 1909, a Royal Commission reported 'there are approximately 9,000,000 acres in the United Kingdom (3,000,000 of them in Ireland) which could be planted without substantially encroaching on agricultural land ... permanent employment would be provided on the basis of one man per 100 acres planted',[3] and the Commissioners 'recorded the following opinion on the subject of unemployed labour in relation to afforestation: that a national scheme of afforestation would contribute to the solution of the unemployed problem...'

Advocates of forestry as a means of dealing with unemployment continued to press the case up to the outbreak of the Second World War. In 1936 the British government increased the Forestry Fund with the intention of increasing the planting programme by 50 per cent, so providing work for the unemployed. Since the end of the war the emphasis on forestry in this regard has been somewhat less.

In Ireland forestry as a means of employment has always, correctly, had its advocates. The principal reasons are simple and obvious. Until very recently Ireland had insufficient forest to meet the social and economic requirements of the community, therefore afforestation was a communal necessity. That of itself generates additional employment, irrespective of any indirect employment resulting from a well-managed and viable industry. The tricky question of whether emphasis should be given to any social role of forestry related to unemployment, or if forestry policy should be based solely on economic principles, would not resolve itself. Connolly believed these aspects not to be mutually exclusive. In presenting his estimate for the year 1935-6 he argued strongly in support of this approach.

In his analysis of policy since the foundation of the State, Gray says: 'Policy declarations were accompanied by references to the advantages to the national economy of the elimination of an adverse trade balance in timber and timber products, the advantages of self-sufficiency in timber and the incidental social gains by way of provision of employment, etc.'[4] As far back as 1885 Howitz had pointed to the West as suitable for afforestation and employment,[5] and O'Rahilly cites Bulmer Hobson: 'The forest alone is capable of giving an economic future to the Irish speaking districts.'[6]

In 1931 Hobson wrote: 'The actual growing of timber would give a volume of employment equal to ten times that now given by non-arable land... Given a forest area of 2,000,000 acres, the labour required for forestry operations alone would effect a revolutionary change in the conditions of employment in a great part of rural Ireland.'[7]

Henry, one of the great authorities on forestry in Ireland at this period, who advocated a cost-effective approach, wrote in 1924 that the results of 'a large-scale' experiment carried out in Great Britain in 1921-2 'show that afforestation is an effective method of giving work to the unemployed. The great advantage of forestry work is that it is mainly done in winter. This is the crux of the problem.'[8]

Conditions today are very different from those of 1933, let alone 1924, and a modern assessment of the employment value of forestry remains to be done. Even if the results were not so positively encouraging, afforestation demonstrably offers the prospect of both direct employment and significant down-stream employment. The British Forestry Commission had sanctioned fourteen unemployment schemes for Ireland, comprising 1057 acres for planting and 220 for making ready, before they ceased to finance forestry operations in Ireland in 1922.

Henry wrote: '[I] am of the opinion that money spent on bonuses to private owners for planting under supervison, would be a ready way to abate unemployment in rural districts...'[9] Mackey, as usual, had no doubts on the matter: 'Two million acres of modern scientific forest will give direct forest employment to a minimum of 25,000 men.'[10]

On the other hand, Mackey is scathing of some parliamentary representatives who dealt with forestry. Citing the Dáil debates of 1929 he comments: 'The approach of some deputies showed that they had not made themselves acquainted with the modern science of forestry. Mr Heffernan (Minister for Posts and Telegraphs) said... "The question of timber imports was not cognate to a debate on forestry".'[11]

Seán MacBride, as recently as 1985, put the matter in context thus: 'Of all public works forestry has to have a labour content. For every £1000 spent on forestry 80 per cent of it is in labour, whereas in roadwork you only get about a 30 per cent labour content. It is a way of providing employment on the spot without having to disperse the population.'[12]

Practical achievements in forestry by the State between 1922 and 1946 may be quantified. It is harder to identify any realistic forward planning on which forest policy might have been based in the immediate aftermath of the war. As usual, funds were scarce and forestry was not given priority.

In the aftermath of the Emergency, restoring the nation to a level of peace-time progress and prosperity was the first priority. There was an obvious and understandable reluctance to invest capital in a long-term project like forestry, still politically unpopular in rural areas, where Fianna Fáil was strongest. Consequently the approach to forestry tended to be gradualist and was aptly summarized: 'Forestry is not an economic panacea for our present difficulties. Putting young trees in the ground will not put more food on the tables.'[13]

By 1946 significant changes had taken place. The status of private forestry declined, and the role envisaged for it by the 1908 Committee as a major feature of a national programme of afforestation had evaporated. On the other hand State forestry, in spite of vicissitudes, had made considerable progress, but was still very underdeveloped. Acquisition was at the rate of about 5000 acres a year and something in the region of 143,000 acres had been acquired. Almost 2000 people were employed, and over £300,000 was expended in 1946 by the Forestry Division of the Department of Lands.[14]

Other teething problems remained, among them administrative difficulties, lack of finance and lack of political will. No policy had emerged to implement either the recommendations of the 1908 Report, or any alternative. Such as it was, policy consisted of an indeterminate planting programme with a vague target of about 500,000 acres, to be achieved at some equally vague date in the vague future.

While much had been achieved by 1946, the whole of Irish forestry was considerably less than the sum of its parts. After the Emergency unemployment was a major problem; men demobilized from the army did not always easily find work. For want of capital, forestry, which might have provided temporary employment and laid up an investment that would continue into the future and mature profitably at regular intervals, remained neglected. The industrious and concerned civil servants who administered forestry were in an unenviable situation. Themselves the products of conservative bureaucracy, they were hampered by being merely a sub-division of the Department of Lands, with whom they had a fundamental conflict of interest. Even were this not the case their ability to adopt the necessary commercial outlook is doubtful, as later events suggest. Without decisive intervention from powerful government sources it seemed as if forestry was to continue, policyless and on a catch-as-catch-can basis, its progress determined mainly by the strength and commitment of whatever individual happened to be in charge.

If the importance of forestry to a nation extends beyond immediate economic considerations, as most of those engaged in forestry would argue, one is faced with the proposition that in Ireland an important resource remained for too long inadequately developed. Yet, as H. J. Gray points out, 'declarations of Forest policy over the entire period from 1922 to 1958 were not made subject to any specific qualification as to acceptable level of financial return'.[15] What he does not point out is the very small scale involved prior to 1936, or that not until 1948 did a forest programme based on policy in any real sense emerge. At the time, of course, Gray was a civil servant with great responsibility in the Forestry Division.

Areas were spoken of in terms of economic planting and acquisition, but it appears to have been in a narrow, rather than a general economic sense. Afforestation was pursued in the interest of maximum economic return, but within a strictly limited sphere qualified by social and political considerations. The term 'economic forestry' lost its meaning because of lack of definition of standards. Normally it was spoken of as a project *per se* without reference to any other possible use for the land resource utilised thereby.[16]

There is no doubt that more might have been achieved under a clear and pursued policy, if that is what is meant. The underlying reasons were more complex than suggested here. Forest programming had become subordinate to short-term, mainly political, interests and little effort was made to realize the national forest programme with the result, since the terms of reference were inappropriate, that forestry was uneconomic almost by definition. Such projects as were in themselves economic tended to be local and fractured. The attempt by Connolly to develop forestry in Gaeltacht areas 'even at expenditure slightly above that which was considered the economic standard of the moment'[17] was out of step with both long- and short-term programming.

It is instructive to observe what was happening in England during the same period – 1939 to 1946. Even before the war the importance of woodlands was evident to those responsible in Britain; 'the possibility of another war had already been anticipated... and some preliminary steps taken to meet this contingency'.[18] Once hostilities commenced the Forestry Commission went on a wartime footing and was divided into management and timber control divisions. More significantly, as early as 1941, within two years of the outbreak of the war, and in what came to be called 'England's Darkest Hour', considerable attention was being devoted to the question of post-war forestry. A report by the Royal English Forestry Society was put forward in November of that year, together with submissions from other sources, and in 1943 the Commission issued its own report on post-war forest policy.[19]

It proposed that there should be five million acres of forest in the country in the future which would be systematically managed and developed – three million acres of land to be afforested and two million of existing woodland. This acreage would take fifty years to achieve. The report also laid down the following five essentials for British forestry:

Firstly, the recognition by the government of the importance of growing timber in this country.

Secondly, the need for continuity in the national forestry policy including the financial aspect.

Thirdly, the existence of a Forest Authority for formulating and implementing government policy.

Fourthly, the maintenance of a unified Forest Service of highly qualified personnel.

Fifthly, the provision of adequate services for research, education and information.[20]

The report also considered private forestry and stated several principles considered essential, including provision for State acquisition and control as well as financial assistance. This led to a supplementary report on private forestry the following year.[21] It gave considerable attention to the question of financial assistance for private woodlands, out of which was to come the Forestry Act, 1947.[22]

The report on post-war forest policy 'not only brought about a reinvigoration of British forestry but also provided a firm foundation for private forestry through the medium of the Dedication Scheme (introduced under the new Act). This gave owners a new incentive to replant their woods and to manage them in accordance with the rules of good forestry'.[23] An Order in 1944 consolidated regulations on timber-felling, acquisition and disposal, and this procedure continued as required until the end of the war in May 1945.

Within six weeks of the end of the war in Europe Britain introduced a new Forestry Act.[24] A census made between 1947 and 1949 showed that 484,000 acres of forest were either 'felled since September 1939' or '"Devastated", that is to say that all worthwhile timber had been cut'.[25]

In Ireland there was little evidence that forestry, or its implications for the future of the country, was receiving similar attention. But a change of government in 1948 brought with it a profound and controversial change of policy. A similar improvement had occurred at the time of the previous change of administration in 1932, when Fianna Fáil first came to office. In each case the impetus given to forestry resulted from the enthusiasm of one individual, a further indication, if one were needed, that a practical, sustained policy did not exist.[26] The individuals in question were Senator Joseph Connolly, whose contribution has already been considered, and Seán MacBride of the Clann na Poblachta Party, who became Minister for External Relations in the first Inter-Party Government (1948-52).

It was now twenty-seven years since the Treaty and twenty-five since the subsequent civil war had ended. United resistance to economic pressure by Britain had been more or less successful during the 'Economic War' of the 1930s. The 1937 Constitution defined the basis of nationhood, and neutrality had been successfully maintained during the Second World War; independence and national sovereignty were accepted facts for a new generation that had matured with no experience of alien government. The subtle shift in political emphasis, that led from idealistic to bread-and-butter issues, now began.

Clann na Poblachta was founded essentially on the principles MacBride had been promoting since his break with de Valera in 1945. The appeal was to youth, to the developing floating vote and to those who, though nominal supporters of Fianna Fáil or Fine Gael and unwilling to change sides, nevertheless were sufficiently disillusioned with established political attitudes to contemplate a *via media*. Clann na Poblachta seemed to offer one. The first Inter-Party Government (in which the dominant party was Fine Gael, supported by Labour, Clann na Poblachta and Clann na Talmhan) took office as the first government of the State not formed from a single party.

MacBride's approach to forestry was characteristically enthusiastic and committed and illuminated with a broad general knowledge of the subject. He had been de Valera's secretary, and had broken with him on this question. He also made agreement on policy a condition of his participation in the new government. MacBride's commitment was such that, according to Viney, his 'interest in afforestation struck some civil servants of the day as "almost fanatical"'.[27]

In a sense MacBride's proposal was almost that of proceeding from a gradualist policy to one of instant expansion (to over three times the best annual planting rate up to then, as it turned out), without any intervening period of acclimatization for the Division. Advance planning, insofar as it related to profitable marketing, remained – and for a considerable period continued to remain – as elusive as before.

But the 'MacBride Expansion', as it might be called, was a positive move in the direction of a long-term forest programme. It represented the first major change of forest policy at government level since the State had come into being, and it brought ancillary, far-reaching, though seldom acknowledged, consequences in its wake. Obviously an expanded programme of such dimensions would require considerable organization and an expanded administration. This, inevitably, swung the balance in favour of administrative control of the Forestry Division, a matter which had been at issue for some time. Hitherto all the chief officers of the Division had been technical forestry experts.

That situation was now about to change. A transfer of effective power within the Forestry Division of the Department of Lands, so that the administrative and no longer the technical branch held the controlling reins (a development essential – if not universally welcomed – within the Division), was now also heralded.[28]

Although MacBride was the prime mover of the new approach, as Minister for External Affairs he was not directly responsible for its implementation, which fell on the shoulders of the Minister for Lands, Joseph Blowick. That did not alter the extraordinary situation of a minister, with no direct responsibility in the matter, being the power-house and the decisive influence behind the sudden burst of forestry activity that followed. It made it, perhaps, easier for him that at least some of the civil servants with whom he dealt abrasively on the issue were not of his own department, for it is a truism that a Minister falls foul of his own civil servants at his peril. It was with the civil servants of Finance and Lands, not Foreign Affairs, that MacBride crossed swords over forestry.

As already noted, the Department of Lands and the Forestry Division had interests which were mutually exclusive. Inescapably, therefore, Blowick was subject to contradictory advice and pressure. A natural reaction, where such pressures indicated a thorny path, would have been to take the line of least resistance, but behind Blowick loomed the formidable figure of MacBride to prevent such a course. Matters were, however, additionally complicated by the fact that MacBride's policy was strongly opposed, on traditional financial grounds, by the Department of Finance. 'The beginnings of economic planning are to be found in this first Coalition period, predating by seven years T. K. Whitaker's famous memorandum ... which is still popularly believed to be the point of departure for Government Economic Planning.'[29]

MacBride, whose commitment to forestry never wavered (it was to lead him into ever-deepening conflict with officials of the Department of Finance), was a key figure in this development. Another was Patrick McGilligan, described by one of his senior officials as 'a model of the austerity he preached ... who could be persuaded to yield to practical necessity. His tenacity and courage and the classical force and simplicity with which he could express himself are revealed many times'.[30]

Once in office, MacBride set about implementing his ideas on forestry. Two matters required his immediate attention, firstly the funding of an afforestation programme, and secondly the amount – and the rate – to be planted. He faced considerable opposition from the Department of Finance, partly because of its reluctance to provide funds for non-productive capital development programmes like forestry and partly, no doubt, because MacBride's public utterances and declared intentions in office persuaded Department of Finance officials that, as Fanning puts it, 'for the first time since the State's inception, financial orthodoxies were being seriously threatened from within the government'.[31] MacBride himself, with an ingenuousness perhaps more apparent than real, observed: 'I could never understand why the Department of Finance was so hostile to the idea of forestry. They were also opposed to the Shannon Scheme and that didn't make sense either. They said it was nonsense. They said this was Socialism – that it was Communism.'[32]

After eleven months in office the government published a White Paper, *Ireland's Long-Term Recovery Programme*, in January 1949. The Department of Finance was 'apathetic, if not hostile, to the ideas of economic programming and it is noteworthy that responsibility for producing the White Paper rested with the Department of External Affairs under Seán MacBride...'[33] Whitaker, who was Department of Finance representative on an Inter-departmental Committee which prepared Ireland's European (Long-Term) Recovery Programme, refutes this suggestion of hostility.[34] But clearly MacBride, though Minister for External Affairs, affected the policies, and therefore the procedures, of the Department of Finance no less than those of the Department of Lands.

MacBride subscribed in general to the views on currency and credit in the Minority Report presented to the Banking Commission by P. J. O'Loghlen; and such notable critics of the Finance attitude as Bulmer Hobson... which... clearly marked his antagonism towards Finance. MacBride himself acknowledged his sense of 'outrage' at the Finance attitude to the questions of economic planning and investment in Ireland from the moment he entered office.[35]

The practice of direct negotiations by the Department of Finance with British Treasury counterparts was (more or less) successfully halted by the Department of External Affairs under MacBride, which insisted that all such negotiations be conducted under its auspices. Although it was but one aspect of many, and its significance was dwarfed by issues of immediate moment, on no question were divergences of view and policy between the Department and MacBride more clearly seen than on that of forestry.

The Marshall Plan (European Economic Recovery Programme) Funds, the American aid-package to post-war Europe of which Ireland proposed to avail, offer a prime example of reasons for inter-departmental conflict. Through them MacBride won a significant 'victory' for his forestry proposals.

I had discussed in detail our afforestation programme with Paul Hoffman (Director of the Economic Co-operation Administration), Averell Harriman, the European representative in Paris, Dr Carrigan, the ECA representative in Ireland, and the American ambassador to Ireland, Mr George Garrett. They supported fully my views and understood the importance of making Marshall Aid assistance conditional upon our compliance with the afforestation programme which had been agreed upon by the government. I did all this deliberately and systematically to ensure that the Department of Finance senior civil servants would not be able to tamper with afforestation policy decided upon by the government...

The policy adopted in 1948 of planting 10,000 hectares (25,000 acres) of new forests annually to achieve a total productive area of State forests totalling 470,000 hectares by 1990 was re-affirmed 10 years later, in 1958, in the Government White Paper entitled *Five Year Programme for Economic Expansion, 1958*.[36]

The relevant paragraphs of the *White Paper on Economic Recovery* are 48 and 49:

48. During the war years Ireland's stock of soft woods was depleted by 60 per cent; this depletion, taken in conjunction with the country's requirements of timber, renders it imperative to engage in large-scale reafforestation. While this will not bring any immediate results (save in so far as it provides employment) it forms part of Ireland's long-term planning.

49. So far, reafforestation has been carried out only on a very small scale; in recent years, the rate of planting in the State Forests has been approximately 6000 acres per year. It is proposed to step this up to 25,000 acres per year. The aim will be to plant a million acres by annual increments of 25,000 acres. At maturity, and assuming an average yield of 3500 cubic feet of timber per acre on a fifty year rotation, the annual return on a 25,000 acres felling will be of the order of 87.5 million cubic feet or approximately 350,000 standards of sawn timber.

Forestry policy 'was enunciated not by the Department of Lands but by the Department of External Affairs in just two brief paragraphs of a 35 page White Paper...'[37] Fanning points out: 'MacBride, as Ireland's Minister for External Affairs, was a Vice President of the Council of Ministers of the OEEC and was given the responsibility by the OEEC to negotiate with the ECA on behalf of the sixteen countries involved...'[38]

MacBride's differences with the Department of Finance had emerged clearly by December 1948 when he sought cabinet authority to tell George Garrett, the American Minister to Ireland, how the government proposed to use ECA funds, and to say also 'what plan has been evolved which, fitting into the objectives of ECA policy, would be best effected by a grant of US funds, rather than by a loan'.[39] Of several headings he considered reafforestation to be the most important.

Finance's reaction was predictably hostile: 'I do not think the Government should be asked to commit itself at such short notice to (this) ambitious programme,' minuted McElligott (Secretary Department of Finance) to McGilligan, having first complained that he himself had been given no time properly to examine it. If further American aid was in the form of loans, he argued, none of their proposals would facilitate repayment and that burden would fall on the taxpayer since none of the schemes would produce revenue for the Exchequer.

It is no business of the Americans what we do with the loan counterpart funds and they should be told that distinctly.[40]

If this advice had been followed it would have pre-empted MacBride's intentions.

MacBride took the view that the funds must be used for economic development – above all for land reclamation and reafforestation – and he characterized the Finance arguments that this would be inflationary as timid, reactionary and owing too much to the British Treasury's influence. He took the precaution of discussing his afforestation proposals with both Costello and McGilligan before publicizing them, and before Finance got wind of them, citing the Shannon Scheme as an earlier example of the kind of State intervention he now thought necessary.

At this juncture the government authorized External Affairs to seek the observations of all other Ministers on the 'utilization of the funds consequent on the loan arranged under the European Recovery Programme' on the hypothesis that the funds referred to could be used for purposes of national development. 'The decision', wrote Whitaker, 'presents the curious spectacle of the Minister for External Affairs inviting the Minister for Finance to furnish proposals for expenditure.'[41]

Whitaker claimed that none of MacBride's ideas

can be said to pass the test of productivity in the sense of producing revenue for the Exchequer from which the interest and redemption instalments could be met as they fell due... land reclamation, drainage and afforestation... fall far short of yielding an adequate revenue to offset the corresponding debt service... Expansionist ideas, however admirable, which involve adding to the public debt and to future taxation are completely out of touch with the reality of the present financial position...[42]

Whitaker's views were reinforced by his Department, most strongly in a memorandum to the government stating that the 'primary responsibility' of the Fund was 'and should remain' the servicing of the American loan: '...to dissipate the resources of the Fund on non-profit earning objects is simply leaving the taxpayers of the future the responsibility of providing the services of the loan'.

But the government deferred consideration of the Finance memorandum until after the summer recess and instead approved a draft letter which MacBride proposed to send to the ECA mission to Ireland intimating the Government's proposals that 10 per cent of the moneys, or approximately £100,000, should be devoted to the promotion of dollar tourism and that the balance should be devoted to capital expenditure required for the expansion and improvement of... forestry... etc.[43]

Whitaker saw no purpose in what he described as 'wishful speculation about the comparative advantage of afforestation as compared with investment in sterling assets, since private individuals would not invest in afforestation, nor could the State have done so because of shortage of seeds and other materials during the years of sterling accumulation'.[44] That situation, described as threatening to financial orthodoxies from within the government, was further aggravated by the sterling devaluation crisis of 1949 during which MacBride's radical financial views would have 'accelerated the polarisation of opinion in government circles on financial policy and this meant that the Department of Finance's advice was increasingly questioned, and sometimes rejected by the government'.[45]

In a memorandum to fellow ministers concerning the ERP MacBride said that the US State Department had made it clear 'that our plans must be dynamic in their constructiveness... to rebuild our economic structure so as to render us independent of outside help, to ensure full employment and ensure prosperity'.

He elaborated on this as follows:

At my suggestion the Taoiseach convened a meeting of the heads of the Department of Lands and the Department of Finance at which it was made clear that the policy decided upon in regard to afforestation would be implemented. At that period I was given responsibility for the handling of all matters relating to Marshall Aid, the OEEC and the European Economic Recovery Programme. It was felt that if these matters were placed under the control of the Department of Finance any programme of development would be stultified.[46]

Ironically, perhaps as a Department of Finance response, none of the ERP funds to Ireland were devoted to forestry. In the long run, perhaps the main benefit of the ERP to forestry was that, virtually for the first time since the foundation of the State, it encouraged government and bureaucracy alike to devote more attention to the question of a defined and sustained forest policy.

The second matter of importance, which is still somewhat shrouded in mystery, is how the rate of planting proposed – namely 25,000 acres per year – was arrived at. A target figure of an initial 1,000,000 acres had been bandied about in one form or another since State afforestation was first propounded. Connolly, as we have noted, intended to reach an annual planting target of 20,000 acres a year gradually, but by the time of MacBride's proposal only about 6000 acres a year had been achieved. But the new government took office in February 1948 and shortly thereafter publicly announced a planting programme of 25,000 acres a year, which MacBride considered an absolute minimum. The origin of this proposed target rate has been far from clear. Like Minerva it seems to have leaped fully armed for battle into the arena when born. Viney attributes its origin to Meldrum.

The then Director of Forestry, Mr J. A. K. Meldrum, was asked in an informal interview for his views on a maximum planting target. He suggested it ought to be practicable to plant 25,000 acres a year. This figure – never more than an expression of opinion on what was possible – was thereafter enshrined as State policy.[47]

Durand is unclear about it and further clouds a confusing point by distorting the chronology without explanation. Thomas Rea, Assistant Secretary to the Department 1986-8, demonstrates a sequence that explains its adoption.[48] His conclusions, confirmed by those given here, may be summarized:

The sequence was: first the decision to plant 25,000 acres per annum, then the land survey, then the request for FAO advice (culminating in the Cameron Report) on the conclusions of the survey. ...The principal conclusions are as follows:

(1) Before the election (February 1948) which installed the first Inter-Party Government, a policy existed of planting 10,000 acres per annum over a fifty-year period. The intention was to extend the national forest area to a total of 700,000 acres, including the privately owned estates (85,000 acres).

(2) The new government did adopt, in November 1948, a policy of planting 25,000 acres per annum over a forty-year period.

(3) The latter programme represented a new planting, i.e. 25,000 per annum on top of the then existing estate (about 170,000 acres).

Much (but not all) of the evidence for these conclusions appeared in the forestry estimate debates in the Dáil Reports covering the years 1948 to 1959.[49]

There is no clear evidence as to how the figure itself came about. The idea of a million acres of forest had been there for years. It may have been reached in the simplest manner – by dividing 1,000,000 by forty years, i.e. 25,000, a conclusion partly supported by MacBride, who said: 'Following the survey made by the Department of Lands and the FAO survey, the FAO expert and myself discussed the matter and we came up with the figure of 25,000 acres.'[50] But, as Rea shows, the figure existed well before Cameron came on the scene. Equally it was propounded ceaselessly by MacBride, whether or not, as he maintains, it originated with him. So far as his statement above is concerned, one must conclude that MacBride's memory is faulty about the sequence of events.

It is as well to bear in mind the strength of MacBride's position at the time, the force of his personality, his political/legal astuteness and, where necessary, ruthlessness, as well as his commitment to forestry. It is evident from his own statement that he duped the Department of Finance by making afforestation a condition of the ERP Treaty without their consent or knowledge. The figure of 25,000 acres may well, as he states, have been arrived at by him in some such way as he outlined. There does not seem to be any more complex or devious idea involved, although the figure later gave rise to considerable speculation. It is quite clear that the responsible department, Lands, had little to do with it – except to try to implement it once the decision was taken. There is no doubt that in December 1948 the government decided on a planting programme of 25,000 acres per annum until 1,000,000 acres of new plantations had been established. It is uncertain, to say the least, if this decision was preceded by any systematic analysis of the implications for such matters as expenditure, revenue, employment, self-sufficiency in wood, or availability of suitable land.

It was not alone on the question of afforestation that MacBride provoked opposition from the Department of Finance. His entire outlook on economic and financial matters was contrary to established Department of Finance thinking, which in itself was enough to prejudice the views of this key department as far as any major project put forward by him was concerned. When that project was a combination of the unorthodox (MacBride being Minister for External Affairs, not Minister for Lands) and (from their point of view) the unacceptable (his scheme was massively capital-intensive and only distantly – if at all – profitable) then the antagonism of the Department was a foregone conclusion. Afforestation on the scale proposed by MacBride gave rise to one of the strangest and most intense battles between the civil service (notably Finance) and a political head ever made public. MacBride commented:

The opposition put up to the scheme of planting on the scale that I wanted was quite considerable, both from the Department of Finance and elsewhere. They said the scheme was unworkable. First of all that they wouldn't have enough seeds. And that they couldn't get them. I sent a minute to all our missions throughout the world and they got us the seeds we wanted within two months. Then they said they hadn't enough fencing wire. Here, again, I had to resort to our diplomatic missions to ensure that adequate supplies were made available – it was really pitiful. The capital required to start this programme was not very considerable. They tried to block it through the Department of Lands. The Department of Finance was controlling the price of land for forestry. I arranged at cabinet that prices should be fixed after consultation with me and it was a question of upping the price so that the farmers were getting better prices for old land.[51]

Whitaker says that this is both an overstatement and an oversimplification. 'The principal objection was against non-productive investment.'[52]

MacBride had adroitly ensured that his afforestation programme was a condition of securing post-war American financial assistance, but the programme was declared and virtually adopted before the vital question of land availability was determined. A land survey became a matter of urgency; hence the 'flying survey' ordered by Blowick and referred to by Rea, followed by the confirmatory assessment by the Food and Agricultural Organization (FAO). (The importance – often stressed – of a comprehensive land survey as an essential priority to any major afforestation programme will be recalled.) These surveys indicated an availability of something in the region of one-and-a-quarter million acres and, not surprisingly, once the findings became common knowledge, the value of marginal and non-agricultural land promptly escalated.

With some 1,250,000 acres of land suitable for forestry in addition to the 142,425 planted up to then,[53] the reports reached the controversial and far-reaching conclusion that, while much of the land they recommended for forestry was in use as rough mountain grazing for sheep, everyone would be better off all round if it were under forestry.

The FAO survey was conducted in 1950 by Mr D. Roy Cameron, Chief of the FAO's Forestry and Forest Products Working Group for Europe and director in Geneva of the Timber Division of the Economic Commission for Europe (referred to by MacBride as 'Forestry Advisor'). Cameron based his conclusions on findings drawn from Scandinavia. He adduced a profit of £2 an acre derived from forestry as against 16/8d. an acre for sheep-grazing, which was the current figure provided by the Department of Agriculture. More significantly he recommended a two-aspected forestry programme: (1) an economically viable programme involving a total of 470,000 acres of afforestation and, (2) a programme of social forestry mainly for western counties, along the lines and about the same size as that proposed by Connolly some seventeen years earlier.

The Cameron Report, which stated baldly, 'Ireland is the poorest off in forests of any European country,'[54] was not too well received. Besides reviewing forestry developments, Viney claimed, Cameron referred to aspects outside his brief and, worse, used the dreaded word 'social'. 'It seemed "obvious" to Mr Cameron that the forestry programme should be divided into two distinct parts. Half the annual planting of 25,000 acres should be aimed at "commercial" forestry, with a sharp eye on costs and yield. The other half should be "social" forestry, putting employment before profitability.'[55]

A POLICY DEFINED (1944-58)

For complicated reasons – having more to do with Dr Noël Browne's proposals for a broad-based programme of social medicine, with emphasis on what became known as 'The Mother and Child Scheme', though perhaps not entirely divorced from the traditional fiscal outlook that opposed MacBride – the word 'social' was anathema. As in parts of the United States, the word had become a dangerous synonym for 'socialist' and became difficult to employ with accuracy even in a proper context. This was part of the 'fall-out' from the repercussive effects of the 'Mother and Child Scheme' confrontation between Browne and the Hierarchy that coloured attitudes to completely unrelated matters for years afterwards. Accordingly, when the question of 'social' forestry was aired in the Forestry Division it was backed away from and some neutral palliative was sought instead. The inevitable results were the obscuring of genuine requirements and inaction. Viney commented:

It seems that Mr Cameron's suggestion for 'social' forestry was regarded as a dangerous concept. It is one thing to plant a piece of exposed blanket bog with the challenge of making the investment profitable, quite another to see the plantation as a piece of 'social forestry'.

Cameron expected that his social forestry operation would be directed mainly to the western counties. By 1965 the area planted in the west had reached 12,000 acres a year – or almost half the total planting programme. In the past decade, in fact, some 97,000 acres of new plantations have been established in the West... But though much of this planting has been done on the least promising land for maximum yields and has, in fact, provided employment where it was most needed, the concept of 'social forestry' is still steadfastly held taboo; and so far as the top forestry civil servants are concerned, this rejection would seem to be sincere.[56]

Whitaker[57] makes the point that the foregoing is 'an admission of the Department of Finance argument against non-productive investment'.

CHAPTER THIRTEEN

A Critical Decade
(1956-66)

'So I say that we had better be without gold than without forests' – John Evelyn.

AN IMPORTANT POINT, LATER TO BECOME a critical pivot on which partisan attitudes towards forestry turned, was contained in a minority finding of the Commission for Emigration, which issued a report in 1951. In general it supported the Cameron conclusions 'to support ... forestry ... (which) would provide additional employment in rural areas. Only the Bishop of Cork, Dr Lucey, had misgivings about the short-term effects of afforestation. He pointed out that families whose holdings are truncated by the sale of rough grazing to the Forestry Division tend to emigrate on the proceeds.'[1]

Interestingly, in view of the suspect status the word social had acquired, the Bishop of Cork's reservations were not on the 'social' aspects of forestry, which, in fact, he approved of, but were expressed in the following trenchant terms: 'The land should be rented by the State instead of bought... Without such special arrangements, afforestation will denude the countryside of still more population in the immediate future and make the task of getting forestry workers in the future well-nigh impossible.'[2] Accurate or not, this statement did not help to dissipate established approaches to the dual social/commercial dilemma. This continued to be a forestry problem for years, during which a long-term national forest programme on the scale recommended in 1908 was subordinated to other short-term commitments.

In the 1940s and 1950s landowners, cattle and sheep farmers in particular, tended to oppose afforestation because it gave no return, let alone profit, for a very long time. Politicians, besides being influenced by the opposition of farmers and landowners, opposed it for the same reasons and because of the unsecured high capital investment involved. 'The difficulties of afforestation lie in the cost of acquiring the land suitable for growing forest trees, and in the expense of planting.'[3] Henry did not include the expense of extraction and of roads, or labour costs, which were, eventually, to predominate.

The desirability of co-operative afforestation, with the State as a funding partner, was advanced, but with little confidence in the outcome. Connolly proposed it in a general way in 1935 and again in 1944,[4] and Bishop Lucey made a forceful statement on the subject in 1951:

The long-term demographic possibilities of afforestation are considerable – roughly 130,000 souls. The short-term effects, however, may well be disastrous for the areas planted. The tendency is for the families whose holdings have been truncated by the loss of so much rough grazing to move out with the aid of the lump sum received as compensation. The result is abandoned homesteads and a still thinner rural population. To prevent this evil we recommend that, for the future, land for afforestation should, as far as possible, be rented rather than bought out, that the rent should be a generous one, and that it should be attached to the holding as such, lapsing altogether if the holding ceases for any reason to be a homestead. In this way the owner would have a compensatory source of income for that lost by the forfeiture of grazing rights and an incentive both to transfer these rights willingly and to remain on in the original homestead. Should the owner wish to sell out at any time he would, of course, be free to do so. This will have the added value of increasing his interest.[5]

As early as 1924 it was stated axiomatically that:

Landowners should be invited to lease suitable land for planting to the Government. Produce-sharing schemes have been devised by which the landowner provides the land and the government does the planting, but a simple lease may prove more attractive, and will be less troublesome to the State, as the dearer cost of the purchase of land is not incurred at the outset.[6]

An approach utilizing the existing structure was outlined by McGlynn: 'There's a very good core of men in the Department with expert advice. These men should be used now as the experts and their services should be made available to the farming community.' A commercial Forest Advisory Service for the purpose outlined was finally opened, in 1986. It is strange that co-operative 'tree-farming' was not actively promoted sooner. A possible explanation is fear of the demographic consequences, yet acceleration of rural migration in the 1950s might have been slowed by employment in forestry.

It appeared to be an unending circle. While the idea of co-operative forestry was good in theory, practice was another matter. No practical solution to the problem of how the farmer might survive the first twelve years or so of non-productive (and labour-intensive) new forestry was offered. Although the input might have been offset by improved overall economics and living standards, the lack of which rendered mountain sheep-farming increasingly unappealing, farmers in general remained opposed to afforestation. Since they opposed it, politicians would not support it. Without political support farmers would not change their minds. Back to square one.

Besides mutual self-help between landowner and State (or other trustee) the idea of social forestry also incorporates the proposition that the commercial consideration of producing commercial timber quickly and easily as the sole criterion for afforestation grotesquely distorts the role and function of forestry. It is held, with considerable conviction, that forestry has an essential environmental – and therefore social – role antecedent to commercial management and not less important to the community as a whole. This idea, though not expressly stated, permeates the views of Henry, Mackey and Hobson. The complexities of balancing commercial and social aspects are well illustrated in the case of Coolattin woods in Wicklow, part of which were clear-felled in the 1980s, the remainder being reprieved from a similar fate by last minute political intervention originating with the Taoiseach, C. J. Haughey.

The imponderables in forecasting financial returns for forestry are immense, as a glance at the following projections and actual costs indicate. Table 4 shows the relative acreages and percentages of land under forest in 1901 and 1951 – *viz* 268,000 acres and 305,00 acres (1.6 per cent and 1.8 per cent respectively of the total).

Table 4[7]

Extent of Land under Crops and Pasture, etc., 1901 and 1951

Use of land	Total land area		Distribution of total land area		Distribution of agricultural land	
	1901	1951	1901	1951	1901	1951
	Acres (thousands)				Percentages	
Agricultural land:						
Corn, roots, green crops	1739	1693	10.2	10.0	13.9	14.6
Flax and fruit	12	24	0.1	0.1	0.1	0.2
Hay	1758	1936	10.4	11.4	14.0	16.7
Pasture	9011	7935	53.0	46.6	72.0	68.5
Total	12,520	11,588	73.7	68.1	100.0	100.0
Non-agric. land:						
Woods and plantations	268	305	1.6	1.8		
Other lands (roads, land under towns, barren mountain and the smaller lakes, rivers and tideways).	4199	5131	24.7	30.1		
Total land	16,987	17,024	100.0	100.0		

These are not very impressive increases in fifty years. The following sentence occurred in the Report: 'The area of non-agricultural land has increased by roughly one million acres – 4,199 to 5,131; 24.7% to 30.1% – since the beginning of the present century, much of this being marginal land in mountainous districts which has gone out of cultivation.'

Estimates based on cost-benefit-return calculations with fixed variables, applicable to short-term agriculture or industry, are not relevant to forestry. The priority rating for forestry relates to contemporary levels of affluence/employment; hardship/unemployment. Relevant to forestry planning are the demand and market price for timber and demographic, soil and land conservation aspects, extremely difficult to quantify over a long term. Availability of land was another important factor. It became possible to put forward realistic and profitable proposals along the lines advanced by Bishop Lucey in 1954 without the likelihood of substantial loss to the Exchequer. It remained to convince the farmer.

It is interesting to see how Henry Gray seeks to skate around the three-legged question – policy/or not; social/or not; economic/or not. In one of the few attempts from within the service to define an on-going policy he states, 'I think it is safe to say that the conclusions reached by the 1917 (Acland) committee influenced the subsequent shape of forest management and *development policy* in this country.' This suggests, though it may not have been intended, that forestry policy since the State came into being adhered to the principles and recommendations of the 1917 committee, whereas we have seen that this was very far from being the case. Gray continues: '...declarations of forest policy over the entire period from 1922 to 1958 were not made subject to any specific qualifications as to acceptable levels of financial return'.[8]

Later, however, he says:

Policy declarations were accompanied by references to the advantages to the national economy of the elimination of an adverse trade balance in timber and timber products... throughout, prime emphasis was placed on commercial or economic objectives, rather than the social advantages of afforestation, though with constant recognition of the merits on social grounds of afforestation of comparatively unattractive sites in impoverished areas where there was a reasonable prospect of growing timber, but a questionable outlook for the crop.

As chief administrator of the Forestry Division Gray seems to have been trying to define a policy where none really existed. Where did matters really stand as the nation hovered on the brink of the industrial explosion of the 1960s?

(1) Far from the Acland Committee shaping forestry progress in this country, State forestry had diverged from the recommendations of

the Acland Report and had assumed dimensions far beyond anything contemplated either in it or in the 1908 Committee Report.

(2) Private forestry, instead of being the mainstay of the forestry programme, was neglected and diminished, reduced (by an estimated 60 per cent) to about 40,000 acres in 1950.[9]

(3) 'Social' forestry, which had been resisted, was becoming 'respectable' (and financially interesting) again, in another guise.

(4) MacBride's programme introduced day-to-day activities that could no longer masquerade as acceptable substitutes for policy, hence the emergence of an initial long-term plan and the switch of control of forestry from the technical to the administrative branches.

(5) Lack of suitable land for afforestation was a problem complicated by the demand for acquisition (at minimal prices) on a vastly increased scale in the 1950s.

(6) Landowners were still alienated.

(7) Some forests had matured, but many of them were uneconomic.

Gray was writing at a time when the mood was expansionist and optimistic. The 'rising tide' was industry-oriented and forestry, no less than any other development potential prospect, required positive presentation. Gray was now the man in the saddle, committed to putting forestry on the map. The favoured 'map' of the period was positive and profit-oriented. 'Social' forestry was not acceptable, but for new reasons; it was neither positive nor contributory. In effect forest policy was made by the forestry administration and Gray was doing, in his way, what Reinhardt and Forbes had done before him, trying to satisfy his political overlords as to the validity of the forest programme in current terms, and then getting on with the job as best he could. It was also necessary to acknowledge the work done in the name of 'social' forestry, and explain substantial tracts of uneconomic forest. Notwithstanding such rationalizations a serious attempt was made by Henry Gray to impose a positive discipline and philosophy on the expanded forestry structure.

McGlynn's view of the situation that had developed is characteristically uncluttered by sentiment or prejudice:

Not alone could we have planted 25,000 acres a year, we could have planted 45,000. We would now have a great forest industry. The whole economy of the Scandinavian countries is centred around forestry. There's at least 3,000,000 acres that should be under forest. We could base our economy in the same way on forestry. One great thing wrong with our forests to-day is that they're far too scattered. And in areas that are far too small and inaccessible. It was never an economic proposition to take over some of these areas, and go to the expense of planting them. The growing of forest trees as an economic proposition, that is real forestry.[10]

In the early 1950s the immediate questions had been those of land availability and policy.

Expensive and well-equipped modern plants cannot be laid down unless they are assured of an ample and continued supply of raw material. Further, the larger the area planted the less it will cost per acre to bring to a state of profitable production...Another consideration of vital importance is that any plan of afforestation must be continuous over a definite and long period. It cannot be taken up this year and dropped next year and then resumed again. In consequence, a policy must be laid down for a sufficient period, and once adopted must be adhered to. This continuity of policy is an essential condition of success.[11]

The foreword to the Cameron Report contains the following caution:

The proposed Irish Afforestation Programme is a very large and complex undertaking. A thorough analysis of all its details and implications, social and technical, would have required much greater time for study on the ground than was afforded to (Cameron). The conclusions and recommendations in this report must, therefore, be considered with this limitation kept constantly in mind.

The Cameron Report summarized its proposals in thirteen recommendations, some of them, for the period, far-reaching – notably that a new and separate Department of Forestry should be created.[12] The report attached importance to the question of what it called 'forest consciousness among the people of Ireland'. It recommended an appeal being made 'to the Church' for support, correctly pointing out 'that there will be innate conservatism, suspicion and hostility to be overcome. It may be taken for granted that the afforestation programme will not endure unless it is "sold" to the people of Ireland.'[13] It strongly recommended that the Irish afforestation programme should be divided into two categories, a commercial forestry programme 'designed to meet minimum requirements for sawn softwood in times of emergency...' and the delicate idea of 'social' forestry.

The Report stated:

The Forestry Advisor is of the view that operations under the social forestry programme should be planned from the outset ...The establishment of (such) forests would bring into productive use lands now lying idle and would provide regular employment during that part of the year when rural employment is scarce to local farm populations ...The objective of this programme should be to produce sufficient forest produce to pay operating costs, but not capital investment charges.

Nowhere, in this quite detailed consideration of the importance of social forestry, does the report mention co-operative enterprise.

The views expressed in the Cameron Report were endorsed in the Report on Emigration.[14]

In spite of MacBride's personal encouragement to the contrary, the old problem of planting as an end in itself deflected attention from long-term considerations. Although planting in 1950 and 1951 increased to almost 15,000 acres – close to double what it had been – the supportive steps recommended by Cameron did not follow.

The problem remained the daunting, perennial one of trying to make acceptable economic sense out of an enterprise riddled with intermittent variables and with a cost-benefit cycle of about fifty years.

In orthodox industry costs are judged by the margin of profit they yield...But cost and profit in forestry are separated by anything from 15 to 40 years...The entire weight of developmental cost was new capital investment... Up to 1948 we were aiming to create a home supply of raw timber sufficient to meet home needs. But the raising of the planting target to one million acres at 25,000 acres a year accepted the challenge of a potential surplus for export.[15]

Besides, while the national interest would, in theory, be well served by the coming on-stream of an export surplus of timber, neither forestry administration nor marketing resources were equipped to deal with the existing potential, let alone quadruple it. In the absence of adequate resources even the national genius for improvization was flummoxed by a programme requiring orthodoxy of planning, order and method rather than enthusiastic reaction, however brilliant.

A policy of maximum production rather than of optimum financial return was a natural consequence (of silvicultural management techniques guided by the urge to correct in the shortest possible time the critical dearth of native sawlog supplies)...This was of little immediate financial significance in the past as the area of woodland at stages where ultimate policy objectives influenced management techniques was not significant in extent, but it did tend to retard discernment of the importance of initiation of research work directed towards the definition of management techniques geared to an objective of maximum economic return.[16]

In addition to being very acquisition-intensive, a planting programme that had become an end in itself had also, to some extent, suppressed the need to plan a market strategy. But now, at last, the question of marketing began to claim the attention of State officials whose responsibility it was.

'The present State afforestation programme is aimed at planting 1,250,000 acres of first-rotation commercial timber at a rate of 25,000 acres per annum on a fifty-year rotational basis. The acreage is to be spread over all counties. Recent surveys have shown that in the counties Donegal, Galway, Mayo and Clare, slightly more than half a million acres are suitable for planting, and in the counties Cork, Waterford, Tipperary, Kerry and Wicklow, somewhat more than 400,000 acres. In the remaining seventeen counties the plantable area is approximately 300,000 acres. This programme, if carried into effect, would provide a suitable forest reserve...'[17]

Between 1950 and 1958 acquisition by the State amounted to 159,061 acres, compared with a total of 210,000 acres (of which 142,425 were planted) between 1922 and 1950, so that in the eight years before 1958 the State acquired an acreage equal to 76 per cent of all it had acquired in the preceding twenty-eight years. Gray states:

A notional investment value as at 31 March 1958, may be got by adding to the £14 million interest at the Exchequer lending rate (a minimum of 2.5% to a maximum of 6.75%. In the period of more critical importance, i.e. from 1950-51 onwards, range limits had been 3.25% and 3.75%). This gives a figure of £21.3 million.[18]

The Report of the Review Group on Forestry[19] offers a perspective in Table 5, which demonstrates gross investment in State commercial forestry.

Table 5

Gross Investment in State Commercial Forestry

(ten-yearly intervals)

Year	Actual £m	In 1983 terms £m
1923	0.03	0.63
1933	0.10	2.62
1943	0.23	3.43
1953	1.35	14.00
1963	3.59	29.16
1973	8.63	36.74
1983	45.04	45.04

Some of the earlier State plantations came to maturity in the 1950s and, together with thinnings from the considerable number of younger plantations, reinforced the need for attention to marketing. Since domestic requirements for all but exotics and hardwoods were by then potentially capable of being filled from native resources, and of providing a surplus, this meant overseas as well as domestic marketing. Given the extraordinary dearth of native timber supplies in 1922, the set-backs during the Emergency, and the dificulties that had to be surmounted, it was a considerable achievement. Nonetheless the requirement now was for an equally major achievement in a different sphere, and those who had already achieved much seemed inadequately aware of what that involved.

A new situation, also affected by the basic problems (want of policy; inadequate marketing; planting for its own sake), now arose. It was not only that small, uneconomic and inaccessible plots (exceptionally as small as half an acre, more usually five to ten) were planted as part of the equivocal and intermittent 'social/not social' outlook, but also that when these plantations reached maturity, timber extraction was also uneconomic. Such plots were not cost-effective in themselves or in what they produced, and contributed little, if anything, in social terms. Revision was undertaken, with the purpose of improving the cost-benefit balance in the face of spiralling costs and accommodating it to government policy outlined in a White Paper.[20] One problem was to try to introduce commercial management techniques into a bureaucratic system. Changes of government and of political outlook in 1951, 1954 and 1957 did not noticeably affect the tendency of forest 'policy' to evolve reactively. The prime consideration remained economic viability.

Thomas Derrig, Minister for Lands in 1951, in general followed procedures already established and put in hand by the first Inter-Party Government (MacBride's plan). Only a couple of years earlier his colleague in Opposition, Seán Moylan, commenting on that proposal, had stated in the Dáil: 'I fully approve that policy, and anything this party (Fianna Fáil) can do to help will be done.'[21] However, Derrig reduced the planting *target*. According to Durand[22] his dilemma was 'whether to continue an increasing planting programme and thereby disrupt essential tending of existing stands, or whether to cut back on the planting programme, despite possible political repercussions'. In fact any cut-back was not very evident. Planting for 1951-2 was up 60 per cent on the previous year and acquisition fell by only 12 per cent. Acquisition increased by some 3.5 per cent in each of the two following years, while planting dropped back by about 14 per cent, and levelled off. Ironically it was not until 1959-60, after two more changes of government, that the 25,000-acre target was reached for the first time – under a Fianna Fáil administration.

In fact Derrig increased planting and acquisition considerably on previous years. To suggest that he would be faced with 'political repercussions' by a cut-back is hardly realistic in view of existing political attitudes to forestry – particularly among the Fianna Fáil rural 'grass-roots'. Durand's comment does, however, seem appropriate to a point of view held by officials concerned that the attempt to reach MacBride's target quickly might diversify available manpower and financial resources to the disruption of existing programmes, and is, no doubt, legitimate in that context. But it hardly applies to Derrig.

During these and the following years emphasis on the 'variable constants' hardened critically as some State forests approached mature felling and the huge programme began to impose the awareness of real economic viability and its imperatives on those engaged in it. Implementing the policy of 25,000 acres a year required co-ordinated action – increased staff, doubling the price payable for land, additional foresters, new heavy machinery, expansion of nurseries, etc.

By the 1950s the reality (advanced from the beginning) was increasingly evident: the capital needed to make forestry profitable could not be diluted, and however desirable 'social' forestry might be, it was wasteful and tended to be uneconomic *unless* undertaken co-operatively.

A CRITICAL DECADE (1956-66)

The office of Director of Forestry was discontinued in 1952. It was a logical development following the expansion and shift in organizational structure, the most important aspect of which was the attention given to management and administration. In 1956 Seamus Mac Piarais became assistant secretary with responsibility for forestry, but forestry was left largely in the hands of the dynamic Gray. A system of regional divisions was introduced to strengthen the central role of the inspectorate which came under the control of two inspectors general, one responsible for land acquisition and research, the other for forest management and services, including engineering, nurseries and so on. This structure was later further rationalized with the creation of the single post of chief inspector.

Economic worth seems to have been on the basis of a notional fixed value subject neither to price fluctuation nor to the devices of commercial expertise. Profit, an important end of afforestation, does not seem to have been given practical consideration. This fundamentally non-bureaucratic word tended to be camouflaged in jargon, such as 'long-term economic viability', thus neatly raising it to the level of theory and removing it from market-place commercial thinking. Consideration of the cyclic formula '(1) sow – (2) nurture – (3) harvest the crop – (4) sell for a profit – (5) cultivate and secure the new crop' does not seem to have gone beyond step three, or, if it did, to have skipped step four altogether, except in a theoretic sense. That this seems so is surely due to the basic conflict between essential bureaucratic principles of accountability and essential commercial ones of profitability, and to shortage of land.

The problem of obtaining sufficient land was intensified because of the increased planting programme. In turn this exacerbated the problem of being offered inadequate and uneconomic plots of land, so that small, scattered parcels of inaccessible land which would never be economic were planted, willy-nilly, injecting a counter-productive, unplanned and unprofitable ingredient of social forestry at a time when the guiding principle was supposed to be economic viability. Governments, while committed to the increased planting programme, would not use powers of compulsory acquisition to obtain suitable land for forestry, irrespective of how it was otherwise being used (or not). The burden of trying to deal with this catch-22 situation fell on the shoulders of the permanent staff. In spite of legislation intended to make the acquisition of land for forestry easier and more attractive to the seller,[23] acquiring land became increasingly difficult and more expensive from the 1950s onwards. Government reluctance to use compulsory powers was described by H. M. FitzPatrick[24] as a 'Micawber-like biding of time and patient waiting for something to turn up.'

The change of government in 1954 had brought the second Inter-Party Government to power. MacBride was not a member of this government. Blowick again became Minister for Lands. Planting was slightly increased from 12,449 acres in 1953-4 to 13,845 acres in 1954-5 (13.5 per cent), with a similar increase the following year, and an increase of 16 per cent to 17,407 acres in 1955-6. Acquisition fell, however, from 20,358 to 17,416 in 1954-5, down 15 per cent. It remained at slightly over 17,000 acres in the following year and went up to 18,026 in 1956-7. The magic figure of 25,000 acres was reached in 1959-60, when acquisition also reached an all-time high of 28,217 acres, after another change of government.

The land shortage problem is clear. A land-bank – known in forestry circles as 'plantable reserve' – seldom exceeded the area planted and sometimes fell behind it. 'The success of a State forestry programme demands a continuing three years' reserve of plantable lands. Forestry nursery production must be planned at least that far in advance and planting stock cannot be held over.'[25]

The new government was as undisposed towards social forestry as the outgoing and the dilemma remained – the constraints and limitations of a forestry service required to expand, coupled with the enjoinder to be economical about it. The charge to be economic was placed on a bureaucratic and not on the commercially structured body which was now badly needed; commercial requirements were greater than before, and, rightly or wrongly, it was held that the substantial granting of large tracts of land to afforestation might be socially disruptive. In short, governments, understandably reluctant to take unnecessary political risks, were not disposed to engage in structured long-term forestry, but would go along with a shorter-term programme, provided it was profitable. As far as forestry is concerned, that is a contradiction in terms.

Speaking in the Dáil in 1956, Blowick stated:

The immediate effect of an over-accelerated transference of land from uneconomic agricultural usage to forest cropping as a result of a demand for rapid planting, might well be a serious blow to local social well-being, since the forest development which would ultimately increase the level of unemployment could not in its early stages be expected to absorb the full diversion of manpower from agricultural pursuits.[26]

This really means that administrations were not willing to grasp this particular nettle. Nevertheless the social aspect of forestry, while never conducted on the planned scale recommended in the Cameron Report, would not go away. The reason is that, like it or not, one way or another, it is an ingredient of forestry. The provision of employment, whether for its own or for forestry's sake, has always been considered an asset and locally important. Blowick summed up his view in 1956: 'I want to put it on the record of the House that if the forestry technicians were left to themselves, they would plant only on that ground on which it was absolutely certain that first-class commercial timber will grow.'

The contrary approach had been courageously challenged by Professor Tom Clear in an article in the *Irish Times* the previous December, when he stated that such a policy would lead to forestry never being anything more than a relief scheme. The issue became a *cause célèbre* in the mid-1950s.

Clear spoke emphatically: 'The difference between the profitability of afforestation with softwoods on good forest land and on poor forest land [was] remarkably large and ... the incapacity of poorest western blanket bog, even though ploughed and manured, to merit consideration even if available at no cost...'[27] Interestingly Clear, in the tradition of his mentor and predecessor Forbes, advocated private forestry as an alternative to uneconomic 'social' forestry.

This indecisive situation moved forward rather haphazardly. In general permanent staffs were not in favour of social forestry and governments favoured a 'reserved' form of 'social' forestry aimed at Gaeltacht sustenance rather than anything else. Thus 'social' forestry was not social forestry at all in the sense that Cameron used it, but had become one of several inadequate schemes for trying to preserve Gaeltachta. Just as the whole idea began to be seen as hopelessly wasteful, new techniques for deep ploughing hitherto unplantable mountain land, and new fertilizers promising viable growth, together with new strains capable of thriving on poor and exposed sites, transformed everything. Such schemes were reprieved and continued now that they seemed to have economic prospects. According to the Annual Reports of the Forestry Division, 1953 to 1957, twelve of the thirteen forests established in that period were in (western) congested areas.[28]

The border-line between economic forestry programming and uneconomic 'social' forestry now tended to become blurred. Between 1950 and 1958 acquisitions, planting, receipts and expenditure increased dramatically.

Forest labour accounts for the great bulk of forestry expenditure. The intensive drive to reduce costs initiated in 1956 had been primarily aimed at a reduction in labour costs and in 1958 a major further stride forward in this field was in hand with the introduction of work study techniques as the basis for an incentive system of payment for forest labour. Allowance was made in the 1958 financial calculations for the anticipated economies of this scheme.[29]

A few years later Viney stated: 'Today [1966] the total investment in State forestry amounts to more than £36M. – more than half of it spent since 1958. And if interest at Exchequer lending rates over the years were added to that capital, the value of the investment rises above £56 million.'[30]

With the change of government in 1958 Erskine Childers became Minister for Lands. He brought to bear on the accumulated problems an acute and realistic appreciation of the by now very complicated matter of forestry, and from the outset proceeded to the goal of establishing the forestry service as a cost-benefit and capital productive – profitable – enterprise.

> In national economic terms forestry expenditure is ... immediately inflationary. It puts millions in wages into circulation without a corresponding increase in the supply of goods on which the money may be spent... With 450,000 acres planted the Forestry Division should be employing 5600 men in the forests excluding its technical staff... the jobs in the wood-processing industries must be added on too, and these are immediately productive.[31]

Childers acknowledged that the deforestation of previous centuries on a scale unprecedented in Europe made it necessary to undertake afforestation on a scale disproportionate to our resources. He proposed a new definitive policy. He asked for: 'reliable predictions of the supply of raw material... A State whose forests are just beginning to produce the first... of what will be an avalanche of timber must have the industries ready to take it. Inevitably it will encourage the industries to set up slightly in advance of the abundant supply and hope it can keep them fed through the first lean years.'

His introduction of the forestry estimate (June 1958) gave rise to a significant debate. A realization that forestry had, so to speak, come of age pervaded the discussion, which considered long-term as well as current problems. In the course of his speech Childers stated that planting had reached 22,500 acres that year and that the long-awaited planting target of 25,000 acres would be achieved the following year. He pointed out that 'present policy provides for the planting of 1,000,000 acres'.

The key matters considered were the success of the planting programme; the introduction of an incentive bonus scheme for forest workers; the importance of private planting (the planting grant had been doubled to £20 an acre), and the thorny question of sheep versus forestry on mountain land; on which point Childers favoured sheep, presumably as part of government, as distinct from Department, policy. He envisaged a financial break-even in 1974 with an eventual return of 5.25 per cent on capital invested and employment in forestry – aided by incentive bonus schemes for forest workers – rising to 20,000. He pointed out that to date the State had spent £14m on forestry of which £10m had been spent in the previous ten years.[32]

During the debate Childers said: 'Our capacity to acquire land upon which we will have the certainty... of growing timber as part of a national investment policy is, and must remain, the key factor in determining the future progress of afforestation.'

In all 1958 was a formative year for forestry, not least because the first detailed reappraisal of planting policy was undertaken (it advocated continuance of the 25,000 acre – later to be translated into 10,000 hectare – programme). A detailed analysis of forestry and forest products was undertaken and published.[33]

The Report of the Review Group on Forestry demonstrates the realism that began to permeate forestry thinking with the onset of a definitive policy and the approach of financial viability. It makes the following important point: 'Forestry is an investment where payments and receipts are spread over periods of up to fifty years or more. There is no definitive method of valuing a forest estate.'[34]

This summary illustrates the volatile position in the late 1950s:

Afforestation has been almost wholly a State responsibility, the contribution made by private enterprise, even with State aid, being very small. In the years preceding 1929, the scale of private planting did not, it is believed, exceed 200 acres a year. State aid was introduced in 1929, the planting grant being fixed at £4 an acre; this was increased in 1943 to £10 an acre and to £20 in 1958. In the period 1930 to 1942 the average annual rate of private planting was about 130 acres. In the period 1943 to 1955 the average was about 250 acres... reaching 424 acres in 1957-58. The higher rate of planting since 1943, which has been concentrated in the post-war years, appears to have been due, in the main, to the re-planting conditions attaching to licences for felling, which was carried out on a large scale during the war...[35]

Steps were taken to 'bring home to the agricultural community the advantages of private planting and the facilities which the Department of Lands can put at their disposal in this connection...Under proper management (small pockets of land) too small for State development, would, in fact, be capable of giving higher timber yields per acre than can be secured from the type of land acquired for State forestry.'[36]

Kiln-drying facilities were introduced on a growing scale and native timber was increasingly used for many purposes for which only imported timber previously had been regarded as suitable.

More significantly, perhaps, the establishment of a State-financed Forest Products Development Board was suggested in 1958, but was not pursued at the time.

The period was one of intense activity and enthusiasm one way or another. The leap proposed and set in motion by MacBride was a universal stimulant, even as it introduced new problems. But because these were new obstacles on the way ahead, clearing them amounted to progress rather than, as before, simply going in circles from one difficulty to the next. The succession of Childers and Gray following MacBride was instrumental in maintaining momentum. A Research

Branch, established in the Forestry Division in 1957, was an important source of advice and guidance on the expanded programme. Further research programmes were undertaken with An Foras Talúntais (the Agriculture and Food Research Institute) and the Institute for Industrial Research and Standards before the end of the decade. Production of saw-log more than doubled, from 1.1 million cubic ft in 1958-9 to 2.6 million cubic ft in 1962-3, when the area under State forestry was 405,000 acres, and some 5600 people were directly employed in State forestry.[37] Things seemed to be on the move at last.

CHAPTER FOURTEEN

The Rising Tide
(1968-76)

'It is my belief that no other internal problem is of greater moment than the rehabilitation of our forests' – Col. Greeley, Chief Forester, United States of America, 1933.

ADVANCES IN FORESTRY COINCIDED WITH new departures in the political and social spheres in Ireland in the 1960s. This explosion of energy was boosted by the introduction of a semi-state commercial television service and the expansion of commercial radio.

A new electorate, matured since independence, were restless and dissatisfied with electoral platforms that rested on worn-out civil-war platitudes. The industrial surge now vitalizing the workforce and the economy attracted home many skilled emigrants; indeed this was government policy. These returning emigrants, many of them skilled technicians and most of them trade unionists, brought with them a familiarity with, and expectancies of, a higher standard of living, and this coloured many of the events of the next decade

The adventurous word 'progress' was in the air, and forestry became part of that adventure in a positive way. State forests were opened to the public on the basis that it was to the public they belonged; 'conservation' and 'environment' became important words. A plan to awaken public consciousness of forests in this context culminated in the European Conservation programme of 1970, conducted by the Forest and Wildlife Service. A natural development was to regularize the position in respect of the abundant forest-associated wildlife, and Gray initiated steps in this direction, resulting in the Wildlife Act of 1976.

How was this point reached?

In spite of unemployment and emigration reaching depressing levels in the late 1950s, the social climate was stimulated by a paradoxical optimism. A representative of the unemployed, put forward as a candidate in the 1958 general election, had been elected in Dublin. The mood of the public was for positive action. The electorate looked to the Fianna Fáil administration – in particular the new, pragmatic Taoiseach, Seán Lemass – with confidence. The social, economic and advisory institutions of the State had all taken steps to diagnose the malaise affecting the country, and were prepared to take it by the throat.

An active campaign was initiated in 1956 to reduce operating costs and focus attention on the impact of silvicultural management objectives on ultimate final yield. This increase in emphasis on economic issues formally became part of Government forest policy under a government Five-Year Programme for Economic Expansion that provided for continuance of the 10,000 hectare planting target, but emphasized the need for ensuring that the country's developing forest industry was based on sound economic lines and the importance of securing maximum efficiency in the forest and on satisfactory industrial outlets capable of paying remunerative prices for wood. The emphasis on maximizing the economic return dating from 1958 constituted a significant change from the earlier policy which largely concentrated on maximum production.[1] On Saturday, 22 November 1958, Economic Development was published.[2]

Those responsible for forestry felt a new sense of realism within their grasp. The proposition that had hovered like some enticing, but elusive, dryad (what else?), that 'the usual purpose in planting a forest is so that eventually the trees can be cut down and sold, making a profit on the the whole enterprize',[3] seemed feasible. The work of thirty-six years, if not exactly bearing fruit, was at least budding. Prospects for the future were envisaged in these unsentimental terms:

Forestry could be considered economic if the sale of crops yielded enough to repay the capital cost, with compound interest. The long production cycle and current high interest rates make this a difficult test. With some reservations regarding western peat areas, climatic and soil conditions in Ireland favour rapid and sustained growth of trees; rural wages are comparatively moderate and there is a reasonable expectation that satisfactory markets will be available to absorb the output in due course. Besides, forestry offers the attractions of providing, in rural areas, a substantial amount of employment, yielding tangible future wealth. This useful employment is provided where there is little economic activity at present and emigration is rife. Forestry fits harmoniously into the agricultural pattern of our economy and, like agriculture, it is virtually independent of imported raw materials for its development. Provided, therefore, the financial implications were not prohibitive, the continuance of a forestry programme on the basis of an annual planting of 25,000 acres would have much to commend it.[4]

The last sentence is an interesting example of bureaucratic understatement. Provided calculations were not overly affected by imponderables, the sooner the programme planting target was achieved, the quicker break-even point would be reached. As far as early planting was concerned the initial fifty-year rotation period would be reached in the 1970s. But, and herein lay the crunch as seen by the Department of Finance, there had been no visible return from any of the £10.3 m. of capital investment put into State forestry up to that time.

In four detailed paragraphs[5] the past performance, current trends and savings, and forecast expectancies for State forestry were examined and brought to conclusions satisfying enough for the Department of Finance to recommend continuance of the programme. These have been admirably summarized as follows:

> At that time estimates based on past experience of costs and revenues indicated that the moneys invested in a forestry programme would provide a return of no more than 2.5 per cent per annum compound interest.
>
> An active campaign to reduce costs was achieving significant savings. At the same time market developments were expected to result in better prices for wood and, consequently, increased revenues.
>
> The effect of these two factors would be an increase in the rate of return from the (then) present 2.5 per cent to 5.25 per cent in future.[6]

It was, of course, a notional equation, but it reconciled the existing commitment of the State to afforestation with the Finance requirement to show reasonable economic justification.

> The excess of expenditure over revenue in 1958/59, inclusive of Allied Services... is estimated at £1,914,000, of which £1,110,000 is reckoned as capital. A gradual decline of the excess to £1,1225,000 in 1971/72, with accelerated decline thereafter producing a revenue excess (£286,000) for the first time in 1974/75 is anticipated. Over the period 1958/59 – 1974/75 the total excess of expenditure over income will have exceeded £23 m. It has been calculated that, on a 50-year rotation, net revenue will increase progressively after 1974-5 until virtual peak at approximately £17 m. is reached in 2009/10 with the final crop of the first annual 25,000 acres... If the assumptions and costings prove to be over-optimistic, net revenue would not accrue until much later and would increase more slowly.[7]

The comfortable era of allowing forestry to grow happily like some outsize garden allotment secure in the shadow of a market garden, ended when the weakness of the market garden, in the shape of private forestry and imported timber, was exposed by the Emergency. An annual planting programme of 25,000 acres meant an end target not only capable of meeting domestic requirements – the hitherto loosely accepted objective – but of having a surplus for export, bringing with it a demand for positive policy.

Up to 1958 it was not possible even to define an acceptable level of financial return. The gap between outlay and return was too great and there was no precedent guideline, added to which was the enhanced capital cost of starting from scratch, with all that that implied. The question of profitability, while it evidently existed, remained, somewhat like the foresters themselves, in the dark. Part of the difficulty was that foresters could more readily calculate costs than returns.

In most commercial activities viability may be judged against the market, the known returns and by comparison with competitors. Forestry had none of these indices. A comparison of sorts might be made with forestry in Britain so far as administrative efficiency, the cost/effectiveness of some specific operations and productivity were concerned. But that was all. The main recommendation of the 1907 Committee was for a National Scheme of Afforestation administered by the State. Under that patronage private, local authority and State forestry were to work in conjunction. That idea was 'lost' in the tumult of Independence and what emerged was virtually a unilateral State endeavour. The effects of the demand for mature timber during the Emergency years rendered private forestry economically unappealing. The support of the European Community for, and the interest of financial institutions in, private forestry, coming just when State forests began to mature in significant quantity, helped to force an overdue policy rethink on this point. Reconsideration of the original concept of 'A National Scheme of Afforestation' might indicate solutions to many problems arising from the complexities of EC funding for forestry, the peculiar structure of forestry in Ireland, and the traditional, if declining, antipathy of the farming community to forestry. But such considerations were still some way off.

All other aspects being equal for the landowner (which they seldom are) the dilemma often came down to trees or sheep. Forestry is, overall, more profitable. But sheep were immediately profitable to the landowner and forestry was not. The solution should have been clear. The Review Group Report said:

Undoubtedly there are difficulties in carrying out a programme of afforestation which requires an extensive area of suitable land in large blocks. There is first the fact that afforestation is a long-term undertaking requiring considerable capital outlay, and the owners of land, either through preference or economic necessity, may be unwilling to embark on a programme which of its nature cannot yield an immediate profit. Private planting should therefore be encouraged. Secondly, afforestation may involve giving up a short-term undertaking. Much of the land suitable for afforestation is used for sheep-raising and there has always been a conflict between these alternative uses for hill-land and mountain-land. Sheep-raising is not only immediately profitable but, being a flexible undertaking, can be adjusted to changing conditions more readily than afforestation. It is, therefore, more favoured by the owners of land. Thirdly, where, because of considerations such as those indicated, the State undertakes afforestation through the compulsory acquisition of land, there are obvious difficulties if this creates local resentment. It is, however, generally accepted that, despite the real obstacles involved in a large afforestation programme, it would be possible to extend greatly the present area of forested land.[8]

A comparison made in the 1930s between return from an acre of land under sheep and an acre under forest is startling even today:

'Forestry gives throughout a coniferous rotation 50 to 100 cubic feet (depending on soil, altitude and species) of timber per acre *per annum* as compared to two to three pounds of mutton and less than one pound of wool'. The yield of Sitka spruce quality-class 3 at 40 years is 140 and a half cubic feet per acre *per annum*. At ruling prices in August, 1932, the cash value of an acre under sheep was 3s. 5d. per acre per annum; the cash value of an acre under sitka spruce (taken at only quality class 3) at London current prices was £21.1s.6d. per acre per annum; of quality class 1, £25.15s per acre per annum.⁹

It was a situation that could not be resolved overnight. Mountain sheep farming is even more traditional than was the rural community's antipathy to forestry. The idea of substituting the one for the other was riddled with difficulty and, in the event, the half-hearted approach by the forest authorities to the hill farmers was, with some reason, merely enough to excite derision. A compelling factor was the reluctance of the younger generation of farmers to engage in the enormous drudgery of mountain sheep farming for what was often a loss-making return in terms of labour and man-hours. But the traditional attitude to sheep farming combined with the inadequate approach by the forestry authorities deprived forestry of suitable land when it was most needed.

As recently as 1983 the official view was expressed as follows with notable absence of special pleading:

The restrictions on developing land for forestry are now severe. Availability of money for agricultural development from government, EEC and lending institutions, and the desire by individuals to retain, reclaim or acquire land has meant that only land with the severest use restrictions has been transferred to forestry.

The main alternative to forestry on such land as is available is, at present, sheep production. The different bases normally used in calculating economic returns for agriculture and forestry make direct comparisons very difficult. However one published comparison for Leitrim drumlin and high-elevation blanket bog (Aughty series) is shown here (Table 6).

Table 6

Estimated net Annual Revenue (3 per ha.) from Agriculture and Forestry

	Labour not Charged			Labour Charged		
	Agri–culture	Forestry	Difference in favour	Agri–culture	Forestry	Difference in favour
Drumlin	31.2	60.1	28.9	4.7	49.5	44.8
High-level blanket bog	6.4	20.0	13.6	1.8	13.7	11.9

In such a table the position of low level blanket bog would be intermediate, although probably closer to the high level blanket bog. Subsidies and EEC price supports in agriculture would tend to increase the difference between agriculture and forestry.

The overall indication is that income from unreclaimed peats, peaty soils, and gleys under forestry would be better than from livestock production.

A number of other sources indicate benefits on economic grounds from forestry versus other land-uses.[10]

After the writhings and convulsions of the previous forty-year larval stage, having weathered opposition, discontent and the Second World War, and the pupal interregnum of the decade 1949-58, the imago of Irish forestry emerged in the 1960s. If it did not precisely resemble what had been hoped for, it was at least fully-fledged and complete with policy-objectives and what appeared to be a calculable future on which might be based the already overdue marketing link and downstream industries.

Writing in the *Irish Times* Jack Whaley commented:

On hill farms carrying less than two sheep per acre, the gross margin per acre is given as £24.60, representing a gross profit of less than £50 per acre (per annum). Forestry could treble this return. It is estimated that between 2.5 and three million acres of such land are available for planting. Most of this land would probably be best utilized for company forests, but individuals with holdings of a reasonable size could themselves convert to forestry with expert guidance.

The extent to which the economics of forestry had improved may be judged by the following quote from the Cameron Report:

The profit returns per acre obtainable from hill grazing of sheep as calculated by the Irish Department of Agriculture in July 1950 were 16/8d. per acre. The equivalent financial returns under forestry could be obtained even if the total profit element were reduced to less than 10 per cent.

...A realistic assessment of the situation demands recognition of the fact that the lands required for the production of timber on an economic basis are to a large extent lands of value for hill grazing although carrying capacity may often be quite low... in many instances the hill grazing is necessary as a complement to valley agricultural lands. Its withdrawal from that use will destroy the economic soundness of many existing farm units.[11]

The major stumbling-block to convincing the farmers was the general official approach to them, still tending to demonstrate that trees were more profitable than sheep over a forty-five year cycle without reference to annual income. Lack of response by farmers to this proposition was sometimes held to be an indication of their ultra-conservatism. The fact was, of course, that while they could follow the long-term economic logic of the official view, the proposals put no immediate bread on their tables. Considerable expenditure and labour (and the risk of crop failure) were involved, and there was no income whatever (except for a small State grant of £90 in three instalments) during the first twenty years for an average yield class (Class 16) of Sitka spruce. There was no profit at all until the twenty-fifth year (and when it came it was calculated as being £11), and no significant profit until the forty-fifth year.

Farmers' reluctance to undertake forestry was fuelled, perhaps even manipulated, by an unrealistic view of potential land values, in the early 1970s, following Ireland's accession to the EEC. During this period a combination of prejudiced agrarian hostility to forestry and inadequate encouragements prevented farmers from seriously contemplating how they might best benefit from a practical alternative to uneconomic sheep-farming. While small, patchwork State afforestation is uneconomic and wasteful it does not follow that similar type 'tree-farming' in today's favourable forest climate, particularly if carried out on a co-operative basis, would also be uneconomic. 'Tree-farming' is considered to be the most profitable use of marginal land. The EC is subsidizing such activity in a significant manner. 'While there must be State involvement to provide the capital programme and a land use policy the actual land, the management and day-to-day administration should be left to the farmer. But they need available the expertise of the forestry department.'[12]

State/private enterprise co-operation deserved greater consideration much earlier on. The great weakness of the Cameron Report, possibly influenced by the then intense American hostility to any suggestion of 'socialism' (even with a small S), is that it did not encourage this idea, which many of the financial institutions are now adopting. It is all the more surprising in view of Cameron's unequivocal statement in the Foreword that: 'National interest would be better served by the dedication of these areas to forestry purposes, rather than to continue as rough mountain grazing, and so lead to the creation of a source of wealth in the form of commercially productive timber.' Agricultural co-operatives were understood and accepted in rural Ireland and the financing involved would hardly have been greater than the financing of the acquisition programme that followed. The necessary structural reorganization would not have presented the problems it does today, when the structure is so much larger.

Cameron reinforces his reference to absence of policy in legislation by quoting paragraphs 48 and 49 of the White Paper[13] as an indication of new policy, commenting: 'This programme is a very great expansion of an Irish afforestation scheme.'[14] Durand comments:

> Areas were spoken of in terms of economic planting or acquisition, but it appears to have been in a narrow, rather than a general economic sense. Afforestation was pursued in the interests of maximum economic return, but within a strictly limited sphere qualified by social and political considerations... The term 'economic forestry' lost its meaning because of lack of definition of standards. Normally

it was spoken of as a project *per se* without reference to any other possible use for the land resource utilized thereby.[15]

This judgment may seem harsh in view of the restrictions imposed by circumstances, but as a comment in isolation it is accurate enough, although it might have been more accurate to say that the term 'economic forestry' acquired a new, if restricted, definition.

The Cameron Report was submitted in 1951. From then onwards, Cameron's 'three major difficulties' of Irish afforestation (even when modified by time and circumstances) endured.

First, and most important, has been the problem of acquiring plantable land *in contiguous areas* of sufficient size to provide economic forest management. The second difficulty has been the procuring of adequate supplies of forest tree seed and the establishment of nurseries for the production of planting stock. The third major difficulty 'has been inherent in the social and economic structure of the country. Lands suitable for the economic production of forest crops are in part lands also suitable for hill grazing.'[16]

However, optimism at the end of the 1950s was given an additional fillip when, with thinnings from State forests beginning to be available in sufficient quantity, pulpwood industries were established in Dublin, Kildare, Clare and Waterford to expand existing timber end-use facilities which were mainly saw-log.[17]

By the beginning of the 1960s the aim of economic self-sufficiency in timber with an export surplus was on target. A fully completed rotation with profit-bearing volumes of timber in increasing production was targetted from the mid-1970s until optimum would be reached between 2005 and 2015. The confidence of those engaged in State forestry in the late 1960s and early 1970s had in it, perhaps, something of the idealism of the forestry enthusiasts of the 1920s, reinforced with a lifetime's experience of practical afforestation. The Second Programme for Economic Expansion stated that:

Ireland has the lowest wooded area in the EEC with only 6% of its total land area under forest compared to, for example, Germany which has 29%, France which has 27%, the UK which has 9%, and the average of 21% for the EEC. However, Ireland has a superior growth rate to that of other EEC countries. Its forest area per capita is on a par with the EEC average.[18]

The State planting rate had increased progressively from 9372 acres to 26,069, and the area planted from a total, since 1922, of little more than 3000 acres acres planted, to 269,897 acres planted by 1960. But even in this period of confidence, when goals had been set and a clear policy was emerging, new clouds darkened the horizon.

The forecast for forestry in the Second Programme had been based on a rotation profit of £650 an acre over fifty years, figures supplied by the Department of Lands. Whitaker indicated that this could equal

5 per cent on total outlay, concluding: 'if the cost assumptions are realized and there is, in fact, a market at the expected prices for the timber available, a 50-year sawlog rotation could be secured without loss if the compound interest did not exceed 5% of all outgoings'. A further comment observed:

> Net outlay on State forestry was estimated in 1958 to average £1.9 million annually during the first programme: approximately £1 million of this would be regarded as capital expenditure for budgetary purposes. Annual expenditures, in fact, exceeded these estimates. The total net annual outlay rose from £1.7 million in 1958/59 to £2.8 million in 1963/64 and averaged £2.3 million over the period; the amount classified as capital expenditure rose from £1.1 million in 1958/59 to £1.8 million in 1963/64, nearly double the programme estimate for the year. The rise in wage rates since 1958 was the main reason for this excess; between 1958 and 1963 forestry wage-rates increased by 30%... Costs in other sectors also rose and, in addition, the development of new forest roads was more rapid than was contemplated in 1958.[19]

During the first half of the 1960s a new dynamic seemed to penetrate social, commercial, industrial and administrative activity under Lemass and Whitaker. Whitaker defined the constants in these terms: 'to allow social services on non-productive forms of expenditure priority over productive projects would cause a misdirection of resources and increase the difficulties of development by raising production costs'.[20]

Such an approach did not deter those directing the forestry programme, now bolstered with clearer policy objectives. 'The final result ... was to show the possibility of bearing a compound interest rate of 5.25 per cent. (The exercise was pursued a step further by plotting the 'per acre' figures of expenditure and income year by year against the actual planting figures for the previous fifty years and an assumed future fixed planting rate of 25,000 acres per annum, using the notional investment figure of £21.3 million already mentioned as a starting point, with some necessary transitional modifications. In this exercise, tests were made with various rates of interest from 1958 onwards. It emerged that at an interest rate of 5.5 per cent the undertaking would be solvent, though capital input would be necessary up to the year 2009-10. The slight difference – 0.25 per cent – in the results from the two calculations reflects the effects of lower interest rates, higher money values and [more pronounced] lower wage levels in the pre-1958 period.) Such an attractive financial return on the State's forest investment coupled with the provision of raw material for substantial remunerative manufacturing development would be a very satisfactory outcome...'[21]

The general opinion was that a forestry policy on a rotation period of fifty years would produce real national wealth. Predictions seemed reasonable and the future looked good. 'By the end of 1963-64 the area under State forests had increased to approximately 405,000 acres.' By 1967 planting had fallen again, to 20,000 acres, and expenditure had increased to over £44 million, while receipts, far from moving towards the forecast break-even point, were still in the region of £600,000. Accumulating adverse factors now came to a head for want of a mid-rotation marketing outlet. They were compounded by the unforeseen effects of the financial crisis of 1971 and 1972, in turn followed by rises in costs and the galloping inflation of the 1973 oil crisis. However, economies did succeed in reducing net capital gross expenditure.

The outlook, while perhaps not as rosy as it had seemed, was still promising, according to the best-informed forecasts. Grudging acknowledgment was even given to social forestry.[22]

> Because of the substantial employment it provides in areas which are largely bereft of other economic activity, a continuance of the present planting programme would be warranted, notwithstanding the heavy financial burden on the Exchequer for some years to come, if there were a reasonable expectation that the financial outcome envisaged by the Department of Lands would be realized in practice. This would involve, in effect, treating the 'non-capital' expenditure in the Forestry Vote as a social service until such time as the level of forestry revenue would make demands on the Exchequer for this purpose no longer necessary... This course is justifiable not only because of the social advantages of afforestation but also because it would automatically come to an end in about fifteen years.[23]

Social forestry had now assumed another definition. No longer uneconomic forestry, to be undertaken for employment purposes in Gaeltacht or other depressed areas, it was now, according to this Finance definition, commercially oriented forestry, the time that it took to becoming economically viable simply a greater variable. It was a significant change of attitude. Yet social forestry *per se*, within the terms of reference propounded by Whitaker, remained *prima facie* an even poorer relation of that departmental poor relation, forestry. Two developments which might, in a sense, be considered alternatives, now emerged. The first was the prospect of viable forestry in western counties, and the second was the conservation and wildlife programme, which also, almost by definition and certainly by popular decision, included responsibility for public amenities connected with afforestation.

By the mid to late 1960s forestry at last seemed to be moving smoothly towards becoming an efficient and adequately managed national resource. Most of the labour pains seemed to be over and the continued evolution of this great national resource and industry looked assured. The essential infrastructure, the tradition and continuity of management and maintenance, common to most other European countries but sadly lacking in the Ireland of 1900, had developed with the forests. The vital initial rotation 'hump' enabling forestry to be economically self-sustaining, while not overcome, was within visible reach. The aftermath of the critically destructive seventeenth and eighteenth centuries had been contained. The foundation for forest thinking and experimentation that emerged in the nineteenth, limited and peripheral though it had been, was built on.

The self-sustaining rotational 'hump' in effect meant a two-rotation period on a forest base large enough to sustain profit-returning spin-off enterprises – sawmilling, lumbering, processing and so on. By the late 1960s that point was apparently close. Suddenly new imperatives presented themselves. The problem of forecasting becomes doubly acute if the forestry infrastructure is not supported by co-ordinated downstream activities such as processing, milling and marketing at the appropriate time. Where such ancillary development and interactive interest and tradition already exists, the problems are less acute and more easily managed. If such developments are not established, setting them up requires additional speculative capital. In a world where money-values alter rapidly this may be compared to casting bread on the waters and hoping for a loaf in return. In such circumstances commercial forestry planning is difficult until the harvesting of the first main-crop is in sight.

Several downstream activities had been expectantly launched, only to find their hopes of viability dashed, largely due to the 1970s recesssion. Another serious difficulty also began to surface. Given the problems at Independence and those consequential on so vast a State undertaking, the requirement, roughly midway through the rotation cycle *en route* to the 'hump', to switch from a bureaucratic approach – essential to the initiation of such a scheme – to a mercantile dynamic tends, as we have seen, to be obscured. But the longer it remains suppressed the more difficult the eventual switch becomes.

There are three requirements enabling this change to take place: the consent of the government of the day, the willing transfer of control by the bureaucracy, and an enlightened, if necessarily capital-intensive, marketing and processing policy. These issues were identified, and a national strategy was prepared to deal with the situation (some day) as the 1960s drew to a close.

The possibility of a semi-State forestry body had been proposed in the 1930s by Connolly.[24] The idea was favoured by de Valera when a resolution to this effect was proposed at a Fianna Fáil Ard Fheis in the 1950s,[25] and it had also been favoured by MacBride in the 1940s, but it was not actively pursued on the grounds that 'since transfer of land ownership to the Public Sector on an extensive scale and the devotion of large and increasing investments of capital without prospect of immediate return were involved, that answerability to the Dáil was most desirable and that furthermore the staff was working efficiently and capably under the existing systems'.[26]

What this means is that the responsible authorities were fudging on precisely the two essential points concerned, namely the need for a mercantile dynamic on a sustained basis within the terms of the overall objective, and a transfer of power and authority out of bureaucratic control to enable this to be achieved.

Then occurred the unexpected recession of the early 1970s. An analytic study of forestry for the Department made the following recommendation:

> An important element in the study of afforestation is the formulation and evaluation of alternative planting programmes. For reasons explained... the existing estate presents the more immediate problem for management, hence the Committee wish to proceed to this aspect straight away... Present afforestation policy is to plant 10,000 hectares per annum on land on which the average yield class is not less than 14 cubic metres per hectare, but does not fix a minimum yield class, for areas to be planted. Possible variations of this policy might be:
> (a) to increase or decrease the planting rate of 10,000 hectares per annum,
> (b) to vary the size of the area to be planted from year to year,
> (c) to alter the average yield class of land to be be planted,
> (d) to fix a minimum yield class for land to be planted.
> In seeking to identify the best combinations, alternative programmes should be specified, as to planting area and yield class, and evaluated.[27]

The question as to whether State forests should have remained under the direct control of a government department or not became more and more a live issue. One of the most curious grounds advanced was that it should remain because administration was costing more than it should, rather than because of the need for commercial enterprise.

The period of galloping inflation that began in 1973-4 was disastrous for the forestry industry, not only in this country, but throughout Europe. It was particularly inopportune for Ireland. It not alone adversely affected forestry with a slump in market prices, increased labour costs and wildly escalating land prices, but the effects of these factors on top of the general recession were instrumental in depressing – and finally ending – the activities of the down-stream chip and particle board industries that had been set up on the assumption of adequate and economic supplies of pulpwood from the State forests.[28]

A clearer picture emerges by comparing the projection with the reality. The projection was as follows:

> Taking into account the capital (i.e. borrowed State monies) of £6.3 million already invested in forestry prior to and including 1957-58, the financing of the present programme envisages investment of (borrowed) State capital rising to an aggregate of £21,618,000 by 1973-4 at the rate of approximately £1 million per

annum. After 1973/74 new capital will be provided from forestry revenue. ...The peak charge on Exchequer revenue, including interest on the capital investment, would be £1.38 million, reached in 1968/69. By 1982/83 net forestry revenue would suffice to meet all the interest on capital and in the following year would enable a start to be made on capital repayments.[29]

In actual fact, forestry gross expenditure grew from £433,000 in 1948 to £16,000,000 in 1977, while income increased in the same period from £100,000 to £4,500,000.

It is obvious that the income position in fact altered from being one of 23 per cent of gross expenditure to one of 28 per cent of gross expenditure only, and not, as predicted, producing a revenue excess (of £286,000) for the first time in 1974-5.

CHAPTER FIFTEEN

A Decade of Decision
(1976-86)

'Would you tell the degree of civilization, of social and political advancement which this or that nation has attained? Look to its laws. Regard how far present benefits are sacrificed to the future generations' – Anon.

WHILE IRELAND WAS ONCE A RICHLY FORESTED COUNTRY, it was not so in the sense of Norway, France or Germany where forests have been managed and cultivated for hundreds of years. A comparable system of cyclic management is about to begin (1991) with the first major crop-yield of forests planted since 1950 coming on stream and clear-felling and rotation management getting underway. In time the country may again be a forest country, but this time in the managed sense, with a forest cover of between 12 and 18 per cent.

In 1976 the Land Commission – in effect all that remained of the Department of Lands – was transferred to the Department of Agriculture, and the Fisheries Division of that department was joined with the Forestry Division of the old Department of Lands to form a new Department of Fisheries and Forestry.[1] For the first time in its career the Forestry Division was not part of a department vying with it for acquisition of land. Almost inevitably this occurred when the problems associated with land acquisition by the Forestry Division had begun to ease.

The recession of the early 1970s was a blow to progress. Prices of imported timber products, especially chipboard and softwood sawn timber, fell, and dumping of chipboard on the EEC market depressed domestic prices. While forecast production targets were maintained by manufacturers, losses resulted. This led to forced closures among domestic fibre and particle board plants and in the pulpwood industry. The sawmilling industry was also affected.

In spite of such set-backs, long-term prospects for forestry were optimistic. Demand for timber within the European Community could not be met from its own resources and Ireland had the largest and most rapidly expanding forest area per capita in Europe, as well as one of the most favourable climates for forest growth, 'almost double the EC average and 19% higher than Denmark's, its closest competitor'.[2]

A number of reports and surveys of forestry with its increasing technical, research, development and management aspects, were undertaken, notably in 1958, 1967 and 1974.[3] They did not always reach similar conclusions.

A 1981 Industrial Development Authority (IDA) Report, whose authors included representatives of the Forest and Wildlife Service, stated that:

> The 1974 recession... slowed the growth rate of many markets and created overcapacity in their related industries. It is now estimated that supply (of European timber plus available imports) will exceed demand until near the end of the century...Although Irish timber supplies are growing rapidly, in terms of most modern processing plants, the timber supply is not large; one or two plants on the scale of new international sawmills or pulpwood mills would be capable of absorbing all of the forest output in 1985.[4]

The accuracy of these statements is not universally accepted. They do not accord with developments since then and Convery's somewhat earlier report contradicts the second point.[5]

But while forecasts varied considerably and the criteria on which the statistical findings were based were far from being consistent, the net effect of each survey was to confirm the validity of the 1948 planting programme. The basic policy of State afforestation (planting 10,000 hectares, as it became, a year) remained more or less constant.

> In 1972/4 the financial, economic and social implications of the afforestation programme were examined in a cost-benefit analysis supervised by a steering committee comprised of senior administrative and technical officers of the Forest and Wildlife Service and two representatives of the Department of Finance. The committee concluded that the rates of return (financial and social) on the investment were sufficient to justify continuance of present afforestation policy in the main.[6]

Convery divided the progress of State forestry into three phases: 1922-50, 'the Era of Consolidation', during which 'a body of experience and expertise was developed which facilitated entry to the next phase' 1950-70, 'the Westward Expansion', when the combination of technical developments, the MacBride decision and social requirement (though Convery omitted the second and third points) resulted in considerable forest expansion in the West; and 1970-8, 'the Age of Amenity', coinciding with European Conservation Year and the opening of the forests to the public.

Over significance may be given to the title of his third phase during which maturity and end-product use were the pressing criteria, and which gets ample provision in the text of the report. But it was during this period that the wonderful forest parks and the national park of Gougane Barra were planned and opened.

Convery's Report emphasized what was increasingly obvious; that forestry, though in a period of temporary difficulty, was an attractive proposition worth the attention of commercial investors. Thus fortified with a sequence of encouraging reports, variable though they may have been in some respects, a 1980 departmental report *The Case for Forestry* (revised in 1983) set out to examine the progress up to then, the structure of existing forest estates, and current financial and social implications for State forestry. The Report put forward six possible planting policies and based an analysis on each. The main findings confirmed Convery's; it was evident, large areas of forest having matured, that the requirement was for marketing at professional, commercial and competitive levels, at home and abroad.

A few years later (1980-3) a survey provided an analysis of the industrial situation *vis-à-vis* product output and end-use. The commercial market-place could no longer be ignored. The need for change in dynamic from the bureaucratic to the mercantile was understood. The findings of various reports had one thing in common; the conclusion that this step was necessary. How to achieve it, and what was to follow then, became increasingly compelling questions.

Native timber can be classified into two commercial categories, one for saw-log industries and one for pulpwood industries (a small quantity is also used for manufacturing purposes). The trade statistics for 1978 showed that we were as yet some distance from the self-sufficiency aimed at (Table 7).

Table 7

Imports and Exports of Wood and Wood Products 1978

Code No.	Trade Statistics Classification	Imports Value (£000)	Imports Quantity (metric tonnes)	Exports Value (£000)	Exports Quantity (metric tonnes)
24	wood raw material	44,862	305,575	2742	103,714
25	pulp and waste paper	5667	35,842	2147	38,140
63	wood manufactures	23,668	74,108	7477	32,584
64	paper and paperboard and articles of paper and paperboard	93,238	248,052	25,471	59,928
	TOTAL	170,435	663,577	37,837	234,928

A DECADE OF DECISION (1976-86)

A study of imports from 1966 to 1973[7] showed that as much as 86 per cent 'of the value of all timber products are of softwoods and therefore readily replaceable by species which can be grown successfully in Ireland'. The report also indicated that of the remaining 14 per cent much might also be replaceable. But an assessment of the per capita consumption figure for what is termed 'Wood raw material equivalent' (WRME) is given as a per capita consumption of 0.4967 cubic metres a year. This is low by European standards. The English equivalent is 0.8 cubic metres a year, also low by EC standards. Irish consumption is expected to rise to about that level by the year 2000.[8]

A confidential 'Analytic Study of State Forest Undertaking – Interim Report on Afforestation Programme, 1974', is extensively cited in chapter 4 of *The Case for Forestry*. That report concluded that the financial yield from State forestry, net of inflation and allowing for certain future imponderables, ranged from 3.8 per cent to 5.6 per cent. Cost-benefit analysis techniques were also used to evaluate and incorporate social benefits associated with the afforestation programme and concluded that the social rate of return ranged from 5.25 per cent to 7.25 per cent. 'This part of the study quantified the benefit arising under job creation as well as the secondary effects arising from the purchase of materials and the purchase and hire of machinery.'[9]

The attractiveness of wood industries was also emphasized, an important aspect being potential savings on foreign currency through exports or import substitutions. *The Case for Forestry* indicated that there would be a shortfall of at least 8 per cent world-wide in the total demand for wood by the year 2000 and that by the year 2025 this shortfall would be 32 per cent. 'These conclusions assume first a doubling of the yield from existing forests and, second, an end to the destruction of natural forests. If these conditions are not met, the supply deficit will come earlier and it will be more acute, thereby increasing pressure on the price of timber.'[10]

Maturation of some of the planting undertaken during the 'MacBride expansion' coupled with increasingly optimistic long-term market forecasts produced an important change in the attitude to forestry from hitherto uninterested financial institutions. These now saw a promising investment prospect in forestry, due, mainly, to four unrelated factors. Firstly the maturation, for the first time in the history of the State, of large forest tracts; secondly the shortened rotation cycle of some species (35-45 years) and thirdly the increasing value of timber world-wide (due to shortage). Some financial institutions reinforced their initial interest with substantial investments. A fourth significant factor appears to be that the shorter cycle tends to match the rotation period of established, minimal-risk capital investment funds (notably pension trusts).

Forecasts of a serious world-wide timber shortage in the short term, becoming progressively more acute, together with compelling forecasts of dramatic rises in the price of timber (allied, no doubt, to relatively modest current land prices) were undoubtedly strong motivating factors in the financial institutions' interest in forestry. State forestry was an established enterprise with the benefit of eighty years' or more management experience. Unless the unforeseen happened, forestry and forest products seemed set to be a major national industry by the turn of the century.

> The affluent society consumes vast quantities of paper and you have an increasing population in the world all using paper. The underdeveloped areas are becoming developed and are also chasing the market for their share. Paper is particularly needed for educational purposes. FAO and UNO predict a crisis point about the year 2000 which will create a very serious problem which will hit the Third World countries most. It will hit them worst at the educational side. It will be too expensive to use copy-books, school-books and so on. The demand for timber will keep on. To prepare ourselves for this situation we should be planting 50,000 acres of timber every year and we should be ready to come on stream by the year 2010 or 2020 with good supplies of timber pulp for paper and we should develop and build up our secondary industries to deal with it.

Within the EC the demand for timber is expected to increase twice as fast as wood production. It is not unreasonably concluded[11] that Convery's finding that there is 'a substantial long-term market for forest products adjacent to Ireland'[12] was correct. The CAS Report makes the case that wood prices can be expected to rise 'by at least 30% in the period 1980 to 2000'.[13]

Acquisition of land for forestry, while it is no longer dominant, remains a problem since there is a clear relationship between the forecast profitability of forestry and the rising price of available land.[14] 'The failure to fully maintain the planting programme set in 1948 has been due to a number of difficulties. These include the rapid increase in the market-price of land and the inability of the Forest and Wildlife Service price-levels to attract sufficient land from that which is available for afforestation.'[15]

The following illustrates the problem:

> The price which foresters were allowed to pay for land was influenced more by the demands of agriculture than by the potential for forestry. This type of land use policy ensured that the major areas of forest are on the mountains and bogs rather than the more productive forest soils. Where the Forest Service succeeded in acquiring and planting the latter, the resulting plantations are the most productive in the Northern Hemisphere. It is reliably estimated that there are over 2 m. hectares of such soils, marginal or sub-marginal for agriculture, in private ownership.[16]

The involvement of financial and commercial institutions in forestry introduced some new thinking, but no great shift of emphasis, until well into the 1980s when the idea of co-operative 'tree-farming' was more closely examined. Then the Forestry Division opened its thinking to possible alternative ways forward.

The approach adopted – and given considerable stimulus by the EEC Western Development Programme – was not only to try to convince farmers that it is economically more prudent to put marginal land under trees than to run sheep on it, but also to provide an income in the long establishment period. It was an approach telegraphed enthusiastically by such forestry experts as McGlynn: 'Farmers should be encouraged to co-operate with the State in planting their own land for forestry. We have a million acres in forestry in State ownership at the moment and I would be glad to see the next million acres in private ownership.'[17]

The absence of an integrated land policy, one aspect of which would be the development of economically profitable forestry in private hands, is held to have been the greatest single impediment to this approach up to the 1990s. The recommendations of the 1907 Committee were very similar, but were swallowed up and tranformed in the convulsions of the 1919-22 period.

Until very recently agriculture had such precedence over forestry that a land policy giving due and appropriate consideration to the latter was virtually impossible. Forestry was not – and was not held to be – an integral part of any agricultural programme, and it was seen as inimical to farming interests. The argument that forestry is not an integral part of any agricultural programme is clearly ill-founded in the light of experience elsewhere; the second argument – from which the first may in part derive – is a mish-mash of attitudes, the origins of which have already been considered. Agricultural policy has generally been traditional and conservative. Flexibility, when it occurred – with sugar beet, for instance – was usually supportive of established practice.

By 1980 the point could validly be made that 'wood is a renewable national resource and will make its contribution to economic development in perpetuity'.[18] To which might also be added – thanks to the work and efforts of those concerned with forestry over the past sixty years.

A further programme, aimed at convincing farmers that hostility to forestry is against their own interests, was therefore initiated in the early 1980s. Steps were being taken by local and commercial interests by 1985. In July 1986 the Irish Farmers' Association proposed a national forestry policy which would involve leasing/partnership arrangements between farmers and investment companies or the State. Farmers should be paid a lease income per annum on top of headage payments in disadvantaged areas and the Western Package measures should be extended to the whole country. The Irish Creamery Milk Suppliers' Association (ICMSA) and other agricultural organizations agreed. Suddenly there was an explosion of interest from this vital sector. Fascinatingly, much of it hinged on co-operative effort.

A Leitrim plantation of 3.5 acres planted in 1949 was sold standing for £20,000, or £5714 per acre – £158 a year for 36 years. In return for a proportion of the revenue from a farmer's (new) plantation, where necessary, companies also offered an annual income. Co-operatives offered to fence, drain and plant land free of cost to the landowner, supply management and technical knowledge and make marketing arrangements. An undertaking established by separate co-operatives, called the Western Forestry Co-operative, launched a programme of private afforestation for their collective members on land unworkable for traditional agriculture, but with high potential for forestry. There were alleged to be 1,000,000 acres of such land in the western catchment area they cater for. The new undertaking was launched: 'as the areas suitable for afforestation on individual farms are generally small, a co-ordinated approach ... is essential'.[19]

Similar schemes were developed by the ICOS in conjunction with the Agricultural Institute, and by the Erris (Co. Mayo) Co-operative which invested £55,000 in special machinery for the work.

Prejudice was being overtaken by practical consideration. Maurice Harvey of the ICMSA said in January 1986,

The most practical incentive which could transform dream into reality would be provision for an income for the farmers while they waited for their forests to mature... Until a scheme is devised to guarantee farmers this income private forestry will expand only slowly and 5,000,000 acres of wasteland which could be producing valuable timber would remain barren.[20]

Macra na Feirme also sponsored community forestry. Cathal Shanahan, national president of Macra na Feirme, emphasized the need for 'an integrated approach to the development of private forestry'.[21] Jack Gardner, of University College Dublin, cited as an example of profitability the case of a Leitrim farmer who planted ten acres in 1950 and sold the timber standing in 1986 for £8000. Trees have been suggested as a source of pension funding for farmers' retirement.[22]

Local endeavour is not national enterprise and policy, but it does much to develop the climate of public opinion necessary to support and sustain one. It is generally agreed that there was a growing tree-awareness in Ireland in the 1980s.

Michael Bulfin, a forestry expert at the Agricultural Institute, writing in *Farm, Food and Research*, 1986, made the following points:

Trees will give between 28% and 117% better return than cattle on three to five acres of wet mineral lowland soil. The lower figure is based on statistics from forestry processes. The higher or increased prices experts think realistic.

The fact that forestry pays, that the EEC imports about IR£14 billion worth of timber a year and that development in several farm enterprises has been halted has provided a considerable impetus to private forestry development in recent times. Certain support systems are necessary.

The involvement of commercial and industrial enterprise, home and overseas, was another major development of the 1980s.

In 1981 the EEC launched a campaign to encourage more private forestry. The depletion of forestry resources in the Community means that it produces only 40.6 per cent of its requirements of timber (making it the Community's largest deficit second to that of energy). The British market alone accounts for £4 billion of the EEC £14 billion timber imports. £300 m. was made available to help in depressed areas and £18 m. of this was offered for the encouragement of private forestry in Ireland. This, in general, is what was known as The Western Package. But the stubborn resistance to trees and forestry was so engrained in rural Ireland that after four years only one-fifth of this amount had been taken up. The negative response produced a positive official reaction and the result was the campaign launched in 1985 with emphasis on the Western Package and grants of up to 85 per cent. In 1988 the scope of the package was broadened with further Community backing. The Western Package Scheme had a planting target of 24,000 hectares (59,000 acres) over the ten years of its life to 1991. In the first four years only 1,110 hectares (2700 acres) was planted under the scheme out of a potential of at least 1,000,000 hectares (2,500,000 acres). Substantial additional grants were introduced in 1990.

The campaign emphasized the inflation-proof aspect of forestry as an investment combining tax-free capital gains on plantation maturity, and with an estimated maturity value of £12,000 per hectare. The essential points were that farmers could develop directly or lease, with tax concessions available in certain cases; that financial institutions offered attractive long-term funding; and that there was a safe tax-free return. Planting to marketing – the integrated cycle of tree-farming – also received attention, especially in relation to the employment potential involved. The argument that processing should be where the trees are is not new. Among the many general reasons for encouraging tree-farming amongst farmers – crop and investment value, land-use, economic stability in the long term, and so on – are some management specifics such as, for instance, the enhanced market flexibility that would result when the producer on the ground controls and negotiates his own crop, unlike the paper-decision system operated from a distance which is an unavoidable part of an organization as huge and as inflexible as State forestry. On 24

September 1985 *The Irish Press* turned its attention to the question of timber-processing, and proclaimed: 'A national timber mill would be a more realistic proposal than a national smelter or a State oil refinery, two projects frequently proposed in recent years.' Clearly, however, the first twenty years or so were a powerful disincentive and some form of annual payment to farmers during that time was essential. Headage payments seemed obvious.[23]

On 8 July 1986 it was announced that compensatory headage payments would be paid to farmers who afforested all or part of their lands. In an additional effort to stimulate interest, it was proposed that full-time courses in forestry be considered for agricultural colleges. These announcements were well received and on headage payments it was stated that it 'was a most significant step to encourage farmers to participate in forestry and ensure that some level of income would be provided during the early years of tree-growth'.[24]

The stimulated commercial interest, however, created new problems reminiscent of the situation that existed between the large landowners and the farmer/smallholder community when the 1907 Report was in preparation. Maurice Harvey, chairman of the ICMSA western development committee,[25] condemned 'tree-barons' anxious to 'grab' large acreages of farming land for planting. 'There is no merit in this type of farming. These entrepreneurs have no interest whatsoever in rural development. Their sole aim is to maximize their profits... A healthy balance between agriculture and forestry could be achieved through community afforestation projects.' Corporate interest and agricultural expectation now clashed over land use.

The need to guard against the danger of exploitation of the maturing forests had, of course, to be guarded against, but was never articulated. That danger surfaced, briefly, in the mid-1980s when the first significantly sizeable crops were approaching maturity. In 1985 the idea of 'selling off' forests to commercial interests was floated – by financial interests who began to see profits in trees – and tested in the air by the government. The resulting outcry, however, quickly raised a storm of political, even national, indignation, and the self-destructive idea was hastily dropped.[26]

Diarmaid Bradley of the Investment Department of Allied Irish Banks, which have a considerable acreage of forest, says 'Forestry is an ideal vehicle for those institutions such as pension funds who do not seek short-term return...,'[27] a point of view modified by Ray Gallagher, ICOS executive, who commented pragmatically:

There is no great Sahara desert out there waiting for afforestation. In the Drumlin belt, there are small pockets of land on small farms that can be turned to tree production and the idea is to pull these together as much as possible. ...Only by co-operative farming can forestry succeed in swinging the pendulum which has been blocked by farmer resistance, notably since 1981. ...The co-operative movement can deliver a thriving private afforestation programme which would have a massive social and economic impact on the communities involved, with increased farm incomes, and increased job opportunities in remote areas.[28]

Following a specific government statement of policy in 1984, that the nation's forests should be developed to the maximum national advantage, a significant examination of the Forest and Wildlife Service was undertaken. 'It was indicated that the overriding objective in the exploitation of this State-owned natural resource should be the maximisation of the benefit accruing to the State on behalf of the community.'[29] At the same time the government announced its intention of establishing a review group to report on the best means of doing so. Under the chairmanship of Paddy O'Keeffe, editor of the *Farmer's Journal*, the Review Group consisted of representatives from the academic and trade union sectors, from the timber industry and the Irish Co-operative movement, officials from the Departments of Fisheries and Forestry, Finance, Industry, Trade, Commerce and Tourism and from the Department of the Public Service. The first meeting of the group was held on 22 November 1984. Addressing it, the Minister for Fisheries and Forestry, Patrick O'Toole, said: 'What is required now is the most objective possible assessment (of the Forest and Wildlife Service), with the national interests taking precedence over any sectoral considerations.'

The group's terms of reference were:

(1) To examine the structure, organization and operation of the Forest and Wildlife Service of the Department of Fisheries and Forestry.
(2) To consider what changes, if any, were necessary.
(3) To make specific recommendations on such changes.
(4) To submit a report to the Minister for Fisheries and Forestry by 1 June 1985.

The group decided on commercial criteria as the basis of their approach, and their report was published in November 1985. It made a number of important proposals and recommendations, based on the following:

(1) That the Forest and Wildlife Service (FWS) has no clear mandate to operate commercially.
(2) That the FWS, as currently integrated into a government department, has a management structure that hinders it from being run as a profit-conscious business.
(3) That there is no clear distinction between commercial and non-commercial activities.

(4) That there are external constraints such as political and sectoral pressures which have adversely affected the FWS profitability and performance.

The main conclusion was that the Forest and Wildlife Service required to be restructured, and that State forestry should be run on commercial lines.

Having reached this conclusion, and recommended its implementation, the group then advised that this should be done within the civil service. They recommended that the body they proposed should have the status of a Commission, for which new legislation would be required.[30] The group recognized the inadequacy of the marketing aspects of the existing State programme, but contented themselves with precisely the sort of compromise that brought about the problem in the first place. They proposed a commercial body to be called National Forest Enterprise, combining functions analagous with those of the British Forestry Commission and an Irish State-sponsored body.[31]

As so often before the nub of the problem was the incompatibility of civil service bureaucratic thinking and practice with commercial thinking and requirements. The merit of the Report was its willingness to see the problems clearly. If the solution they proposed was one that could hardly suffice, that is understandable, taking account of the long initiatory, sometimes extremely difficult, almost parental role up to this of the Forest Service of the Civil Service. The reality, however, was that it simply was not feasible for an accountable body such as a civil service department to have the necessary flexibility and financial independence required for commercial dealing.

Many of the limitations arose, not from lack of will within the Forestry Division, but from the nature of bureaucracy itself. Commercial enterprise and thinking is not (and should not be) a civil service function. The civil service observes that 'accountability for public funds' is their over-riding principle. There is, of course, the problem that this argument may be used as a smoke-screen to conceal inadequacies and lack of initiative. That was not the case here and it is important to recognize that fact.

Besides development of a forest programme, a process of forest management evolution also occurred. Some downstream industries were established,[32] perhaps too soon, with the result that when, in the 1980s, they might have operated in the security of an overall supply and marketing infrastructure to help to absorb temporary crises and to enable them to operate viably 'in the worst of times', they were not there.

The Report proposed that National Forest Enterprise (NFE) should adopt a leading role in development, exploitation and the important question of land policy, observing also that 'much of the increased forestry vote was taken up by inflation in wages and salaries'.[33]

From 1983 the government confirmed a planting policy of 10,000 hectares per annum as 'an overall policy objective... pending an improvement in the land reserve situation in relation to State-owned lands, and an acceleration of private forestry, an annual State planting programme, including reforestation, of the order of 7500 hectares – to be augmented as much as possible by private planting – should be accepted as the most practical short-term policy for the next five years'.[34] The government also stated that 'for the purposes of achieving this objective, reforestation and planting of privately-owned land should be taken into account'. In this fashion the government appeared to keep most options open without making any significant change in continuing policy.

A change of government in 1987 was quickly followed by the establishment of a new State Forestry Board, which would 'contribute in a practical manner to continued forest expansion and, through aggressive and imaginative marketing, optimise the financial return on the State's forest investment'.[35] It is therefore important to examine the main, and in some cases radical, points made in the Report in the context of that decision and the enabling legislation that followed.

On forest management the Review Group said: 'We consider that a return of 3% over inflation is a desirable and achievable objective.'[36] The group concluded 'that the State forests should be run commercially... in future any non-commercial activity should be independently recorded and funded separately'. The group outlined a proposal for a new accounting system.[37] Marketing was given priority. Following the Report and in a largely inadequate attempt to improve the position with regard to marketing and sales, the FWS in 1986 introduced a national auction system for timber sales after criticisms were directed at the tendering system, which was also continued.

Although the group observed that 'there is potential in co-operative forestry development... we support the approach recently adopted by the Irish co-operative organisations' society in promoting farmer involvement in forestry', no recommendation was made in respect of co-operative forestry. The Review Group seemed to take the view that this should be confined to small-scale farm forestry, and as such was a relatively minor category of private forestry, more the concern of private enterprise than of the State.

But the Report also stated that 'the development of a vigorous private sector would help to reduce the imbalance in commercial forestry which is, at present, a virtual State monopoly'. They recommended financial support, access to information, taxation, administration and that their proposed NFE should include a division which would, on behalf of the industry as a whole, co-ordinate matters relating to private forestry. This raised questions previously touched on, for instance in relation to the precedent established by Michael Moran when, as Minister for Lands, he indicated that he would not permit land to be used for forestry investment.

The value of the fixed and variable assets comprising a forest estate is a fluctuating factor calculable only in retrospect. Equally clearly, flexibility and the capacity to refocus policy effectively and rapidly in the light of changing circumstances, are essential. The social value is another matter, and the combined net and social worths of this natural resource are important community assets.

The emphasis of the Report was economic.

The fundamental aim... must be the creation of wealth. In the process of wealth creation there will be job opportunities, and well-conceived policies will bring maximum benefits to local communities and to the economy at large. The potential might be considered under a number of headings:
(a) Production potential.
(b) Land use policy.
(c) Market opportunities.
(d) Silviculture and research.
(e) Maximizing added value.

The following point is worth noting:

If private funds approximating to those invested in commercial properties over the past decade were invested in afforestation, a large proportion of this area could be planted... our marketing strategy would be continuously reviewed and that strategy should be reflected in our forest management policy, in our research effort and in our type of wood industry development which we encourage.[38]

The involvement of the commercial and finance interests, which now saw in forestry a prospect for long-term investment of low-risk capital, introduced a new edge into the market-place, and new tensions. 1986, for instance, saw accusation and counter-accusation between the inadequate timber-marketing element of State forestry and sawmillers on the question of tendering. That system was seen by the millers as prejudicial to their interests, since buyers from Northern Ireland (benefiting from cheaper fuel and certain subsidies) could tender lower than themselves. Those responsible for State marketing countered that the millers were engaging in price-fixing. This unpleasant, unnecessary, and avoidable dispute achieved an uneasy compromise without being fully resolved with the introduction of timber-auctioning in 1986. State contracts to supply pulpwood to commercial factories were also being criticized, on the grounds that details of the contracts were not made public.

The controversy in relation to the sale and tendering of timber also led to reductions in staff on a fairly substantial scale in some sawmills.

It was alleged that the Co. Wicklow timber industry had been placed in serious doubt by lucrative contracts going across the Border. The point was also made that Northern Ireland sawmills could undercut southern millers tendering for saw-logs from State forests, with resulting redundancy and unemployment in southern mills. This unfortunate development highlights one of the difficulties encountered when the essential switch to a competent commercial-marketing structure does not take place in time. The situation should not have arisen and should have been resolved sooner. In 1986 an official committee was established to look into the problem, which was a result of the fact that production was in the main a State enterprise, whereas mills and processing were in the main private enterprises.

'The apparent inability of the Department to accurately forecast the volumes of timber to be released to the market in any given period is the single most disruptive aspect of the current situation.'[39] While this is an over-simplification, it underlines the fact that the differences between State and mercantile enterprises involved with the same raw material, plus inadequate co-operation, created a problem that would not otherwise have arisen. 'It is noteworthy that Smurfits, one of the biggest industrial concerns in the country, is also one of the largest private investors in the Irish forest industry, concentrating on the utilization side. It had ploughed £10 m. of its shareholders' funds into the business through its Woodfab offshoot.'[40]

The Smurfit expansion in sawmilling in 1985, in spite of the recession and lay-offs in the timber industry, was pursuing a strategy already outlined. 'By 1990 only four large mills will be in operation, with the bulk of the smaller mills wiped out. By this stage, however, the amount of timber coming on stream from the State forests will be truly massive and Woodfab (Smurfit-owned)... will be waiting to clean up on this harvest ... the company is ideally placed ... at Fermoy and Aughrim.'[41]

The general attention forestry was getting at this time reawakened some of that enthusiasm for it evident in the period between the First and Second World Wars. But it also illuminated dangers to be observed elsewhere. One example, perhaps, of the enthusiasm is a Trees for Ireland scheme, jointly fostered by the Irish-American Cultural Institute and the Forest and Wildlife Service, which encourages Irish-Americans to contribute $10 per tree for planting in the land of their origins. A similar idea, subscribed to by the American Jewish community for Israel, was very successful. The success of the American-Irish venture, inaugurated in September 1986 with contributions of $50,000, remains to be seen. An example of possible danger developed in Scotland, where there was powerful criticism of the activities of private forestry companies buying land for afforestation.⁴² The Ramblers' Association publication condemned subsidized afforestation in Scotland on precisely the grounds that were also advanced to demonstrate its economic viability, namely, that large coniferous plantations are easily planted and grow quickly. The *Theft of the Hills* Report accuses private forestry companies of being 'strident and ruthless advocates of planting all possible land ... who... scramble to find land for their investors to buy and plant. ...As a result,' says Tomkins, 'prices of hill land used for forestry have risen by an average of 15 per cent a year between 1950 and 1980.' The report claimed that in the ten years between 1976 and 1986 the area of Scotland under coniferous plantations had increased by nearly 500,000 acres, and also claimed that such land was 'ugly, inhibited access to the countryside because the trees are thickly planted, and destroys important wildlife habitat, while remaining an attractive investment and encouraged by the government'.

During the summer of 1988 concern became more evident on this score. There were a number of letters to the newspapers and these were synthesized in September of that year:

> Roads which once were open to airy infinities of moorland are becoming walled in by impenetrable pine and spruce... The Connemara conifers have broken the skyline bang on cue, for they have sprung up, like dragons' teeth, from decisions made twenty years ago. In Britain ... the dark tide of conifers enveloping the uplands prompted fiercer and fiercer opposition from countryside lobbies and nature conservancy groups. This spring, the UK government declared an end to massed conifer planting on the English uplands.⁴³

But it was the Union of Professional and Technical Civil Servants that articulated this concern most strongly in a document entitled *Ireland's Forestry, A Review*. It urged a halt to planting conifers on western bogs, arguing that general environmental considerations gave the bogs a conservation priority that took precedence over conifer planting. The point was also made that hill streams draining conifer plantations are alleged to be more acidic and contain more aluminium than similar streams on contiguous moorlands with similar soils and rock foundations.

Much of the strongest rural-based hostility to local afforestation proceeded from Leitrim, the area in the country generally considered to be that most suitable for afforestation. It is a combination clash of social, cultural, traditional, and economic values. Perhaps unfortunately its resolution is, for the moment, seen only in inflated financial terms.

Leitrim is an area of small and essentially poor farms where bread-and-butter economics are generally the norm. Consequently the involvement of what are classed as forestry speculators (State or commercial, but mainly the latter) is seen by those opposed to forestry to contribute little to the county, and will 'undermine the economic prospects and the fabric of Leitrim society'.⁴⁴ The Irish Farmers' Association in the county was critical of the idea of afforestation, claiming that it was a question of trees or people. It opposed the industrial development of forestry, while prepared to consider 'an integrated policy which would include agriculture, industry, tourism as well as forestry for the future development of Leitrim',⁴⁵ but did not enunciate details of such a generous concept.

On the reluctance of farmers to consider tree-farming seriously and the hostility that has been engendered by the participation in it of major industrial and commercial interests, Joyce says:

> Attempts to generate enthusiasm among farmers to engage in afforestation have, in the main, been unsuccessful. Even the forestry development scheme contained in the Western Package has met with a luke-warm response. After four years in operation, less than 1000 hectares of the 24,000 hectares target have been afforested. Most of this has been by investment agencies who have purchased the land prior to planting. In sharp contrast to the farming community, the financial institutions have not been slow to recognise the potential of forestry.⁴⁶

The controversy became a confrontation of interests between those who concentrated on the maximum mid-term profitability involved, and those who sought to protect the community investment in forestry to date from the prospect of exploitation. Ironically, as a result of such self-interest, both sides were coming to a closer recognition of the reality – namely the need for both technical and mercantile expertise in managed forestry that is up and running. Joyce also made the following pertinent observation: 'A more positive aspect of the controversy is that now, for the first time, a tree-grower can see a market for his plantation at all stages of development. This should be an encouragement to owners of marginal agricultural land to engage in a forestry enterprise.'⁴⁷

Also in Leitrim, a new venture, Crann, promoting the growth of broadleaves, was founded in 1986 by an Australian woman, Jan Anderson, and an Irishman, Ciaran McGinley, with the help and encouragement of John (Lord) Kilbracken, thus neatly – and presumably accidentally – reviving one perspective of the 1908 Report. The voluntary Tree Council of Ireland also dates from about this time. Crann leased, at a peppercorn rent, twenty acres from Kilbracken's estate in Killegar, and turned them into new forest land with an additional hectare for a nursery. The aim of this non profit-making charitable trust was to create a modern broadleaf forest and train young people in its cultivation, much as Avondale was originally intended to do on a broader basis.

These developments, summarizing what was most significant to forestry in the 1980s – namely the imminence of the long-awaited prospect of economic self-sufficiency and foreseeable profitability – were given focus and purpose with the announcement by the government in 1988 that it proposed to establish an Irish Forestry Board.

Coillte Teoranta was established on 22 December 1988, under the Forestry Act, 1988, to:

– undertake the full range of commercial forestry activities;
– operate forestry on a profitable basis;
– commercially exploit all available and potential opportunities for growth;
– develop and diversify the forest industry in Ireland;
– pursue an aggressive promotions and marketing programme for Irish timber and forest products.

It was also recommended that import substitutes by Irish timber should be increased, downstream industries encouraged and, perhaps most significantly, that the annual planting target should be increased to 13,000 hectares, with the emphasis on private forestry.

A continuing difficulty has been the need to provide calculations which would satisfy current requirements of the Department of Finance, while simultaneously endeavouring to move towards the eventual goal of crop rotation viability. These criteria are mutually exclusive and were responsible for many problems, in particular for the plethora of contradictory calculations produced from time to time.

One recalls the Department of Finance 'conditions' regarding forestry as expressed in *Economic Development*. The problem was, of course, that Finance, quite properly, had one set of criteria which Whitaker clearly expressed, to the effect that public funds must be spent profitably, or at least not unprofitably. On the other hand, the Forestry Division laboured under the genuine difficulty of being unable to demonstrate the profitability of their undertaking in any but notional and very variable terms. Costs tend to be estimated over a period at a constant rate of compound interest, while capital stock is valued at current prices, inflation or no. The conflict between the two sets of principles may be roughly summarized: the Department of Finance required a clear indication of progressive profitability for the funding provided for forestry over any given period, while the Forestry Division was in a position to say only that timber prices, in general, keep ahead of inflation and that, leaving aside any social and environmental benefits that might also tend to accrue, there appeared to be an acceptable profit at the end of the line. The forestry case, whether it is acceptable or not to the funding Department, was probably best stated as follows: 'A comparative review of the real financial rate of return in other government programmes would probably show forestry investment in a relatively favourable light.'[48]

Maintaining a viable industry 'in the worst of times' is a basic industrial criterion, as yet questionable in the case of forestry. It is to be hoped that Coillte will be able to eliminate any doubt, by being in control of its own budget. Appropriate industrial integration for competitive commercial activity is vital, and therefore harvesting and marketing should not become vulnerable by being dependent on an external processing structure.

The future can expect to see several important developments in forestry in Ireland. In another eighty to one hundred years a further 500,000 hectares of forest is expected to have been planted. This will bring the country a step nearer to the economic independence conferred by a major exploitable resource. Most of this will be due to a resurgence of private forestry of which there was a marked increase between 1985 and 1990, more in the industrial sector than in the tree-farming one, but evident in both. Much of this development is expected to be in the West.

Employment will probably rise, both directly and indirectly, either in the forest and timber industries or because of the activity they generate. It is likely to be most noticeable in the down-stream areas affected by harvesting, processing, marketing and exporting. Usage will diversify and expand. It is anticipated that wood will become an important alternative fuel, especially as turf supplies diminish. The strong probability is that much of the 350,000 hectares of low-level bog will be used for forestry. Most satisfying for the man or woman-in-the-street – apart from the amenity aspect – is that saw-log production will reach, and maintain, significant levels before the new millenium.

The volcanic strides made in the five years 1985-90 are demonstrated clearly in the *National Development Plan*, 1989-1993, in which forestry objectives are set out. Essentially what occurred is that 'since 1987 the government have given forestry a particularly high priority as an area for development.'[49] These include doubling 1988 planting levels from 15,000 hectares a year to 30,000 hectares in 1993; building 750 km of new forest roads to facilitate 'economic exploitation of existing forests'; increasing harvesting and marketing programmes from 1.5 million cubic metres to two million by 1993; and implementing recreational and tourism programmes. Resulting from this and from EC policies (with which Irish policy was required to conform to qualify for EC forest funding) and the funds accordingly made available, a national forest policy of substance and purpose emerged and was outlined in *Operational Forestry Programme*, 1989-1993. The OFP is a 79-page document in which some of the more radical proposals concern the growth and projected performance of the private forestry sector (see note 2, chapter 17). With the exception of a definitive land-use policy – which can hardly be much longer delayed – this document appears to forecast the sudden coming-of-age of forestry and forest planning and activities. Policy, while possibly as yet not fully developed, is no longer based simply on acquisition and planting. Marketing, environmental, social, demographic and other aspects of national importance are given due consideration and place.

It is here hardly out of place to point to a remarkable achievement, due entirely to the work and dedication of those who devoted their energies and skills to State, and to a lesser extent private, afforestation, sometimes in the face of considerable and ill-judged opposition.

As we know the essential recommendation of the 1907 Committee was for 'A National Scheme of Afforestation'. The target to be reached was 1,000,000 acres in eighty years. It is often forgotten that the target was reached inside the distance, in seventy-eight years. Admittedly there were some profound sea-changes and unforeseen hiccups along the way. Who, for instance, of that committee, could have foreseen that when the target was reached it would not be under the umbrella of a British national scheme for all Ireland, but be the State scheme of an independent nation? That the target would preponderantly consist of exotic, American species? Or that private forestry, at that time the *only* forestry, would have shrunk to less than half its original size and then have been so stimulated by massive EC, government and private-enterprise capital injection that in the space of four years it had recovered to the extent of being 46 per cent of total national planting in 1989?

CHAPTER SIXTEEN

Timber and Species

'There is not any other wood that does so well agree with the glue as spruce' – John Evelyn, *Sylva*.

THE MOST SIGNIFICANT FACT in respect of species, apart from the question of broadleaf *v.* conifers, is that since the recommendations of the 1907 Committee, which were orthodox and traditional in respect of species, the composition of Irish forests has altered, from being composed mainly of European species, to having a preponderance of species from North America. The progress in forest development since 1922 also led to a situation in which: 'The predominance of the State in Ireland as a forest owner is unique in the western world.'[1]

Since 1958, which seems to have been the critical year, an important aspect of the planting programme has been the increased use of Sitka spruce and Lodgepole pine (*Pinus contorta*), both of which are 'capable of high production on difficult sites'.[2]

Between 1958 and 1985 the percentage of Sitka spruce almost doubled, from being 26 per cent of the total in 1958 to being over 49 per cent of the total. Lodgepole pine remained fairly constant with an increase from 22 per cent to about 26 per cent. The use of these two species is reflected in a corresponding decline in other species, illustrated in Table 8.

Table 8[3]

Percentage Distribution of Species in State Forests

Species	pre-1958	1984	Dec. 1985
Sitka spruce	26	49	49.4
Lodgepole pine	22	26	25.6
Norway spruce	16	8	7.4
Scots pine	15	4	4.4
Other conifers	13	9	8.6
Broadleaves	8	4	4.67

In 1983 something less than 5 per cent, 328,500 hectares, of the total land area was under forest, some 85 per cent of it in State forest and 15 per cent in private woodland. Of the total 90 per cent of State forests consisted of conifers, whereas conifers comprised only 20 per cent of private woodlands – so that private forests had a slightly larger hardwood content than State forestry (32,360 hectares as against 17,000 hectares for State forest, approximately).[4]

Broadleaves are sometimes (wrongly) seen as the traditional forest tree and from time to time the protest is heard that these more attractive, and often more spectacular, deciduous trees with their spreading branches and variegated seasonal colours, in marked contrast to the serried shades of sombre green and shadow presented by blocks of conifers, were overlooked by State afforestation. In Britain broadleaved forests were cultivated for hundreds of years for economic and strategic purposes until they became redundant in the last century and brought economic disaster for many. But in this country where managed oak forests were few, the native oak was virtually wiped out.

The reasons for the low level (8 per cent) of hardwoods in State forests may be summed up in the words Sitka spruce (*Picea sitchensis*). 'The Sitka spruce, which likes humidity and high rainfall on moist mineral soils, was first planted as a single specimen on private estates in about 1834.'[5] A native of the west coast of America, Sitka spruce is the most important tree in Ireland. It is well suited to our climate and soil and produces one of the highest yield classes in the country, which means in Europe. The timber is suitable for building and construction work generally, and it provides the raw material for manufactured boards and paper.

When I refer in this writing to hardwood I mean broad-leaved timber, and when I refer to softwoods I mean conifers. The entire forest area of the world is estimated at 7.5 billion (a thousand million) acres, of which more than half is hardwood. The value of this hardwood is not, however, in the same proportion. If a hundred be taken as a unit, the value of the world hardwood supply to meet the demands of humanity would be less than 10, the value of softwood over 90. With the exception perhaps of beech, the hardwoods instead of increasing the fertility, exhaust the soil... You may mature crops of prosperous commercial oak and other hardwoods in from 80 to 120 years – On the other hand, in this climate with its humidity you can mature Tideland (Sitka) spruce and Douglas fir for lumber in under 40 years, but Tideland spruce may be cut for pulp within a generation... Within 25 years... the maincrop of existing new forest – can be working full circle.[6]

In the 1930s the problem was to introduce a new and unorthodox perspective that ran counter both to traditional forest appearance and to species. While the economic advisability of a switch from broadleaves to conifers was generally accepted by foresters, the matter of species was far from being agreed. The conservative view was not confined to Ireland. In countries where the lumber industry flourished on a foundation of huge natural forests even more acute problems existed.[7] These were met head-on in the United States in 1932.

We must get over the old hardwood idea of viewing trees as containing so many logs. We must adhere to the view of looking at a tree as a national product, which not merely develops wood substance, but grows cellulose. If we grasp it in this light we get a vista of the great diversity of products now obtainable from forest growth and capable of being put to commercial use.[8]

In 1924 Professor Ostenfeld made the same point in respect of Europe: 'Apart from fuel, usually obtained from hardwoods, an overwhelming proportion of the timber now required for industrial purposes is produced by coniferous species.'[9]

Here it was argued that indigenous species should be replanted under a managed system. There was a groundswell in favour of species such as Scots pine, Norway spruce and Douglas fir.

A statistic from 1927 has considerable implications for a world faced simultaneously with a universal timber shortage and a rapidly increasing paper-demand from the Third World: '36,000 cords of wood are required to produce this magazine for a year – a cord = 4.74 cubic yards; a spruce forest at 40 years is taken as yielding 6,660 cubic feet of pulpwood to the acre. 692 acres will yield 36,000 cords.'[10]

Paper consumption is increasing with unbelievable rapidity. It is now questionable if even the United States or the USSR can provide sustained pulpwood supplies commensurate with increasing world demand as new paper markets are being created, so that, today, a thriving timber-growing industry is as basic as agriculture. 'Upon timber crops are founded permanent local industries. Timber growing and its dependent industries supply the livelihood for a population... which adds to the national strength.'[11]

The value of softwood as an industrial timber far exceeds the structural value of hardwoods, with which in any case several softwoods can compete. The relationship of hardwoods to modern life is that of a luxury, even exotic, timber, whereas softwoods are the comprehensive universal wood. 'Nations today (are) concentrating on softwoods... because the lumber value of this tree (the Sitka spruce) almost dwindles beside its yield of the precious fibres from which the wood-pulp known as sulphite is obtained... cellulose.'[12]

Such views as these were the expression of informed forestry opinion over the previous sixty years, especially in Great Britain, where the lethargy and indifference of the second half of the nineteenth century, following the collapse of the oak market, was jolted by the sharp realities of the First World War when the vital strategic importance of home-grown timber re-emerged. But by then the requirement was for commercial softwood, and so it has remained.

It is important to establish two points on which rests another one taken for granted; namely that the purpose of modern forest management is to plant, husband, reap and profitably market the timber product. (The social aspect of forestry is not, it is argued, primarily a management function.)[13] These two points are: (1) The principle of sustained yield, which may be defined as the steady supply of the required timber to the optimum production capacity of the forest,[14] and (2) The fact that this is achieved 'through the concept of the Normal Forest'.[15] The concept is an ideal one rarely, if ever, achieved in practice due to the variables involved, but it is the goal at which forest management aims. It consists of a uniform forest area with a single forest tree species and equal areas of every age from one year to the number of years in the rotation – 35 years for a 35-year rotation, and so on. Every year one area would be felled and replanted.

The importance of these points as the basis of sound forest management is put bluntly into context: 'The Irish State Forest Estate, taken as a whole, is at present very far from the condition of a Normal Forest.'[16]

Interestingly the 1908 Report – far-seeing in so many respects though it was – remained conservative in the matter of species, in spite of Augustine Henry's advocacy of softwoods and the introduction of new fast-growing species. The bulk of the evidence in the Report favours Scots pine, European larch and Norway spruce, no doubt because 'modern' European forestry methods at that time favoured the use of European conifers. It was not until after the foundation of the State that the successful American mix was introduced.

In the mid-1920s the first experimental plots began to show that Sitka spruce was a potential economic mainstay of State afforestation. It now accounts for nearly half of all State plantations, and since 1958 it has dominated every other species. Lodgepole pine also increased. These figures reflect decisions based on study and experimentation during the early years to determine the most suitable species for this country.

Sitka spruce had already stood the test of time as a sound commercial timber when State afforestation was first undertaken. It performed better than any other species even under difficult conditions – in Connemara producing very satisfactory values (70 Hoppus feet *per* class 1 tree) in a plantation within seven miles of the sea. The timber is not exceptionally prone to disease. Species selection was made in accordance with principles of established experimentation with seed and soil types.

As we have already seen, potentially profitable forest expansion in the poorer westerly areas was made possible in the 1950s with the development of new techniques and fertilizers as well as the use of the hardy Lodgepole pine. But, of course, there were development problems and in 1957 it was pointed out[17] that, due to the very short life-span of State afforestation and the limitations of piecemeal acquisition, age-class distribution of Irish forests was very abnormal, more than 50 per cent of the stocked area being in the 1-10 year age class (much of which was also, of course, due to the accelerated planting programme then underway for some years). The balance included 20 per cent in each of the 11-20 and 21-30 year groups, and the remaining 10 per cent of over 30 years was made up equally of acquired woods and early State plantations.

The abnormality was held to be most marked in the case of *Pinus contorta*, or Lodgepole pine (preferred for phonetic reasons), accounting for 75 per cent of the 1-10 year old class. This was a dependence on a relatively unknown species even more marked than the position of Sitka today.[18]

In some ways, perhaps, Lodgepole pine provided the most interesting experiment by State foresters. Called *Pinus contorta* because of the twist in its spines, and Lodgepole because of its use by Indian nations for their lodges, it was little known as a cultivated crop even in America before the 1950s. During the 1930s its use in conjunction with new planting techniques on wet and peaty ground began. The results were encouraging enough to ensure that it achieved and retained its place as the second 'bag' crop of the State afforestation programme, being especially suitable for peaty soils which other species reject. 'Before 1950 very little was known of *contorta* pine elsewhere. Its silviculture had been largely ignored in its native habitat on the west coast of North America. It appears that Ireland... has the greatest area of *contorta* pine plantations, but it looks as if Britain will soon take the lead in this respect.'[19]

Following the Irish pioneer work Lodgepole pine was adopted as a planting species in Britain in the 1950s. Today the most intensive cultivation of this species in Europe is in British State forests, which now have the greatest area of Lodgepole pine plantation in Europe – a development which may be said to have been scientifically pioneered in this country. One cannot but regret that similar experimentation did not precede Knockboy.

A native of the west coast of America (like Sitka spruce), Lodgepole's distribution is greater, giving rise to many varieties of the species. This 'caused a lot of trouble when it was first used... In 1960 this problem... was solved, and now all seed comes from the coast of the States of Washington and Oregon, or better still, from home-grown trees – The timber is not yet well established, but it is promising, particularly in ornamental woodwork. It is also acceptable as pulpwood.'[20]

In general Sitka tends to be planted on better soil and *contorta* on less promising land. Between them Sitka spruce and Lodgepole pine appear to offer the optimum return from a wide spectrum of soils, from the most promising to the least.

A curious and complicated feature of the progress of State forestry has been the one we began with; namely, title. By outside standards Irish agricultural holdings are small, usually between twenty and forty acres. Few exceed two hundred acres. Mountainous regions and tracts of marginal land suitable for afforestation often consist of scattered plots. These often involve common grazing rights. Accordingly acquisition was unduly complicated and vexed, commonages not unusually being held by as many as forty separate titles together,[21] any one of which might frustrate the acquisition and all of which required to be satisfied before title might be acquired.

One consequence of this is the patchwork and – in terms of planned commercial forest management – uneconomic character of many Irish forests complained of by McGlynn and others. That is not the dominant situation today, of course, but in 1958 'the typical forest now consists of many detached plantations and is quite unlike the traditional picture of a forest as a large unbroken tract of woodland. The acquisition process has been likened to the compilation of a jig-saw puzzle on a gigantic scale'.[22]

Tallow forest in Waterford (then about 2000 hectares) was cited as having been pieced together from 63 acquisitions none of which was a square mile (640 acres) in extent. 'The number of separate woodland blocks is now (1958) 35, scattered over a district of 100 square miles and occupying 5% of the land surface.'[23] The problem of acquisition consisted, therefore, not only of getting land, much less suitable land, it was also one of getting it in parcels large enough to be worthwhile for the purpose. At the same time the question of species was still under experimentation.

Perhaps the most dramatic event in the formative period of species selection during the 1930s was the temporary fall from grace of Douglas fir (*Pseudotsuga menziesii*), which had been looked on as a species of great promise. It was popular in the 1930s, but proved more sensitive to soil and climatic conditions and more susceptible to infection than had been anticipated.

State forestry may claim that besides creating a national forest enterprise, it also pioneered the successful cultivation of American species on a significant commercial scale. In so doing new sites, hitherto considered unsuitable, were opened up and a high output per acre developed. Perhaps most important of all was that experimentation led to the successful reduction of the rotation period from 50+ years to between 35 and 45 years, thus reducing the time-span between investment and return.

At least partly as a result of this research and experimentation 'during this century... the forest estate has been transformed from one of over-mature, under-stocked, mixed forests, privately owned, to one of predominantly young conifers, State owned'.[24] Silviculture and management techniques required adjustment involving, the introduction of and adaptation to the new species, and there were changes also in the type of land being planted.

Afforestation on peatland, an important aspect of State afforestation, became possible at the right time for the reasons outlined. The areas most affected, as we have noted, were in the West, where western blanket bog (a distinct type from the midland raised bog) is the type of soil in question. 'It is on these bogs that the greatest potentiality seems to lie, because midland bogs have formed only a small proportion of our peat plantings in the past, and are unlikely to increase in importance in the future.'[25]

In the midland area the fuel requirement is a priority. The question is whether cutover bogs give a better return under agriculture or forestry. The future use of timber as an alternative commercial fuel is undoubtedly a factor. The likelihood is that about 50 per cent of Bord na Móna (the State Bog Board) bogs will be used for afforestation, about 25 per cent for grassland and that the remainder, being below sea-level, will revert to natural habitat.

Little had been done about peatland afforestation prior to the 1940s. Thereafter peatland afforestation becomes significant, going from 225 hectares in the West in 1950-1 to 3750 in 1960-2, from 10 per cent to 32 per cent of the overall total.[26]

The thrust underlying all this activity in these years was rooted in optimism for the future. Pointing out that less than half the 1976 consumption of wood in the EEC was produced by member states, *The Case for Forestry* report said:

The moist oceanic climate of Ireland, linked to timber production from relatively fast-growing north-west American conifers (Sitka spruce and Lodgepole pine) means that...quite exceptional rates of growth are experienced on certain gley soils in Leitrim, Cavan, Monaghan, Clare, Limerick, Kilkenny and Carlow...a survey in County Leitrim assessed an average yield class for the county of 22 and individual plots in excess of 32... recent research on Lodgepole pine shows that midland peat can produce yield classes of over 20.[27]

There remained the problem of acquiring land. The compilers of *The Case for Forestry* calculated – 'conservatively' – the land potential for forestry in the country to be 3,348,672 hectares (8,362,489 acres), classified as being of limited agricultural use. The report also postulated six alternative planting policies. These were considered against (1) the requirements of wood production for a period to 2040, (2) national self-sufficiency 'in terms of production and consumption of wood and wood-products to the year 2025, including balance of payments and implications',[28] (3) expenditure and revenue implications for the Exchequer to the year 2020, (4) employment consequences to the year 2040, and (5) maintenance of a regular flow of raw material for industries. A forecast to the year 2040, based on forest management principles, and showing the individual effects on wood-production, saw-log, pulp, annual planting and reafforestation, was made for each. 'Certain climatic and biological factors such as wind-blow, diseases and pests may alter the future production. Allowance has been made for these factors reducing the expected production by a percentage.'[29] A management system outlined by D. O'Brien[30] was adopted. The forecasts were based on two sets of data.

Field Data: These come from areas planted up to and including the year 1957 in accordance with the field inventory of State woodlands for all areas in the production stage, as documented in the Inventory of State Forests 1968 by Liam O'Flanagan (senior inspector, FWS, later general manager, operations, Coillte). All stands are classified according to species, stocking density, planting year and yield class. The data were updated to take account of all clear-felling and wind-blown areas up to 1974.

Supplementary Data: These are estimated for stands planted or to be planted from 1958 to 2024 which are not included in the field data of the inventory. They have been divided into six eleven year-periods – areas planted 1958/68; areas planted 1969/79; and areas planted post 1979.[31]

Because plantings from 1980 onwards will have a negligible effect on production in the year 2000, self-sufficiency in that year is expected to be about 97 per cent whichever policy is adopted.[32]

However, by the year 2025 the picture will be considerably altered, as Table 9 shows:

Table 9[33]

Projected Self-Sufficiency in Year 2025

	1978	POLICY 1	2	3	4	5	6
Production (000 sq. m.)							
State Forests only	547	6236	4768	4403	3680	2957	2148
Consumption	1737	5266	5266	5266	5266	5266	5266
Self-Sufficiency	31%	118%	91%	84%	70%	56%	41%

Convery[34] considers the employment forecasts between the years 2000 and 2040 according to these forecasts, and finds significant variations.

In chapter 9, TCFF makes the observation that: '...the concept of *maximum sustainable yield*... requires that the flow of material from the forest be regulated so that it reaches its maximum as soon as possible and... that the level of yield never diminishes'.

The hypothetical effects on forest management terms of each of the six policies are considered, but without any recommendation. A list of fourteen conclusions is provided, the first nine of which reflect the optimism about forestry, supported, and apparently justified, by independent interests. Conclusions 13 and 14 refer to the means by which self-sufficiency in forest products might be achieved, and when.[35] Conclusions 10, 11 and 12 assess the respective merits of the six policies.[36]

In another report an important point in respect of home-grown timber is referred to as follows: 'The fast growing timber (of Ireland) differs in quality from the slower growing softwoods produced in Europe and in natural forests.'[37] The point being made is that while timber in Ireland grows faster than elsewhere, the quality is lower. It is literally a case of quantity, not quality. This need not, however, prove an obstacle once the correct marketing parameters are identified. Apart from Greece Ireland has the smallest area of forest in private ownership in the EC.[38]

The Review Group also referred periodically to what was taking place in forestry in Britain. So it may be seen that the vestiges of the bizarre historical interaction between British and Irish forests remain relevant, though in a very different context. Irish timber-processing capacity at 1984 may be gauged from Table 10.

Table 10[39]

The Processing Sector

	No. of plants	Number employed	Timber Intake sq. m. (000's)	Capacity sq. m. (000's)
Sawmills (minimum throughput 250 sq. m. round timber a year)	150	1200	675 (estimate)	1000
Pulpwood	2	325	400 (estimate)	550

Dr F. C. Hummel, former commissioner of the Forestry Commission in Britain and former head of the Forestry and Environment Division of the Commission of the European Communities, was employed as a consultant 'to examine and report on the quality of silviculture in the forest estate and comment on the State forestry programme'.[40]

A complicated and technical method of estimating the value of State forestry was employed.[41] Depending on the discount value of all future revenue and expenses, given estimated values of between 1 per cent and 3 per cent per annum, a value ranging from £606 m. at a discount rate of 3 per cent to £2569 m. at a discount rate of 1 per cent (assuming a 5 per cent per annum productivity and timber price increase), based on an annual planting target of 7500 hectares, was achieved.[42] The value estimated by Dr Hummel was £810 m. Both evaluations exclude additional assets, equipment, buildings and nursery stock.

The other important finding was that State forestry had achieved profitability of approximately 2 per cent (compound interest) over inflation and also provided 'highly valued recreational and amenity facilities'.[43]

An interesting statistic emerged in regard to employment:

One million pounds allocated by the State to private sector afforestation could produce approximately 200 jobs in planting, while £1m allocated to State planting would, with present cost-structures, result in about 40 extra jobs, but, as indicated earlier, the investment in State forests should also secure an asset giving a yield of 3% or more over inflation.[44]

In general employment of forest workers by the State, which works out at about one forest worker to ten hectares, relates to new planting and plantation establishment. Harvesting, tree-pruning and processing also offer potential for job creation. The establishment of new forests is somewhat more labour-intensive than on-going management and has slightly higher direct employment potential in the shorter term (the shortfall being more than offset in the longer term by indirect and downstream activities).

Irish forest produce and forest lands differ from those of most other countries in that Irish timber production is exclusively from relatively immature and evolving plantation forests. These cannot yet be compared with long-established and managed forests. New forest development in Britain and New Zealand is comparable.

An interesting point emerged from the Review Group deliberations affecting management techniques, with emphatic reference to the once neglected question of marketing. While the origin and development of modern Irish forestry owes much to continental forestry practice, such continuing influence was queried, in particular because of the different and much faster growth rate here. The Review Group expressed it as follows: 'The desirability of such influences is worth reviewing. Innovative thinking in silviculture, species selection and plant breeding should be encouraged , but all of these should reflect an overall marketing strategy.'[45]

Since *production* was, in the main, a State enterprise, whereas *mills* and *processing* were private enterprises, a situation existed that, if it continued once State afforestation became geared to marketing, would become one in which the tail would start to wag the dog.

> The major wood products exporting countries are characterized by a substantial level of integration in the producing units from forest to final product... in Sweden there is a State Forest Industries body (Assi) which is a major wood processor. Assi shares common statutorily defined objectives with the Swedish Forest Service... Vertical integration allows the most economic 'fit' between harvesting and processing: harvesting, wood, specification, scheduling, plant location etc., can all be arranged so as to maximize advantage to the unit as a whole. All forest operations can be conducted in the context of an overall wood growth and processing plan.[46]

An integrated forest structure allowing flexibility in respect of the export market also calls for a horizontally integrated structure allowing maximum efficiency in wood utilization, so that fluctuations in demand can be adequately coped with by effective product control.[47] Convery's outline is rooted in common sense.

The want of an integrated State harvesting/processing/marketing system was also criticized.[48] There were compelling reasons why processing plants were begun, and evident reasons for the failure of the early starters following the recession of the 1970s. The government of the day could not be expected to use public monies on speculative commercial processing ventures when it was already funding an immense speculative forestry venture (which would eventually provide the processors with their raw materials) and which was hard put to satisfy the budgetary criteria of the Department of Finance. Similar arguments apply to the question of direct State processing, which would also unavoidably be faced with the requirement of becoming a commercial enterprise in time. In the event the processors came on stream at an inopportune time and, as a result, failed before they could be revived – at a cost to the State of almost £2 millions from Foir Teoranta, the State Rescue Company.[49] But, in 1990, in line with the remarkable progress made in the previous four years, 'the feasibility of introducing a new mill to utilize the extra supplies of small wood becoming available is being examined... total investment in expanding the wood processing sector could exceed £150 million by 1993'. This, had the negotiations with a Swedish pulpmill not failed, would have been in addition to two existing mills, at Clonmel and Scarriff. Coillte Teo remains committed to the principle of expansion in this area, according to official policy. There are approximately a hundred operating saw-mills in the country.[50] Greater emphasis, and substantial grants, up to £1200 per hectare – have been introduced for broadleaves.

CHAPTER SEVENTEEN

Private Forestry

'The Land Acts, together with the political and social changes of recent years, have finished the private owner so far as economic forestry is concerned' – Bulmer Hobson.

HOBSON WROTE THE ABOVE IN 1931. At the time, it certainly seemed to be the case; and the events of the Emergency seemed to endorse the view.

Time and events, as we have seen in the pattern of general development, brought change and practical financial encouragement to the private woodland owner. The rotation period – that flexible and often imponderable factor governing forest economics so much – shortened. Nevertheless the decline of privately owned woodlands was radical and, at 15 per cent of the total (excluding scrub), Ireland, in 1990, had the lowest private woodland ownership in the EC and Europe. But by the late 1980s private afforestation in a new guise – perhaps in two new guises, co-operative 'tree-farming' for the farmer/landowner, and large-scale afforestation by commercial interests – seemed poised to make important progress.

What of traditional private forestry from which all else stemmed, including the growth of State afforestation to 400,000 hectares in less than seventy years? When the 1907 Committee sat to consider the future of forestry in this country there was not a single acre of State forestry, except for the experimental plots and associated land, in all amounting to about 3700 acres at Avondale, which was then simply a school to train foresters for future employment in private forests.[1] Many of those who sat on, and gave evidence before, the committee were private forest owners who, when they so enthusiastically recommended a scheme of national afforestation, could not have anticipated the adverse developments in their sphere which occurred later. At the time private forestry amounted to some 307,000 acres for the entire country. Since these were largely well-established woodlands in the traditional mould (if, in the main, rather more experimental than those of Britain), they contained, and continue to contain, a much higher proportion of broadleaves than State forests. But by 1978 private forests amounted to only 82,000 acres, of which 40 per cent was scrub.[2]

Gloomy though the situation was there were signs of a resurgence of interest by private forest owners from the end of the Emergency onwards. As we have seen, that interest was accelerated in the 1980s, mainly by the direct financial encouragements available from government and EC sources, and by the interest of financial institutions in forestry as an investment prospect, and by 1990 there was a dramatic change, as we have noted (see note 2).

Essentially there are three classes of private owner, and all three sub-divide. They are the tree farmer, the industrial holder and the traditional forest owner, perhaps the most interesting of the three. The third is at once more vulnerable, because of a traditional and, sometimes, semi-emotional relationship with woodlands, and tends to be deeply committed to the principles of forestry and afforestation.

An important aspect of private forestry is that it still contains (1991) a higher proportion of slower-growing hardwoods than State forests – a situation that must alter as new planting increases. Most of these hardwood or broadleaved woodlands are of mature timber.

The terms 'hardwoods' and 'softwoods' are used as broad classifications in the timber industry. In fact some broadleaves produce softwood and vice versa. As employed loosely by the timber trade, softwoods are produced from conifers, and hardwoods are produced by broadleaved trees. Generally speaking softwood produces long, straight timber, relatively easy to work, used in construction and for pulpwood and paper. It accounts for about 75 per cent of manufacturing wood-use. Hardwood is generally used for furniture, carving, veneers and so on.

The Report of the 1907 Committee observed that

> the independent detailed surveys made for the committee in certain parts of the country go to show that the total figure (for forestry) given in the agricultural statistics of the estimated area, may be taken fairly to represent the actual area. That area was, in 1907, 306,661 acres, or 1.5 per cent of the total area of the country. Whether Ireland's area of 1.5 per cent of woodland is too low for the economic welfare of the country, we will consider when we have examined the actual state of this area at the present time.
>
> We find that, such as it is, this area (under timber) is steadily shrinking. The total area under woods is being reduced; and from the character of the cutting, the quality of the woods left standing is being deteriorated.[3]

The modern concept of State forests could not have existed before the modern concept of statehood, but it can also be argued that the Crown forests of England were, in fact, State forests. The point would be irrelevant were it not for the fact that in encouraging (Irish) private owners to afforestation, good forest management and conservation, when compulsory legislation had proved ineffective, the authorities were exhorting them to contribute to England's maritime requirements.[4]

Private forestry in Ireland during the nineteenth century was, perhaps fortuitously, characterized by a more independent and forward-looking approach than was the case in England, traditional and dominated as its forestry was by naval requirements and profiteering rings.

Before that, and to recapitulate briefly, Ireland was held by such as Phillip Cottingham (who in 1608 reported on the availability of timber for the navy, Lord Chichester endorsing the report), Robert Payne, Sir Henry Sidney and others, to contain a great supply 'of noble oaks fit to construct ships to carry Britain's thunder o'er the deep'.[5] But this timber was being used for pipe-staves to such an extent that in 1698 an Act of Parliament stated that 'the timber is utterly destroyed'.[6] Moreover, the cost of Irish timber (in 1630) was only a fraction of what it was in England. As a result, it was claimed, the cost of producing iron in Ireland was only one-tenth what it was in England. The argument (perhaps somewhat over-stated) was that it was not the shortage but the cost of timber in England that led to the destruction of Irish forests by English speculators.[7]

On the whole legislation was preventive and lacking in any compulsion. While it purported to be enacted in the common interest, most legislation of the period distinguished in favour of landowners.[8]

The year 1700 can be regarded as the low water mark of native Irish woods, and brings us to the most important development in the history of Irish forestry of the past, and of considerable importance in regard to its future. This is the creation and laying out of demesnes in all parts of the country, but more especially in those districts and on those types of soil which in the ordinary way would be out of reach of the economic planter.[9]

But any suggestion that there was a planned movement to reafforest the country in the overall interests of the nation would be misleading. What is referred to is the general interest taken by the 'gentry' and, in particular, by the landed variety (*parvenu* or not), in silviculture. This had become a universal interest of the leisured classes, and was by no means confined to Britain (or Ireland). Its motives were partly intellectual or scientific in accordance with the fashion of the period. The encouragement of the RDS and Grattan's parliament helped progress. But afforestation tended to be cosmetic – screening, avenues and so forth – rather than seriously commercial, until quite late. Nonetheless the enthusiasm which scattered acorns and beech-nuts with increasing – if modest – profusion on the fertile land, coupled with pressure from the timber traders, could have but one eventual result, namely the need to harvest the mature crop, thus leading to the commercial forestry outlook of the early nineteenth century and the establishment of the climate of thought in which an incipient industry could develop.

That interest in silviculture was an important factor contributing to the growth of commercial forestry in the early nineteenth century, when the demand for shipbuilding materials was high. The spectacular changes that occurred in mid-century provided a jumping-off point that led to the support, first, for Knockboy and then for the proposals of the 1908 Report, from private forest owners.

If early-nineteenth-century political and social events – Catholic Emancipation, the Famine, the Fenian Risings – affected forestry development, they were as nothing compared with the changes wrought in naval requirements by the American Civil War followed by the continuing Land Agitations in the second half of the century, leading to the Land Acts at the turn of the century.

Following the Act of Union Ireland was deemed to be part of the United Kingdom of Great Britain and Ireland. The difficulties created by the Land Acts and why a tenant 'as soon as he gets a conveyance executed to himself proceeds to cut down and sell any trees that may be on his holding' have already been explored.[10] 'Trees, in the eyes of our farmers, are meant to be felled; they are a useless encumbrance on the land.'[11]

The problem became so acute with the forestry acreage remaining virtually static (at about 340,000 acres) between 1800 and 1885, that the 1887 Recess Committee on Forestry convened to enquire into the question. Some of the evidence given was depressing. 'The present state of things had utterly killed landlords' improvements.'[12] Powerscourt also believed that the government should begin planting 'rather than rely on private owners', which is eventually what occurred – though hardly as envisaged. In the meantime, following on the Land Acts, private woodlands took a severe battering.

The failure of Knockboy was a major set-back. Only in 1903 did the State decide to intervene directly in the conservation of existing woodlands. Avondale was purchased in 1904. But it was not until 1909 that the State intervened to control felling – even to a limited extent – by statute.[13] When the State programme was inaugurated at Knockboy private forestry, the only established system of managed forestry in the country, was barely a hundred years old itself. It was intended that the forestry school at Avondale would provide foresters for private forests, not for State schemes, which is what happened.

As the role of State afforestation increased, that of private forestry began to shrink and alter, largely due to changing political and social circumstances, and to the blueprint of the 1907 Committee Report being significantly altered or misinterpreted in execution. The resulting weakening of interest by landowners was subsequently aggravated by the effects of the First World War on stocks and, more decisively, by the trauma accompanying the emergence of the Irish Free State.

'The condition of private woodlands was to deteriorate so much that the State eventually intervened to remedy the defects.'[14] State encouragement to private planters (by way of technical advice and financial assistance) went side by side with statutory restrictions on indiscriminate fellings. From 1922 there was some encouragement from the State for private planting, but it was unevenly directed and was seldom more than lukewarm. The 1928 Forestry Act provided a system of grants for private afforestation similar to one earlier introduced in Great Britain. By the 1950s subsidies and free advice in respect of planting co-operation were readily available, but private forestry was nevertheless declining at a rate of about 500 acres a year. On the whole the farming community remained unimpressed and unconvinced.[15]

The private forest acreage referred to indicated a staggering decline from the 306,000 acres of private forestry in 1907, mentioned in the 1908 Report.[16]

Of importance is the fact that many of those who considered, framed, subscribed to and implemented this policy were private foresters acting, as they saw it, in the best interests of forestry as a whole.

Although no data are available, the large woodland estates appear to belong primarily to the descendants of the old Ascendancy landlord class. These owners have survived depressions, land wars, the implementation of the Land Acts, the workings of the Land Commission and the imposition of inheritance taxes. Tenacity, an enduring commitment to the land and, in some cases, considerable frugality, characterize this group. Trees are important as both symbols of good stewardship and as links with the past; they represent a patrimony and a tradition to be handed on to one's heirs. While return on investment considerations *per se* are not predominant concerns with these owners, they nevertheless have sufficient business acumen to ensure that the forest yields some revenue. Aesthetic, amenity and wildlife considerations will typically have an important bearing on woodland management.[17]

This point is important to any understanding of the events which shaped forestry matters from 1920 onwards. Those who introduced and executed the idea of State forestry not only represented the only forestry, they also held a virtual monopoly of accumulated silvicultural knowledge and experience in the country. It can be argued that without private forestry there could *be* no forestry policy. At the time the proposal was put forward private owners and their support were essential to the success of the national scheme; without them it must fail or be altered. The Cameron Report contributes another aspect which partly reflects an accepted wisdom of the 1940s rather than a fully informed judgment.

The system of land tenure prevailing prior to the turn of the century was inimical to the increase of forest lands. Land ownership was in very large holdings, often with absentee landlords. Tenant farmers renting small areas on an insecure basis had no inducement to plant trees. Furthermore, in the early days it was often the case that voluntary improvements of tenant holdings were followed by rental increases.

On the other hand, the demesnes surrounding the landowners' residences were uniformly kept in part in forest. Such forests were primarily maintained for sport and amenity purposes. They did, however, serve as a restricted source of supply for the most essential wood needs of the rural population. The need for wood on the farm in Ireland has always been very much less than any other European country, because over most of the country the farmers' fuel needs are met more simply and with less labour by burning turf. Only a small percentage of the farmers in the 26 counties use wood for fuel.[18]

The question may legitimately be asked: what course might forestry have taken in Ireland if, after the foundation of the State in 1922, the original 'tandem' concept had been pursued with continued support and involvement from traditional landowners?

When the government of the new State came to power in 1922 the position was clear. The circumstances – civil war, rioting and pogroms in the North, tensions with Great Britain – gave forestry a low government priority. Circumstances became such as to alienate most if not all private forest-owners. The reasons for the metamorphosis of defined tandem policy into one in which State forestry became the over-riding principle thus become plainer, as do those for the early decline of existing private forestry as a major aspect of national forest policy.

Private forest owners and new administration were, by background, opposed to each other. 'Only 3000 acres had been planted by the State before the first Irish government came into being in 1921.'[19] In spite of the difficulties experienced by the neophyte State, it would be unrealistic to minimize those facing private forest owners, whether of their own making or not. The reality is that, whether in private or in public hands, forests are a renewable resource of environmental as well as economic value to the community. In a land as denuded of tree-cover as the Ireland of the 1920s these considerations were of considerable proportionate importance, but were not given the attention they deserved. Apart from existing woodland owners, little or no – certainly insufficient – encouragement (for the reasons already stated) was given to new owners or to tree-farming.

In 1924 Augustine Henry wrote: 'In Ireland, the main trouble lies, not so much in the reckless felling of the timber, as in leaving the ground unplanted. Such land becomes waste, covered with weeds and scrub, worthless to the owner, and diminishing the resources of the country.' He called for the introduction of controlling legislation, including the following: 'Some notice should be taken of the barbarous manner in which street trees are mutilated in Dublin and other towns in Ireland... Corporation should be compelled to seek the advice of the Forest Officer....'[20]

Henry was a strong advocate of small farm forestry and may, unwittingly, have contributed to the enduring official view that there was a mutual exclusivity of interest between large and small landowners when it came to private forestry. In reality the question is one of different requirements and differing approaches (mainly managerial – i.e. co-operatives for small, grouped pockets – financial and technical). But they share a common interest and such variations as exist are capable of accommodation within a specific policy of private afforestation.

> Land-owners should be invited to lease suitable land for planting to the Government. Produce-sharing schemes have been devised by which the landowner provides the land and the Government does the planting, but the simple lease may prove more attractive, and will be less troublesome to the State, as the dearer cost of the purchase of land is not incurred at the onset. ...The difficulties of afforestation lie in the cost of acquiring the land suitable for growing forest trees, and in the expense of planting.[21]

Henry was blunt. Since he was writing well within the period of the influence of George Russell (AE), perhaps he might have developed the co-operative aspects, but his idea does not seem to have progressed. Perhaps the combined weight of official opposition, political caution and farming hostility was too much for such a radical proposal.

The 306,000 acres of private forest in the (entire) country were reduced to 296,493 acres[22] during the First World War, and in 1922 there were 248,000 acres of private forestry in the whole State which still 'comprised 84% of the total State woodland of the country'.[23] By 1931 this was reduced to 'under' 200,000 acres and to 108,000 by 1942.[24] By 1950 that had become 94,871 acres, but 'it is reckoned to be a safe estimate that private woodlands in general are only 50% stocked. On the whole it is considered by the Forestry Division that the total area of private woodlands fully and satisfactorily stocked would not exceed 40,000 acres at 31st March, 1950.'[25]

The same paragraph contains the following chilling observation:

> No large increase in the area of private woodland can be expected in view of the virtual elimination of large estates, the predominance of small agricultural holdings and the continuing purchase of woodlands by the forest authority.

Hence the conclusion seems to be that... private planting... cannot be expected to make a significant contribution to the national timber supply.

Of course this was written in a transitionary economic climate and the demise of large estates as a common social phenomenon, though foreshadowed, was not yet wholesale. Neither had the developments which interested financial institutions in investing in forestry and large estates taken place.[26] The general lack of interest in planting by private owners from 1922 to 1929 may be demonstrated by a calculation which purports to show that the 'scale of private planting did not, it is believed, exceed 200 acres a year'[27] for the whole country!

It should not be overlooked that the Second World War would almost certainly have put a greater strain on Irish private woodlands had this country been still part of the United Kingdom, and that benefits-in-kind or direct cash benefits to private forestry were not significantly more beneficial in Britain than in this country.

As we have seen encouragements were introduced under the 1928 Act – including a planting grant of £4 an acre, but with little result. 'In the period 1930 to 1942 the average annual rate of private planting was about 130 acres. In the period 1943 to 1955 the average was about 250 acres.'[28] Because of the nature and variety of the variables involved it is hardly possible to subject private planting to economic analyses of the kind undertaken in relation to State forestry. 'In practice, timber felling on small privately owned woodland areas is timed by economic requirements on the farm rather than by normal considerations of forestry rotation.'[29] This important point demonstrates the difficulty.

The point is also made: 'Private woodlands do, however, constitute a crucial reserve which can be called upon by the owner, when required, to finance improvements in his holding.'[30] That statement, made in 1958 when State policy was well down the road from clarity of objective and firmness of purpose, is a good example of the lack of realism sometimes evident in the official approach to private forestry – especially to 'tree-farming' on a small scale. It is a curious example – so far as consideration of private forests is concerned – of the moribund private forestry philosophy of the first four decades of Statehood. But, oddly enough, it was in that same year, 1958, that Erskine Childers, in a major speech to the Oireachtas, articulated a new and radical (if, alas, short-lived) policy on private forestry.

Rather like the change that affected the superior idea of a national forest programme in tandem, the 'idea' of private forestry also underwent something of a change. Traditionally, of course, private forestry meant 'estate forestry', the forestry of great landowners. Gradually, however, the principle of 'social' forestry was more and more incorporated into the idea. When the Cameron Report was published in 1951, 'social' forestry had begun significantly to supersede that of the large land/forest owner whenever 'private' forestry was referred to. It also seems clear that, from 1922 until his retirement ten years later (but with little success and in the face of considerable opposition), Forbes strove to persevere along the lines recommended by the 1908 Report and the Acland Committee (and, indeed, of the Forestry Act of 1919), so far as private forestry was concerned. He (who better?) was aware of the *fait accompli* position regarding private forestry. 'During the last fifty years a serious decline in planting is shown on private estates and energetic State action becomes necessary if the woodland area is to be maintained.'[31]

It may be argued (and sometimes has been) that Forbes, rather than promoting the encouragement of private owners to plant, was proposing State 'planting' action as a substitute. But this argument is counter to the thrust of his purpose in including the private planting grant in the 1928 Act, as the following makes clear. Recognizing the *de facto* situation Forbes also recognized (and deplored) the implications:

It is to be feared that planting on an extensive scale must either be carried out by the State or remain undone. In a country like Ireland this is a misfortune. The private planter can ... work on small areas... The State can only work in large blocks... without the co-operation of the private planters, therefore, the wooded condition of the country as a whole must diminish.[32]

When, in 1934, Connolly made the first serious proposals for social forestry, the idea of 'private' forestry was well on the way to being transformed from a major activity by large landowners to a minor one by smallholders. The grant scheme, initiated under the 1928 Act, which came into effect in 1931, offered little encouragement to the large estate owners. But, while the new Act might have benefited farmers and smallholders, they were the very ones who still viewed forestry with distrust and suspicion, not least as they became locked in the struggle known as the Economic War with Britain at about the same time, allowing no room for unproved, exotic, uneconomic experiments in forestry, which they disliked anyway. The argument that 'tree-farming' is more profitable than mountain sheepgrazing, if it ignores the question of regular income, becomes preposterous, illogical and unrealistic, which was the case for years, having the effect of reinforcing the doubts of the farming community in general about forestry and those promoting it.

Not surprisingly farmers felt that those putting forward such an argument either don't understand the realities of the situation or that they were trying to fool them. Paradoxically the argument was more successful with large landowners who began to see prospects of a better long-term return for marginal land. Thus the wheel eventually began to turn full circle with 'private' forestry coming to mean large, interested landowners, this time, in the main, corporate institutions. But even farmers who could afford the capital for long-term planning found the approach unsympathetic. Correct though the proposition is, so long as it ignored elementary practicalities it had little validity for those it was mainly supposed to convince.

Farmers may also have been inhibited from time to time by the fact that if they undertook afforestation it could be deemed to increase their incomes, with resulting diminution of any social welfare benefits they might receive. That concern was later invalidated when tree farming was specifically exempted from being considered in this way.

Until recently a farmer of thirty-five years of age with, say, fifty acres (and in the 1930s the average age for such a farmer would probably have been considerably higher) was being asked to consider something like the following:

(1) He plants ten acres in forest, worth £5500 in 50 years.

Age	35	45	55	65	75	85	
10 acres Profits	nil	nil	nil	£1000	£1000	£3500	= £5500.

(2) He runs 15 sheep a year on ten acres getting £90 a year worth £4500 over 50 years.

Age	35	45	55	65	75	85	
10 acres 15 sheep £6 a head (£90 x 10)	£90	£900	£900	£900	£900	£900	= £4590

Labour and capital costs are roughly similar, but a forestry planting grant of £4 per acre (paid in instalments *after* planting) was payable. No contest.

According to this view of things if sheep were an unattractive proposition, trees were less so. Between the 1930s and the late 1970s, except for what might be called the Childers Initiative, the position as between a choice of sheep or trees was, as broadly indicated above.

Given the choice between no income except a planting grant (with no allowance for failure rate or a variable demand for thinnings) and no significant income until the end of the thirtieth year (when he is sixty-five), or a fairly reliable income from mountain grazing, it is not surprising that farmers engaged in sheepfarming were less than enthusiastic about the unknown, the untried and, on the face of it, the impractical, or that the remainder were unimpressed. Yet those who chose to plant were warmly received with, until recently, an interesting caveat.

The farmer who wanted to plant an acre or ten acres was always very welcome and he got a great deal of encouragement. But if an industrialist came in and said he wanted to buy a thousand acres and put it under forest, he was discouraged on the grounds that he was taking over agricultural land. Many such were from other countries, and this became an additional objection. But it would have put land into production.

It would have been better to have encouraged private forestry more and leave it in the hands of the farmers, but encourage them to plant trees on it. Trees can be grown far cheaper that way than can be grown by an army of foresters like myself and others who are highly paid by the State.[33]

While efforts to cultivate an interest in forestry amongst the farming community continued, the neglect of large private forests – by both owners and government, refusing on the one hand to continue the tradition and on the other ignoring it – continued into the 1970s. But there was one period when the large and established private forests were again of crucial importance. Speaking at the inaugural meeting of the Society of Irish Foresters in February 1943, Anderson paid this tribute to private forestry during the Emergency: 'in supplying materials essential for the life of the community...'[34]

In 1948 the Private Planting Grant was increased to £10 an acre, but with little immediate effect on an increase in private planting. In 1956 an Irish Landowners' Convention was held in Dublin to consider that point and a booklet, 'Private Forestry in the Irish Republic', was later issued. It contained the observation that 'very little interest has been shown in private forestry ... there is a tendency to overlook the fact that private woods are still the principal source of mature timber in Ireland ... there has been little re-planting in the past *due to lack of confidence* with the result that many woodland areas are now understocked.'[35]

The booklet outlined – with some special pleading – the plight of forests and of forest-owners. It correctly pointed out that after the Emergency, since the immature State forests could not supply the demand, owners responded to appeals from the Minister of Industry and Commerce to keep the home market supplied with mature timber at controlled prices. It made the (then) important point that taxation and Estate and Succession Duties were discouragements to private forestry, making it financially 'impossible' for owners to provide employment and establish forest industries in rural areas.

The argument was that where capital was locked up for between fifty and one hundred and twenty years (depending on whether hardwoods or softwoods were being grown), forestry should not be taxed on the same quick-turnover basis as industry and trade. Comparisons with the situation in other countries were adduced in support of their case. It was pointed out that in England private forests were de-rated, whereas £1 borrowed in Ireland (to pay rates) at 5 per cent amounted to £209 at the end of a fifty-year rotation period. Death duties were also criticized: 'The greatest incentive to commercially managed private woodlands would be to abolish all death duties on them.'[36]

A point made by the private owners, and subsequently adopted as departmental practice, was that they should be in a position to get technical assistance and advice 'from the Department's trained officials in very much the same way that farmers obtain technical advice through the Department of Agriculture and the County Committee of Agriculture'.[37] They emphasized the lack of central timber marketing facilities and urged the State to deal with this problem. In a reference to the Forestry Debate in the Dáil that year (27 June 1955) they said: 'The emphasis in that debate all seemed to be on State ownership and State acquisition of more land for planting, but nothing was said about the private woodlands which are now supplying the home markets or the necessity for their continuance and re-planting.'[38]

The booklet, described as a 'Memorandum of Private Forestry', contains nine conclusions in the form of recommendations to the Minister for Lands. Of these the most important were those in respect of taxation and rates; the seemingly not unreasonable request that the State 'instead of concentrating exclusively on State forests and the acquisition of more land for planting' should help existing woodlands in private ownership 'which still form the main reservoir which supplies the market with mature native timber...'; the request for technical assistance; for a comprehensive marketing programme, and that owners be allowed to carry out their own thinnings and 'other silvicultural operations... part of the normal routine of good forestry,' which were not freely permitted. This important document was forwarded to the then Minister, Blowick. No steps in the matter became evident until after the change of government the following year, 1957.

Childers, the incoming Minister for Lands, announced (1958) an ambitious programme that included new thinking and a new policy towards private forestry. He intended to implement the State planting target of 25,000 acres a year set by MacBride, not achieved up to then, and said that it was hoped to achieve this target by a joint enterprise undertaking involving private forestry, mainly smallholders and farmers. To encourage this he raised the planting grant from £10 to £20 an acre and announced a major promotional campaign which included free technical advice and lectures.

The bonus-incentive scheme with which this initiative was associated was a major social undertaking aimed at providing 'a new national wage base for forestry workers', in an enterprise where wages and conditions throughout the country varied widely, and was stated by Childers to be 'by far the biggest advance in work-study of its kind... since the inception of the State'.[39]

He also announced that he had asked the Institute for Industrial Research and Standards to establish standards for home-grown timber. He went on to say: 'Our private forestry plantations in 1880 totalled 380,000 acres and today amount to 90,000 acres... Only 1,966 acres have been (privately) planted in the past four years although a grant of £10 an acre is available.'[40] In a powerful policy break-away he stated:

> The emphasis on State planting has been far too great and the whole activities of the Department have been concentrated on State forestry... The Government has agreed that we can make a tremendous effort to interest people in private forestry: apart from making grants available not a penny piece has been spent for over 30 years on private planting promotion as a nationally desirable policy: we start from scratch.[41]

'Again', he might well have added. But this time with less success than with the State programme. The combination of accumulated prejudice and inadequate, even misguided, financial encouragement turned out to be counter-productive rather than anything else, and proved too much for the ideal. The innovation was short-lived.

Dillon's caustic observations on Childers's proposal were, unfortunately, largely borne out since, although the State planting target was quickly enough reached, and though there was a response from private owners, it was neither sufficient nor lasting. The undesirability of a national forest estate mostly in State ownership became a talking point. But lacking an appropriate response from the landowners, no way out of that impasse emerged. It seemed that if the nation was to have woodlands on an adequate scale at all, only the State was going to do anything about it. In a paradoxical illustration of this Childers also made the point that to date (1958) the State had spent £14 million on forestry, £10 millions of it in the previous ten years.

By the mid-1960s we were back again to the official notion that private forestry meant trees on small holdings only.

> The predominance of small-holdings in Ireland precludes any possibility of large-scale afforestation by the farming community. The prime aim of the Government's private forestry policy is to secure the conversion of productive forest use of those small pockets of waste land which cannot be included in the State forest programme. Even within this limited objective progress is slow because of the almost complete absence of any forestry tradition among the farming community... the annual planting under the grants scheme is now about 1,100 acres, compared with less than 500 acres prior to 1958.[42]

An important – and controversial – decision followed the 1958 planting grant increases, and did re-enliven interest in private forestry in one sense. Some estates were bought by non-nationals who believed that purchase of land for planting would be a practical long-term investment. These purchases raised a political storm amongst the agricultural community. Furthermore some of the estates in question bordered State plantations and had been offered to (and refused by) the State. Thus 'the provisions of the increased grant were therefore affecting State acquisition and management in a manner which the Forestry Division considered undesirable'.[43]

What this apparently means is that (a) the political storm over foreigners buying land and (b) the fact that they were able to do so for forestry purposes in competition with the Department, was not to the liking of the Department – in other words, private forestry was all right so long as it was kept in its proper (very subordinate) place. Such thinking would appear to reinforce the view of an absence of strategic or mercantile thinking in official forestry circles at this time. With State forests ready to come on stream within a decade, private forestry pressing for a marketing strategy and rationalization highly desirable, such an attitude is entirely consistent with the 'tunnel-vision' outlook resulting from 'forestry' being considered a virtual State monopoly.

It was because of the initiatives by foreign buyers and the consequent uproar from farming and State forestry sources that the Minister for Lands, Michael Moran, warned in October 1960 that 'if (he) considered any private planting project as inimical to the accretion of further State ownership he would withhold a planting grant'.[44] This declaration, so counter to what Childers had said earlier, was supported by the statement that 'the grant was devised primarily to secure the planting of the many small areas of wasteland on agricultural holdings, and not as an encouragement to non-agricultural interests'.[45]

But the negative attitudes – traditional and economic – altered greatly in the 1980s due mainly to financial encouragement from the European Community, making it possible to bridge the income gap in the first years.

A second important factor in bringing about a change of attitude on the question may have been changes in rural lifestyle. The so-called 'drift from the land' assumed epidemic proportions in places. Coupled with increased mechanization this meant less available manpower. Add to that improved living standards generally, and greater social expectations, which rendered labour-intensive hill sheepfarming unattractive and unprofitable. Younger farmers simply would not get involved in running sheep on mountain land with the laborious, time-consuming work involved for low rewards, when they might more profitably do something else.

By the late 1980s forestry reached another turning-point at which the thrust – however vague and inaccurate from time to time – had always been aimed – namely at maturity of the basic economic crop, replanting and the prospect of cyclic economic viability. The developments brought with them considerable change, both in practical and in attitudinal terms. Forestry was seen in a new light by people and interests who had hitherto scarcely considered it seriously. The prospect of economic viability brought with it hard-nosed assessment. The forest structure was examined along with markets and marketing. The blue-print for a co-ordinated industry began to emerge.

Five main areas of forestry activity were identified:

(1) Well-managed State-owned forests mature and maturing.
(2) Great tracts of land suitable for forest development.
(3) Private co-operative forest development.
(4) Industrial forestry development.
(5) The complex of downstream industrial and employment potential.

From the mid-1980s onwards forestry was accepted as a practical, commercially profit-yielding, proposition. Emphasis was being placed increasingly on its importance to the farming community as a less onerous crop alternative both to mountain sheepfarming and as another marginal land-use, and this was getting a positive response.

The size of the annual planting programme has important implications for forest management and timber surplus. On the principle that you can reap only what you sow and, in forestry, the time period to the productive thinning stage is 20 to 25 years, we are now harvesting thinnings from the large planting programme of the 1960s. The accompanying table gives an indication of the approximate area and age structure of State forests. It shows a preponderance of young crops and the tremendous future potential. It also shows the paucity of material in the 50-year age class when final harvesting might be expected.[46]

Table 11 gives approximate distribution of area by age class in State forests at 1983.

Table 11

Age Class	Area in '000 Hectares	% of Total
1-10	77	26
11-20	89	30
21-30	71	24
31-40	26	8
41-50	22	7
51-60	9	3
60+	6	2
TOTALS	300	100

The shortened crop rotation rate accommodates pension trust funds admirably, but, since forestry investment could not be realized at short notice and buyers were still thin on the ground in the market-place, the finance houses were understandably cautious at first, and initially committed only about 2 per cent of their investment assets in this field – amounting to about £4 million a year. Such investors preferred to buy established forests with short-term maturity. But gradually an interest in forest ownership developed. Financial companies' interest in forestry, rooted in changing trends in the stock markets where much pension money was invested, extended to planting, nurserying and financing farmers on a shared basis.

The Irish Life Insurance Company invested in forestry in Wales where it has 14 per cent of its forestry holding. Over 40 per cent of its fund in forestry was in bare land which was planted in 1986, and 35 per cent of its holding was in trees under five years old, all of which indicated a long-term interest.

As the total forest area approached 425,000 hectares in the mid-1980s (338,000 hectares of State forest, 85,000 hectares – only 57,000 hectares of it productive – of private forestry), the most profound change in forestry since the foundation of the State and its far-reaching consequences affected both State and private forestry, particularly the latter. The overall approach to forestry now adopted by those in positions of influence unknowingly reverted to an approximation of what was proposed by the 1907 Committee. The lone, and sometimes despairing, voices of such as Forbes and Henry, Connolly and, most forcefully, Childers, protesting that private forestry was being neglected to the disadvantage of the nation were, if not recognized, acted on.

Future developments in forestry were seen as being largely in the province of private enterprise, both small and large. The financial returns were seen to be attractive and convincing. The EC Western Package Grant scheme was pivotal so far as this about-turn on private forestry was concerned. The EC concern was not simply for Irish forestry, but was based on a clear understanding of the current forestry position in Europe related to projected timber requirements, as already discussed. Ireland offered considerable scope for rapid forestry development, the main obstacles being the related ones of funding and attitude. It was on these areas that the EC concentrated its drive through the Western Package Grant scheme. The process of conversion was slow.[47]

The Western Package was introduced in 1981 for a ten-year period. It was jointly and equally funded by the EC and the government. At first it applied only to thirteen counties, all but three of them western counties. In January 1988 the scheme was extended in two important ways. Hitherto it had applied only to marginal land; from 1988 it applied to all agricultural land in disadvantaged areas. Furthermore the designated disadvantaged areas now included areas in an additional thirteen counties so that, in effect, every county in the country was in a position to benefit. The scheme was also extended to include part-time farmers, retired farmers and farmer co-operatives. Prior to that only fulltime farmers had qualified. The EC funding was also increased and the costs were now borne in the proportions of EC 70 per cent, government 30 per cent.

Some examples of the break-down of the grant are interesting.

– Planting grant, 80 per cent for farmers and farmer co-operatives, 20 per cent for others, subject to a maximum of £800.
– Road grant, 80 per cent of cost subject to a maximum of £12 per metre.

State grants were also available, but *not* in addition. Compensatory headage allowances were available to farmers substituting forestry for agricultural land where livestock headage payments already existed. Loan or partnership finance was available from (and to) co-operatives to enable farmers to establish woodlands and have an income while doing so. By 1988 the entrenched hostility of the farming community to forestry had considerably dissipated, with the prospect of assured, continuing profitability in forestry at a net 5 per cent base for the foreseeable future exclusive of land values.

Besides providing wood and wood-products forests also protect land from flooding and erosion, ease drainage problems, provide shelter belts and enable marginal land to become productive. In the ecologically alert climate of the 1980s, when grave concern was expressed about the damage caused to the ozone layer and oxygen deficiency alleged to result in large part from mass destruction of tropical rain (and other) forests, such environmental considerations became increasingly important. The claim was intensified that the justification for afforestation did not rest on considerations of direct financial return alone, but on much broader criteria.

While forestry – tree-farming – is unlike other types of farming in that it is essentially an investment and not a cash-flow undertaking, this problem was substantially reduced when it became possible to arrange for an income from the outset. The land, a capital asset with considerable and separate growth value, usually remains the property of the farmer when the crop is felled – unless he decides to dispose of it as well. For taxation purposes the annual growth of trees was viewed as income and was exempt from income tax.

Forestry thus involved investment in *both* land and timber, each of which is profitable, but in different ways. The rate of return compared with the real rate of return from other forms of investment was generally advantageous and ahead of inflation. Both land and timber prices showed a consistent rise and all the indications were that this would continue. A number of important tax concessions were available, notably exemption from income tax on profits or gains from woodlands, and the effect of these was to increase the investment value. The total value at clear-felling of a hectare of Yield Class 20 (which is high: the average is 14-16) was calculated[48] at almost £18,000 if held, and having a disposable value of £1500 to £13,500 calculated at five-year intervals from the fifth year to clear-felling at the fortieth year.

Apart from the comprehensive scheme of tax reliefs and inducements for woodland owners, important grants and financial schemes for private afforestation stimulated rural interest in forest investment. These fell into two categories: Grants, and Loan or partnership finance. The grants were the EC/State-funded Western Package Scheme and Exchequer Grant schemes.

In May 1988 the Minister for Forestry, Michael Smith, who had proved himself to be a very hard-working and energetic minister for forestry, announced the Forestry Bill, re-enabling the establishment of a national commercial State forest company. While it was not made specifically clear in the announcement, a general thrust of the enabling legislation was to encompass the original proposal for a National Forest Programme administered by the State under which both the new State company, Coillte, and private forestry might be encouraged and expected to flourish.

The sudden explosion of 'economic' forestry, besides enthusiasm of various kinds, brought with it criticism too, sharpest, ironically enough, at the political end which had ignored forestry in the difficult years. Some criticisms were realistic and were directed at areas, such as marketing, in which the Forest and Wildlife Service showed itself to be weakest. But, on the whole, external and non-expert criticism tended to be ill-informed and insufficiently considered, or followed arguments put forward earlier and more cogently elsewhere. For instance in its Review of State Expenditure on Forest and Wildlife Service, October 1986, the Dáil Committee on Public Expenditure criticized accounting methods which, it claimed, were 'inadequate and cost-ineffective'. The report recommended that commercial criteria should be applied; that private afforestation should be developed and that amenity, conservation and wildlife should be amalgamated and funded separately from commercial activities. Many of the recommendations in the report – which bears evidence of over-hasty compilation – reflected those already put forward with more restraint and understanding by the Review Group on Forestry. In effect the report of the Dáil Committee skimmed the surface of the forestry situation and made recommendations which had the appearance of pronouncements, but without much in the way of serious indication on procedures or costings. While some criticisms were constructive, there was a good deal of what might be called jumping on the bandwagon.

One of the strongest statements in the Cameron Report was, 'It is noteworthy that there is no overall definition of forestry policy embodied in the legislation in Ireland, such as might be expected in the Forestry Act, 1946.'[49]

That position had changed. While the economics involved in forestry had, by the 1980s, altered for the better, an anti-forestry campaign in some areas, particularly south Leitrim, remained strident and orchestrated. From being directed at the idea of forestry in general, it now focussed on private forestry undertaken by financial institutions. It sometimes led to the destruction of forestry machinery and plants. The confrontation was essentially among local vested interests. The campaign was something of a pipe-dream, tending to focus on a notional (often unrealistic) potential value of land as if calculated for industrial use, and lack of significant industry (other than forestry) in these areas. This condition, attributed to lack of local political endeavour and low voting power rather than to the county's suitability for forestry, was advanced as the reason for lack of alternative industrial development. The emotive pitch embroiled marginal landowners in a dispute from which they appeared to have little to gain.

The question was presented as being that of the possibility of immediate gain being obstructed by unproven long-term forest development of no use to farmers. To a considerable extent this approach was vitiated by subsequent events and the possibility of the farmer being able to keep his land and obtain an annual income from the outset – while participating as a partner in the marketing of the crop.

But values were articulated in terms akin to the false surge of land values after Ireland joined the EEC in 1973. Perhaps similar inflationary values were hoped for. The consequences for farmers (in the late 1970s and early 1980s) of inflated land prices, when the banks foreclosed on loans they had a few years earlier happily encouraged, were not lightly forgotten and the campaign faltered as forest economics began to prove themselves. Nevertheless the statement that 'The question of how a farmer who planted trees survived for the 10-15 years they provided him with no income ... was the reason why there was no interest in forestry'[50] put the farmers' hostility to forestry in the 1970s and 1980s in a nutshell.

Following the development of co-operative tree-farming and of leasing and private forestry by financial institutions, the emphasis on acquisition as the underlying problem – as it was in the context of a forestry programme based on State planting on State land – altered. The emphasis remained, but it focused on acquisition to provide a foundation: on private participation and partnership, not on State monopoly.

The difficulties which beset traditional agriculture, reflected in agriculture's performance in real balance of trade terms, and the impediment of agricultural surpluses within the EC, were also important incentives for forestry reappraisal. Forestry would seem to be an obvious alternative to some traditional types of farming. The question of a common forest policy for the EC has been proposed (and opposed) by member states. There are obvious difficulties, and the example of the CAP is not encouraging at the time of writing. However encouragement of forestry in member states is active and the question of Community action on forestry outside a common policy is very much alive.

It became stated forest policy in 1988 that the national forest estate should be increased in Ireland by another 340,000 hectares at least (to 10 per cent of the land area) and that most of this undertaking should be by private woodland development.[51] If that target is achieved within a similar time-span, the middle of the next century should see this country once more respectably afforested with a well-developed forest industry and infrastructure providing the State with one of its largest industrial bases and areas of employment, and a substantial natural resource.

In *Operational Forestry Programme*, 1989-1993, a detailed outline of the activities and targets proposed is set out, together with annual costs. Particular attention is devoted to encouraging farmers and other potential forest owners from the private sector, for whose benefit an overall forestry scheme, it was stated, would be published. The total investment over the five years was calculated to be some £163 millions.

Perhaps the final word should go to McGlynn, who, like so many of his colleagues, played an active and formative part in the development of modern forestry at a critical time and whose interest and devotion to the cause of forestry, State and private, and to the achievement of a proper balance between them with resulting prosperity for landowners – particularly the hard-hit smallholder – never diminished.

'Every farmer should have his forest. It should be part of every farm.'[52]

CHAPTER EIGHTEEN

Northern Ireland; Biomass

'Habitarunt di, quoque silvas' – Virgil.

THERE IS NO ESSENTIAL DIFFERENCE between the background forest history of Northern Ireland and that of the country as a whole until the foundation of the State in 1922, when Northern Ireland, as part of the United Kingdom, came under the aegis of the new British Forestry Commission. From then on certain differences emerged in several areas, notably in administration and species selection.

In 1910 Forbes bought Ballykelly wood in Co. Derry – a wood of about 200 acres – for the Department of Agriculture and Technical Instruction, paying something in the region of £1800. This was the first State forest in Ulster, and was later, when part of the province became Northern Ireland, to become the main forest showplace in that area. Forests at Knockmany (460 acres), Co. Tyrone, and Castlecaldwell (300 acres), Co. Fermanagh, were acquired in 1911 and 1913 respectively. The first plantings took place in Ballykelly in 1912, when 11 acres of Douglas fir were put down. In 1912 the Forestry Commission leased 3000 acres from the Duke of Abercorn at Baronscourt, Co. Tyrone, which was the first major acquisition in the area.

After the Forestry Commission assumed responsibility for United Kingdom forest affairs in 1919, the position of Ireland was uncertain. Accordingly, in January 1920, the arrangement earlier outlined was arrived at whereby the Forestry staff of DATI were transferred to the Forestry Commission in order to comply with the enabling Forestry Act,[1] but were immediately seconded back to the DATI which would act as agents for the Forestry Commission.

David Stewart, one of the Scottish foresters recruited in 1908, became District Officer in the Northern-Ireland-to-be, and the headquarters of the District was moved to Baronscourt. When Partition took effect on 7 December 1922, the State forest area in Northern Ireland amounted to something less than 4000 acres, run by a staff consisting of Stewart, two foresters, a forest staff of twenty-five and a couple of clerks.

In 1923 the government of Northern Ireland acquired responsibility for forests in the area and this responsibility was vested in the Northern Ireland Ministry of Agriculture. There were about 40,000 acres of private forestry in Northern Ireland at this time. In the same year a Commission was set up to 'consider and report on how the natural industrial resources of Northern Ireland can best be developed' and forestry was the subject of a report later published by the Northern Ireland government. Among those called to give evidence on afforestation with Forbes was Colonel John Sutherland, Forbes's erstwhile opposite number as Assistant Forestry Commissioner for Scotland.[2]

The report, known as the Charlemont Report after its chairman, Lord Charlemont, stated that the forest areas of Northern Ireland could be substantially increased. It recommended that the two largest existing acquisitions, Baronscourt and Newcastle, Co. Down (3000 acres), which had been acquired in 1923, should be developed as quickly as possible to make them economically viable. The report also advised the encouragement of private planting by extending the recently introduced Forestry Commission Planting Grant Scheme of £3 an acre to Northern Ireland, and the provision of plants at cost.

Stewart proceeded to recruit experienced foresters from Scotland and simultaneously moved his own headquarters to Belfast where he could become directly involved in policy. He arranged for forestry students from Northern Ireland to be trained at forestry schools such as Beaulieu, later Benmore, in Scotland. A second forestry inspector, John McAlpine, also a Scot, was appointed in 1929, and remained with the service until 1948 when he was succeeded by Col. F. G. Burgess.

During the 1920s there was an annual planting rate of about 400 acres, which increased to about 1000 acres between 1930 and 1946. During the 1950s, paralleling developments in the south, this was doubled so that by 1954 some 67,000 acres had been acquired, of which 38,000 were planted, and there was a permanent staff of over 1100. By 1930 unemployment in Northern Ireland was over 22 per cent[3] and this aspect of policy was stated to be of considerable importance, particularly so far as short-term winter relief schemes were concerned. These continued during the 1930s. The three legs of Northern Ireland forest policy were:

(1) To reverse deforestation and create home-grown timber resources equal to at least emergency requirements;
(2) to provide productive work in areas of serious unemployment;
(3) to encourage private planting.[4]

The long-term aim in the 1950s was to create a total area of 150,000 acres of productive State forest, with a sustained yield capable of making Northern Ireland self-sufficient in timber consumption.[5]

Because of financial restrictions the planting grant of £2 introduced in 1927 (a year before the Saorstát Éireann 1928 Act and four years before the introduction of the Southern grant of £4 an acre) was considerably less than that recommended by the Charlemont committee. It was not successful. According to Kilpatrick[6] 'a report to the Secretary, J. S. Gordon included the following – "this scheme has not been availed of by many owners as private forestry in Northern Ireland has almost disappeared and is not likely to be revived"'.[7]

The situation of private forestry in Northern Ireland, therefore, appears to have been, if anything, even bleaker than that in Saorstát Éireann at this time. Curiously, while the main scheme of encouraging private planting does not seem to have been very successful right up to the 1950s, a separate scheme of supplying plants at cost to encourage farmers and smallholders to establish shelter belts and to plant marginal land was quite successful, in spite of the fact that varieties were restricted. Such post-war private planting amounted to about 10 per cent of the annual total. It appeared that farmers in Northern Ireland did not share their southern counterparts' prejudices against trees, no doubt due to differing historical perspectives.

By 1927 the State forest area had increased to over 12,000 acres. The region was divided into four forestry areas in counties Antrim, Derry, Down and Tyrone. In a report by Stewart it was proposed that, as part of an overall long-term programme to plant 40,000 acres at a rate of 1000 acres a year for forty years, spruce would be the main crop in Tyrone, Derry and Antrim, and larch and Scots pine in Down, at a rate of some 250 acres a year for the first twenty-five years. Nurseries to supply seedlings were already established at Baronscourt, Rostrevor and Newcastle. This proposal was modified by the Ministry of Finance and limited to a seven year programme with a target of 1000 acres a year, for which £22,000 was allocated to 1935-6. A limit of £6000 expenditure in any one year was provided; since annual expenditure had several times exceeded that figure already, this was a severely limiting factor.

The first woodland census in Northern Ireland was carried out between 1939 and 1940, impelled by strategic considerations and the wartime supply situation. Felling controls were also introduced. Both factors made urgent the matter of a timber census. The results showed a total woodland acreage of 60,000, of which 20,000, representing about 70 per cent of the total area, was found to be scrub or otherwise commercially useless. The remaining 40,000 acres was divided roughly into 25,000 acres of private woodlands and 15,000 acres of State forest. In contrast with the situation in the south, a substantial portion of the Northern Ireland forest estate was composed of acquired mature woodland. Some 5500 acres were felled during the war, and about 4000 acres were planted. At the end of the war there remained about 40,000 acres of effective woodland, now divided roughly half and half between private and State ownership – the largest area of which, by far, was in Co. Derry, with well over 13,000 acres.

A policy revision took place after the war. Training and recruitment had been suspended during the war and in 1946 arrangements were made to qualify professional foresters at British universities, there being no faculty of forestry in Northern Ireland. Vocational training in forestry was, however, provided in Northern Ireland at Greenmount Agricultural College.

A government committee, chaired by Mr Justice Babington and known as the Babington Committee, was set up to consider post-war agricultural policy, including forestry. It produced a report which gave a powerful recommendation to forestry development in Northern Ireland: 'there is an unanswerable case for going ahead with afforestation in a comprehensive way... We suggest that the planting rate be doubled ... to at least 2000 acres per annum over the succeeding fifty years'.[8] The planting grant was increased to £10 an acre, acquisition was increased (including some experimental work on peatland, as in the south) and the work-force was significantly increased.

The Northern Ireland Forestry Act, 1953, was consequential on a New Forestry Act in Britain superseding the 1919 Act. The Northern Ireland Act did not include a provision for the compulsory purchase of land for forestry purposes. 'It is thought that the absence of the power resulted in a better relationship between potential vendors of land for planting and the government.'[9] This is an interesting comment and one that may cast light on some of the social and demographic differences in Northern Ireland, especially at that time. That Act became the instrument for the implementation of the three-point policy outlined earlier, which by and large remained unaltered until the Development Programme 1970 was formulated.

In spite of some of the differences mentioned, forestry in Northern Ireland shared problems in common with what was now the Republic of Ireland, particularly, so far as acquisition was concerned, that of small parcels of land which had to be obtained piecemeal, making the acquisition of forest blocks of economic size a time-consuming and expensive business. The acquisition programme proceeded energetically, however, and between the end of the war and 1954 some 30,000 acres were acquired, bringing the total holding to 67,000 acres, a three-and a half-fold increase on the rate of pre-war acquisition. As in the south, Sitka spruce was the dominant species, but poplars were also given substantial encouragement.

In 1955 Northern Ireland had its own version of Knockboy when the attempt was made – with disastrous results – to afforest 150 acres on Rathlin Island. It is estimated that 'approximately 99 per cent of the area failed and of the remaining 1 per cent lying in the lee of rocky outcrops, the trees survived in the form of bushes'.[10] In 1955, too, the first public forest park was opened at Tullymore Park, Co. Down.

In 1957 the retiring chief forest officer, R. O. Drummond, made recommendations about future forest policy. These included a number of administrative proposals and the development of two regions, east and west, with an effective regional staff in each. This proposal was effected in 1964. Expansion had continued and by 1960 over 1600 men were employed in forestry, between 500 and 900 of them under the 1958 relief scheme. By 1962 there were also 125 foresters in the Northern Ireland forest service.

By the 1950s production from thinnings was a constant feature and marketing policies were devised to cope with this at local level as far as practicable. Up to 16,000 tons of pulpwood was supplied to Bowater Ltd of Birkenhead. The establishment of a chipboard factory at Coleraine in 1959 improved that position. Fencing material was also supplied and in the mid-1950s saw-log became available in increasing quantities. Because of these developments and the prospect of local timber supplies coming increasingly on stream, two important bodies were established in the 1960s, namely the Northern Ireland Home Timber Merchants' Association, the principal aim of which was to negotiate with the Ministry of Agriculture on behalf of timber merchants, and the Ulster Timber Growers' Association, representing private woodland owners.

Pomeroy Forestry School was opened in 1961 to provide training for forest staff and in 1962, largely as a result of a similar situation in Britain brought about by similar post-war circumstances (an excessive number of professional foresters having been recruited), restructuring took place. Two years later some of the Drummond recommendations, particularly the dividing of Northern Ireland into two forest regions, were implemented. In 1962, also, the Abercorn Committee recommended the establishment of a nature conservancy in Northern Ireland. In 1965, as a result of the Amenity Land Act, a wildlife branch similar to that in the Republic was established, with a game farm at Seskinore forest, and wildlife courses were added to the curriculum at Pomeroy.

In 1969 the approaching maturation of State crops brought to a head a number of on-going questions, among them that of social or economic forestry. Another forest policy review was initiated by the new assistant secretary for forestry, Harry Oliver. A draft White Paper met with the perennial political problem (which, it might be thought, was less of a problem in Northern Ireland than elsewhere at the time), that of short-term political considerations versus the lengthy rotation period of the forest crop requiring long-term planning and financing. After some amendments the basic proposals of the White Paper were adopted by Stormont and emerged as policy, subject to five-yearly reviews. Shortly afterwards Direct Rule from Britain was introduced in Northern Ireland.

Following accession to the EEC in 1973, 'more material was crossing the Border than was being produced in Northern Ireland'.[11] This point has already been shown to have created difficulties for timber processors and millers in the Republic; it also, apparently, caused difficulties for producers in Northern Ireland. The essential point was – and, at the time of writing, remains – that because of lower transport costs hauliers and timber merchants can buy more economically in the Republic. A second important point to emerge at this time was the active discouragement of establishing more chipboard factories 'which might utilize small diameter material in the short-term, but would effectively prevent the build-up of raw material to satisfy a pulpmill by the end of the century'.[12]

Private forestry development had now fallen significantly behind general progress and the targets forecast, 300,000 acres of which 75,000 were intended to be private, were considered inadequate in the second respect. Reorganization of the service continued within the new policy initiative. It was realized that management was over-staffed and over-structured. It was accordingly streamlined.

The question of economic or social forestry was also a feature of the Northern Forest Service, if with considerably different emphases than in the south. The question would not resolve itself in a society where up to 30 per cent of the community in forest areas was unemployed. In Northern Ireland this aspect received additional attention with the announcement of the Urban and Rural Improvement Campaign, introduced as a matter of urgency in 1970. Forestry was an obvious area for such relief works and with considerable reluctance and under specific guidelines which tended to divorce the work from established forest requirements – relief works within the compass of the overall programme continued under the aegis of the Forest Service until 1980.

Ironically, because of a hiatus in drafting the enabling statutes that followed the White Paper of 1970, the Forest Service was given virtually a free hand so far as forest parks and amenities were concerned, and those employed under the relief schemes tended to be directed to work in these areas, to the lasting benefit of the Northern Ireland community. Glenariff Forest Park, opened in 1977, and the Augustine Henry Memorial Grove at Portglenone were two of their major achievements. However, the overall relief project was not popular with the permanent forest staff in general, although it provided considerable employment.[13]

In 1971 100,000 acres of State Forestry had been achieved, but in the following year the planting programme suffered a set-back, as it was found that requisitions had not kept pace with the rate of planting.

In the light of political developments which inescapably affected the structure of government agencies, a further review was carried out in 1975. It included conclusions and recommendations for the period to 1980. But, as there was no longer a Ministry to which to submit it, the document containing these became an internal guideline. It generally followed the principles which had already been established in the Republic for forecasting, namely estimating future economic viability by means of formulae, modernizing management and management techniques in line with crop maturation, and new imperatives, particularly in respect of marketing, improved productivity and so on.

More immediately, the requirement to reduce staff numbers was of pressing economic importance. As a first step the relief schemes (in spite of not being directly related to forestry activities) were 'disbanded', and an incentive scheme and one for the reduction of staffs were brought in to help to streamline the newly designated Forest Service. These reductions were carried through, generally, in line with proposals. The results were encouraging and production expanded rapidly in the 1980s.

In 1981 a census of private woodlands was undertaken, which showed that some 230,000 hectares existed, exclusive of hedgerows and areas of less than half a hectare.

* * *

Some space must be given to the subject of global forestry, both natural and cultivated, its importance to man and the environment, as well as what is happening to it. The environmental importance of woodlands and forests has come under considerable scrutiny in recent years, largely because of the deforestation of vast areas of natural forest in South America, Asia and Africa. One viewpoint that has gained considerable currency is that destruction of these natural forests on so enormous a scale as that taking place since the end of the Second World War has a profound effect on the global environment, not simply cosmetically but to the extent of upsetting the balance of the ecosphere to which we belong. While this proposition remains unproven and controversial there is general scientific agreement on the fact that the ozone layer is being steadily depleted and that this will have significant consequences for the earth. It follows that any afforestation programme in this country bears a positive relationship to the proposition.

The questions of the 'greenhouse effect' and the so-called Hanaker theory have received considerable attention in recent years. While it is indisputable that tropical forests are being reduced at a rate that provides legitimate cause for alarm, the same is not true of forests in the temperate, largely industrial, zones (except in the acid-rain belts, especially in Europe). 'When forests in these countries disappear the losses are compensated by reforestation or natural forest expansion into agriculturally marginal or sparsely populated rural areas.'[14]

The greenhouse theory asserts, in effect, that the increase in the level of carbon dioxide in the atmosphere, because of a reduction in tree-cover to absorb it (plus artificial generation caused by human activity), produces a hot-house effect with increased solar radiation in tropical areas. An extension of the theory asserts that consequent evaporation, paradoxically, returns in polar regions causing an increase in the ice-caps, in turn spreading to temperate zones. A new and devastating geological cycle is rapidly induced and a period of glaciation follows. The argument is that the destruction of the earth's tree-cover has reached a critical point in this ecological balance.

However, there are insufficient data to form any positive scientific judgement in the matter. 'There is little argument on the effects of rising carbon dioxide levels. It has even been suggested that high CO^2 levels will prevent the onset of the next glaciation.'[15]

A more common version of the theory is that the temperature of the earth will simply rise by several degrees centigrade with resulting floods. Such theories acknowledge that the main danger resulting from a reduction of the atmospheric ozone layer is in the penetration of increased amounts of ultraviolet rays. Since ozone (O^3) is formed when sunlight strikes oxygen (O^2) in the upper atmosphere, the importance to the human ecological system and to the continuation of the biological world structure as we know it today of dense masses of oxygen-producing vegetation is self-evident.

While there is no way of knowing precisely what the effects of depletion of the ozone layer might be, there is universal agreement that it would result in major damage and distortion to the earth's biological system with calamitous consequences. The reason for the want of precision is lack of knowledge. Although a great deal is known about the ecosystem, much of it acquired in the second half of this century, the principal result of this accumulated knowledge is to demonstrate how little is known. This is especially so with regard to stratospheric chemistry.

However, Dr Coxon quotes from a book by an environmentalist,[16] as follows: 'warnings of an imminent ice-age and other catastrophes are ill-founded and irresponsible. The recent droughts in Africa, floods in Pakistan and tropical storms in Australia all have parallels in the past and do not imply that the global pattern of climate is undergoing a radical and permanent change.' That, however, does not alter the fact that tropical forests are being destroyed.

In the Amazon basin deforestation is taking place at a staggering rate, and much of the detritus is being burned off to add to the CO_2 content. The exposed soil is often leached of nutrients and remains barren. In Vietnam the total defoliation policy of the United States during the war there left much of the country's rain and mangrove forest areas denuded. Rainfall also leached much of the land of nutrients so that it will be difficult, if not impossible, to reconstitute forests in such areas for a very long time.

In the mid-1970s the United Nations Environment Programme and the FAO (Food and Agriculture Organization) jointly developed a programme to monitor the world's tropical forest areas and a programme to acquire statistics was launched in 1978. Forest was divided into four categories:

(1) Closed tree formation, in which the canopy and underlayer obscures the ground.
(2) Open tree formation, which has sufficient light to provide continuous grass cover as well.
(3) Fallows, which is a mixture, normally secondary vegetation following clearances.
(4) Scrub trees and shrubs of various ages and various forms.

Closed tree formations were found to be the largest group with, surprisingly enough and as in each of the groups, the largest percentage in tropical America and the smallest in Africa. Asia contains about 25 per cent in each group.

The survey found that 7.5 million hectares of closed forest (total 1200 million hectares) and 3.8 million hectares of open forests would disappear *each year* between 1981 and 1985. A further 4.4 million hectares would be affected by exploitation of one sort or another *per annum*. The staggering total was 15.7 million hectares annually – an area the size of Ireland and Belgium combined – representing a reduction of 0.8 per cent per year of the total world cover.[17] Because of loss of tree cover, soil erosion, and fuelwoods rural communities in the Third World face severe crises.

The Irish organization Trócaire is engaged in tree-planting and educational programmes throughout the Third World to help to offset these severe and punishing problems: 'The majority of the people in the Third World – over two billion – depend on fuelwood for cooking and heating. Every single day in the developing world hundreds of millions of women and children have to spend hours looking for a few sticks to light a fire with... The solution to the fuelwood crisis involves comprehensive reafforestation, conservation and development programmes.'[18]

The importance of the tropical rainforest lies not only in its timber or ecological contribution. 'One in every four drugs sold in Western chemist shops owes its origin to rainforest species. And over 1,400 forest plants are being investigated as possible cures for cancer.'[19] Ireland shares this burden of responsibility. Most of our hardwood comes from the Ivory Coast, 'a country that has lost 66 per cent of its forests and woodlands in the last twenty-five years'.[20]

Paradoxical as it seems, there is a current drive to establish 'tree-farming' in traditional rainforest developing countries. The proposal is to establish these on deforested lands, where conditions are ideal for quick-growing timber. 'A plantation of eucalyptus or pine can generate ten times more wood than can a similar sized patch of virgin forest.'[21] The solution must be more control of the destruction of natural forest and, where that is unavoidable, the encouragement of replanting and commercial plantations for tree-farming and fuelwood.

Because of the evident damage to its surroundings, the world is rapidly becoming more environmentally aware. In the context of forestry the most acute problem is the destruction of the rainforests of Asia, Africa and South America. But there is also a school of thought, with a powerful lobby, which argues that, while it might be considered that planting a tree is a good thing, planting acre upon acre of conifers on land which had nurtured many species and varieties is at odds with the ecosystem which it has taken thousands of years to develop.'[22]

About one-third of the earth's surface has a suitable climate and environment to support forest growth. What has always been significant is that it is precisely the areas that can support forest growth that also offer the most favourable conditions for man to live in.

Man has tended therefore to concentrate in the forested regions of the earth from the earliest times, and that is where he has made the greatest environmental changes. This process is still going on, but it has, in the main, shifted from the temperate to the tropic and sub-tropic zones.

Ireland is a classic example of the ravages that unchecked exploitation of natural forest can produce. It was to rectify the results of this spoliation that the programme of national afforestation was begun, on a managed basis, in the wake of the 1907 Committee.

Europe, and Ireland in particular, is now less well-endowed with forests than the world in general. 'Global forest cover is about 30 per cent of the land surface area, that of the EC slightly more than 24 per cent' (that of Ireland slightly more than 6 per cent).[23]

Trees are, of course, essential to our very existence, providing the oxygen we breathe. But they also play an important part in safe-guarding the stability of our environment – hence both agriculture and economy.

The EC imports 66 per cent of its timber requirements. The total forest cover of the EC is 'composed of 58 per cent broadleaved, largely deciduous species, the remainder being coniferous'. Individual states vary, 'Italy having 76 per cent broadleaved, while Ireland is almost the converse with 80 per cent coniferous cover. The percentage of State versus private holdings also varies greatly, with Ireland and Greece owning 85 per cent and 66 per cent respectively, while in Germany and France private ownerships account for 74 per cent and 65 per cent of the total forest area.'[24]

The unresolved controversy about acidity levels in thickly planted conifer areas provided a jumping-off ground for a broader environmental issue – the general effects of widespread, high-yield, maximum-revenue-potential conifer-planting. An Taisce, the National Trust for Ireland, is to the forefront in this debate.[25]

Their report raises a number of seemingly valid points, somewhat tempered by special pleading and generality. The report says 'the evidence in Ireland is that current forestry practices are not necessarily based upon ecological principles. Indeed, the reverse is sometimes the case.'[26] No clear evidence is adduced in support of these statements, true or not. But the case for a national land-use policy is stated succinctly.

In essence the report advocates what has been repeatedly maintained throughout this book, the need for a clear, forward-planned land-use policy. Understandably this proposal is seen by An Taisce in emphatic environmental terms: 'The government should consider the adoption of a land use policy in order to minimize the conflict between large-scale forestry and other land uses, including conservation.'[27]

'It is most important (it also emphasizes) that the current debate should not degenerate into pro- *versus* anti-forestry. The debate should centre on how forestry can develop in the future. The conflicts arise because of poor strategic planning and a lack of integration with other rural land uses.'[28] It asks rhetorically 'if it is possible for commercial forestry to be compatible with conservation and sustained development?'[29] And answers 'Yes', provided changes in current emphasis – including more broadleaf forests, fewer conifer forests, especially on blanket bogs, more 'Tree-farming', and an overall land-use policy are introduced.

In its programme published in 1990,[30] the Department of Energy outlined a five-year plan for forestry, recognizing its importance at a number of essential levels, including regional development, general economic importance to the country and environment. In the Foreword the Minister for Energy with responsibility for forestry, Robert Molloy, TD, said: 'The operational programme places emphasis on protecting the environment and the measures are geared towards ensuring that forestry development generally will contribute to environmental improvement.'

The programme itself states[31] that grants will be discouraged for planting schemes which 'are likely to have a negative impact on such features as areas of scientific interest... (that) the encouragement of broadleaved trees will form an important aspect of the forestry programme... (that) afforestation by Coillte Teo on Bord na Móna boglands will be confined to cut-away bogs made available by Bord na Móna.'[32] It announced the establishment of an Environmental Protection Agency, among other environmental measures. The only aspect on which it does not meet the points raised by An Taisce appears to be that of a national land-use policy.

The An Taisce Report welcomed much current (1990) forest policy, especially tree-farming:

The bulk of Ireland's afforestation could be accommodated on marginal farmed land on mineral soils rather than peatlands. Native and naturalized broadleaves can and should become an integral part of Irish plantation forestry, preferably on Bord na Móna cutaway bogs throughout the midlands[33]... A major obstacle to progress is the assumption that large-scale coniferous afforestation of the virgin blanket peats is in the national interest... To our knowledge no comprehensive economic assessment of present forest policy has been carried out.[34]

Coillte has already indicated its intention of moving away from planting on blanket bogs, and of adopting some of the points made in this report.

The claim is that widespread conifer-planting in blanket boglands destroys the original ecosystem and that habitat and wildlife, sometimes rare, tend to alter and be replaced by something else. In reasserting the claim that high acidity levels result from excessive conifer-planting, the report cites the following:

The balance of evidence from numerous studies is that conifers growing on poor soils promote greater surface organic matter accumulation, greater acidity and a higher degree of podzolization... by contrast certain broadleaved trees, notably birch, aspen and holly, can reverse the process found under conifers.[35] However, that has been challenged.[36]

The report acknowledges the significance – both general and ecological – of the EC policy on forestry,[37] in accordance with which the *Operational Forestry Programme, 1989-1993* was drawn up and approved. EC policy is determined on the following broad strategic base:

(1) To increase timber resources.
(2) To provide an alternative land-use, thus reducing food surpluses.
(3) To protect and develop existing woodlands and forests.
(4) To encourage environmental improvements.
(5) To provide employment in rural areas.

The total cost of the Irish forestry programme (to 1997) is estimated to be £163,246,000 of which the EC will contribute 32 per cent, the State 48 per cent and private forestry 20 per cent.[38]

Citing the Community Strategy and Action Programme for the Forestry Sector, Commission of European Communities (COM 88, 1988) – 'Forestry is seen as a real alternative agricultural enterprise, especially since the EC has a net timber deficit, and as an essential element in the reform of the CAP' – the An Taisce report nevertheless goes on to attack the EC forest grants system on the general grounds that 'The introduction of substantial grant assistance from the EC is creating an artificial economic climate for afforestation in areas of great conservation importance, which in general have no natural suitability for industrial timber production'[39] – an argument it is likely to find increasingly hard to sustain.

The following cited point is also contentious.

A major complicating factor has been the various State and EEC farming and forestry incentives which have encouraged the destruction of several sites of conservation interest. This has given rise to the ridiculous situation where one arm of government financially encourages the destruction of a site of conservation interest while another tries to protect it without adequate funding.[40]

It reminds us in reverse of what happened at the turn of the century when woodlands were destroyed under the Land Act purchases, while the forest committee of the DATI had no funds with which to buy or protect them.

But the bottom line of An Taisce's 1990 argument is this: 'the most contentious issue (is) whether endangered semi-natural habitats should be afforested in the first place...' a statement so broad as to leave room for endless argument and debate and one that, paradoxically, rather than reinforce the point they also emphasize, namely that the debate they raise should 'not degenerate into pro- *versus* anti-forestry', would seem more likely to produce the opposite reaction.

The achievement of this country in terms of afforestation has been remarkable. In spite of diverging from the principle enunciated by the 1907 Committee, and in spite of the traumas of the 1920s and the severe economic burdens imposed by the Economic War with Britain, followed by the Second World War, remarkably the target set by the Committee has been achieved within the forecast period, a tribute to those engaged in the undertaking, more particularly to the enduring validity of the proposition, simply expressed, that forests are a good and a necessary thing. The Irish dilemma was to provide them out of virtually nothing.

As always, capital funding was a major problem. Its lack had diverted political and official attention from forestry, sapping worker-morale and self-confidence. Now public and public confidence appear to be positive where it matters.

In December 1990 Minister for Agriculture Michael O'Kennedy, TD, announced an EC-approved parallel Operational Programme for Rural Development for Alternative Farm Enterprises, providing about £10 additional millions for forestry 'for development programme opportunities not covered in the Operational Forestry Programme'.

Simultaneously Molloy detailed how this additional funding would be allocated – including recreational forestry geared to tourism; the environment and local wood production; demonstration farm forests providing educational experience for farmers; forestry business, planning and investment advice; nurseries and training programmes.

As in the eighteenth and nineteenth centuries when it stimulated interest in new species and methods, and in the 1930s when it provided practical training and direction, continental *agacerie* appears to have given the incentive needed to impel those responsible towards bringing initial forestry development to the status of a fully fledged national scheme of afforestation.

During the early research for this book it became obvious that co-operative tree-farming was – at the time – a neglected yet potentially fertile area for forestry development. The degree of change is such that it is now the basis for much forward planning.

In the course of his announcement, moreover, Molloy made an important statement: 'These proposals underline the recognition now being given to the very important role which forestry plays in the balanced development of the rural economy.' It seems likely that the equally long-overdue land-use policy must soon emerge and that Mackey's two criteria for a thriving national forestry industry – (1) financial profit, (2) indirect gain – will, hand-in-hand, take purposeful hold of the future.

Those with responsibility for forestry can look forward to developing what has become an established and viable infrastructure, and towards a time when forestry will be one of the largest and most important national undertakings and a vital part of the life, economic and environmental, of the community as a whole.

Appendices

APPENDIX 1

Disafforestation: Glendalough, Co. Wicklow

The following information was supplied to Mr A. C. Forbes by Mr Liam Price, who researched the territorial boundaries of Co. Wicklow. It was given by Forbes in a paper, entitled 'Some Legendary and Historical References to Irish Woods, and their Significance', subsequently published by the Royal Irish Academy in June 1932.

The Disafforestation charter of 1229 (relating to the area of Glendalough) is printed in full in Gilbert's Crede Mihi (No. xxxvi, page 39); and in Gilbert's Historical and Municipal Documents of Ireland (pages 539,540). It includes (i) Saufkeyvin, (ii) Fertire, (iii) Coillach, (iv) 'the lands belonging to the demesne of the Archbishop and of the Church of Dublin, and which once belonged to the demesne of the Archbishop of Glendalough, and of Thomas, formerly Abbot of Glendalough,' their boundaries being given.

Coillach was the property of the Archbishop of Dublin under a grant of King John (while Earl of Mortain), date between 1181 and 1190. From a note in Alan's Register, folio 21, I think it may be taken that it comprised a territory corresponding roughly to the present parishes of Kilbride and Blessington, in the barony of Lower Talbotstown.

It seems to me that the territory in these districts is indicated more clearly by the boundaries of the demesne lands of the Archbishop, and that the demesne lands disafforested are not intended to include more than is comprised in these districts. For these boundaries see Orpen, *History of Ireland Under the Normans*, vol. iv, pp. 10-13. Ó Brún and Othee evidently lay between the sea and the mountains in South Dublin and North East Wicklow, the boundary perhaps running from Enniskerry to the source of the Vartry River, between Calary and Glasnamullen, and thence following the Vartry River. The next place on the boundary is 'Wykinglo, the land of the Baron of Naas'; this, I believe, included all the parishes adjoining Wicklow, so that this boundary would follow the boundary of the present parish of Derrylossary (which used to be called Kildalough) as far as the Avonmore River at Cronybyrne, and thence it would follow the Avonmore until the next place on the boundary, 'Arclou, the land of Theobald Butler'. This again included the land around Arklow; it is called 'Offynachlis' (i.e., the Irish Ui Enechlais, or Ui Fenechlais), in a list of rents in Alan's Register, folio 163. This probably stretched up to the Avonmore river, somewhere in the neighbourhood of Ballyarthur, and then stretched along the south bank of the Aughrim river up to the district near Aughrim. Thus, all the territory west of the Avonmore and north of the Aughrim river would, if I be right, be included in this disafforestation grant; and this piece of territory, comprising the present parishes of Rathdrum, Knockrath, Ballykine, and Ballinacor, with part of Castlemacadam, was, I believe, originally included in Saufkeyvin. The next place on the boundary is 'Omayl, the land of Philip Fitz Resus,' so that the boundary seems to have followed the Ow and the mountains, following the present boundary of the united Dioceses of Dublin and Glendalough, as far as the parish of Donoughmore, which is in the Glen of Imaal. The places named as bounds all seem to be excluded from the grant, so we may take it 'Omayl' was excluded, and the boundary then goes 'on the west side of the land of Naas, and then to Rathmor, the land of Maurice Fitz Gerald'. This is pretty vague, but I think it probably followed the mountains round the north side of the Glen of Imaal, included most of the parish of Hollywood, and so reached the present boundary of the County Kildare. The parish of Burgage probably was included in the territory of Coillach (Alan's Register, fol. 21).

Fertire I would reckon to comprise Derrylossary parish with that part of Knockrath parish north of the Avonmore, and as much of the Calary Parish as is included in the barony of Ballinacor North; and perhaps it also included Boystown parish, but this is difficult to say: at any rate, all the western and mountainous part of Boystown parish was later, e.g., on the Down Survey Map, called 'The Bishop's Lordship,' as though it was in some special way identified with the see. It must, I think, have been either part of Fertire or else of Coillach.

For a boundary of the territory of Ranelagh, which, to a certain extent, i.e. from Cronyburne along the Avonmore to Kilcarra and thence to Killacloran, and again from Toorboy to Convalla, seems to coincide with the above-suggested boundary of Saufkeyvin, see Chanc. Inquis., Wicklow, Jas. 1, No. 12, dated 1617.

APPENDIX 2

General Deforestation in the Seventeenth Century

The period from 1608 to 1691 denuded Ireland not only of her leaders, but of her trees.

'Trees,' said Sir Jonah Barrington, quoting the saying as the sentiment of the great Irish landlords of his day, according to Hore, 'are an excrescence provided by nature for the payment of debts.'

Swift's view was very different. In the 7th Drapier's Letter he wrote: 'I believe there is not another example in Europe of such a prodigious quantity of excellent timber cut down in so short a time, with so little advantage to the country either in shipping or building.'

The question of the best method of 'dealing with these fastnesses', as Le Fanu puts it, 'was the principal problem which presented itself to those in command of the Queen's forces in Ireland'. From the English standpoint the woods are described in the State Papers as 'a shelter for all ill-disposed' and 'the seat and nursery of rebellion'. The deforestation process which proceeded from about the year 1200, according to some sources (but not on anything approaching the Tudor scale), was hastened by the year-round grazing which the Irish climate permitted and which helped to destroy young, regenerating trees. Neither of these causes, however, affected forests in any fundamental manner, so far as we can judge today. And it is clear that as late as 1600, at least one-eighth of the country was still covered with (woodland) forest, a very substantial amount. It does not suggest the great proportion suggested by Harvey and other writers whose observations (made, it must be recollected, under considerable stress and with great difficulty) incline to the idea of a country entirely covered by forests, bog and mist from which the kernes fought Elizabeth's troops.

'Despite the export of Irish oak over the three previous centuries, it was not until Elizabethan times that the forests became important for military and commercial purposes. There were suggestions that clearance of the forests would remove the shelter used by persons regarded by the authorities as troublesome and undesirable, but it seems that this clearance was in practice affected by the late seventeenth century commercial expansion which followed the period of colonisation. The main uses of wood were in house-building, ship-building and barrel making; as fuel for iron smelting and glass making, and large quantities of oak bark were stripped for use in leather tanning. By the first decades of the eighteenth century Ireland had reached its present condition of being primarily a "timber-importing country"' – *The Forests of Ireland (Forests Past and Present)*, ed. Niall O'Carroll, p.33.

However, in spite of this view during the last Desmond rebellion (1579) Sir Warham St Leger, President of Munster, wrote to Burleigh that the 'scope of the Geraldines' range includes the Great Wood, Aharlogh, Dromfynine, Glanmore, and Glanflesk, which are their chief fortresses' and he proposed to employ a force of 4000 English soldiers, besides the army already in the field under Ormond, to protect the labourers in hewing down and burning these woods. He described 'Arlough': 'It conteyneth in length three miles, in breadth six miles, distant from Limerick, south by east, sixteen miles; situated betwixt two mountains south and north; the south mountain being a marvellous high mountain called Slivegrote.'

Sir George Carew, noting the length and breadth of the fastnesses of Munster about the same time, wrote:

'Glangarriff, in O Sullivan More's country, 4 miles long and 2 broad.
Glanrought, in Desmond, 3 long and 2 broad.
Leanmore, in do. 3 do. and 3 do.
Clenglas (Glen?) and Kilmore, in the co. Limerick, 12 long and 7 broad. ('Between them they formed a mass of wood which, with the exception of Glenconkyne, was the largest forest in Ireland,' according to Falkiner, *op. cit.*, p.156.)
Dromfynine, in the co. Corke, on the Blackwater, 6 do. and 2 do.
Arlogh and Muskryquirke, in Tipperary, 9 do. and 3 do.
Kilhuggy, in Tipperary, bordering on Limerick, 10 do. and 7 do.
Glenflesk, 4 do. and 2 do.'

In 1608, the year of the Flight of the Earls, Sir John Davys writes of the sequestered O'Neill estates: 'From Dungannon we passed into the county of Coleraine, through the Glinnes and woods of Clanconkeyn, where the wild inhabitants did as much wonder to see the lord deputy, as Virgil's ghosts did to see Aeneas alive in hell. But his lordship's (the viceroy's) passing that way was of good importance two ways for his majesty's service; for both himself and all the officers of his army have discovered that unknown fastnes; and also the people of the country, knowing their fastnes to be discovered, will not trust so much therein as heretofore, which trust made them presume to commit so many thefts, murders and rebellions: – for assuredly they presumed more upon our ignorance of their country than upon their own strength.' Davys then, according to Hore, wrote to the English government to suggest that 'the great forest of Glenconkeyn, well nigh as large as the New Forest in Hampshire, and stored with the best timber' should be retained as a reserve for the navy. 'But,' writes Hore, 'as it was important to the peace of Ulster that this vast shelter for rebels and robbers should be destroyed, and more suitable that its oaks, in place of being used in building "wooden walls" for England, should be employed in erecting a town whose walls would prove a "chief fastness and refuge" to colonists in the north of Ireland, the king, in 1609, gave permission to cut down 50,000 oak-trees at 10/= a piece, 100,000 ash trees at 5/=, and 10,000 elms at 6/8d., for the purpose of building Londonderry.'

One planter, however, attracted to Ireland in the seventeenth century by the prospect of acquiring land and fortune from the confiscations was Sir William Petty, who had a more enlightened view of the importance of forestry. In his *Political Anatomy of Ireland* he recommended the 'planting of three millions of timber trees upon the bounds and mears of every denomination of lands in the country'. However, 'so rapid was the consumption,' writes Falkiner, 'that the want of fuel, formerly abundant, began to make itself felt'. Thomas Dinely writing in his *Journal* about the year 1682, remarks on the consequent substitution for the first time of turf for wood firing. 'The wars', he says, 'and their rebellions having destroyed almost all their woods both for timber and firing, their want is supplied by the bogs.'

Arthur Young was also anxious to see an improvement in policy and felt that 'instead of being the destroyers of trees they (the peasantry) might be made preservers of them...', suggesting that the peasantry were enlisted in the destruction in addition to what they used themselves. Young proposed that farmers should be encouraged with premiums to plant and preserve trees and that tenants should be obliged to plant under a special clause in their leases requiring them to plant a given number of trees a year in proportion to the size of their holding.

According to Falkiner the abundance – one might almost say super-abundance – of timber and its derivatives was recorded well into Anglo-Norman times. He cites the *Annals of Ulster* for the year 835 which states that the acorn and nut crop was so great that year 'as to close up the streams, so that they ceased to flow in their usual course'. Allowing for the accepted hyperbole of some of the annalists, it nevertheless indicates the condition of the woodlands at that time. Falkiner continues: 'That this state of things survived to an era well within historical memory is abundantly demonstrated by many authorities. Sir John Davies, a writer whose observations and conclusions, even when we disagree with them, are always suggestive, has noted the degree in which the political system adopted by the Norman colonists of Ireland, and pursued, whether by choice or necessity by the English Government for many centuries, had the effect of preserving this feature. That system was to drive the native population from the plains to the woods; with the result that the Irish territories tended to become ever more and more a succession of forest fastnesses. Had a different plan been adopted, the woods, as Davies points out, would have been wasted by English habitations, as happened just before his own time in the territories of Leix and Offaly, round the new made forts of Maryborough and Philipstown...' This is an interesting example of *post-factum a-posteriori* argument in which Davies, according to Falkiner, attributes Tudor and post-Tudor policies to pre-Tudor times. Falkiner goes on: 'No attempt was made, however, for above three centuries after the arrival of the English in Ireland to encroach to any serious extent upon the native reserves of the Irish inhabitants, though a statute of Edward I, passed in 1296, contained a clause which was designed to provide highways through the country. But the wars of the Bruces which followed within a few years of this enactment, and the subsequent decadence of English power prevented the taking of any effective steps under the statute.

'Neither the English colonels (to say nothing of other ranks) whom Cromwell metamorphosed into Irish landlords, nor the Dutchmen whom William of Orange rewarded with Irish soil, regarded their new forests with much liking; even their successors do not seem to have looked on their woods as ancestral inheritances,' wrote Hore.

The destruction of the woodlands was due in the first place, as we have noted, to deliberate policy. This was extended and encouraged by war and accelerated before and after the Rising of 1641. Arthur Young in his *Tour of Ireland*, noted that in the neighbourhood of Mitchelstown there were 'a hundred thousand acres in which you might take a breathing gallop to find a stick large enough to beat a dog, yet there is not an enclosure without the remnants of trees, many of them large'.

Boate noted in Cromwell's time that in some parts of Ireland a man could travel for days without seeing any trees 'save a few about gentlemen's houses'. Towards the north 'for a distance of sixty miles from Dublin, not a wood worth speaking of was to be seen. For,' he adds, 'the great woods which the maps do represent to us upon the mountains between Dundalk and Newry are quite vanished, there being nothing left of them these many years since but only one tree standing close by the highway, at the very top of one of these mountains, so as it may be seen a great way off, and therefore serveth travellers for a mark.'

It was almost as if the planters attacked the trees in lieu of a human enemy. And perhaps in some way they associated the woods and fastnesses so closely with the Irish they had dispossessed, and who fought back from them, that they had a double, even treble, satisfaction in clearing them – deprivation of refuge, profit and whatever psychological contentment they could derive from doing so.

In the uneasy period following the Cromwellian settlements and to the end of the seventeenth century the settlers 'realized what they could by stripping it (the land) of its feathers; and, subsequently, the vengeful dryads of the departed groves appeared to them in the shapes of "tories" and "rapparees"', wrote Hore.

Thousands of trees were felled. The notorious Sir Valentine Browne 'cut down and destroyed' timber to the value of £20,000 around the lakes of Killarney, according to the trustees of the estates who went on to note the waste and glut on the market caused by cutting the Earl of Clancarty's woods to a value of £27,000. 'So hasty', they wrote, according to Hore, 'have several of the grantees, or their agents, been in this disposition of the forfeited woods, that vast numbers of trees have been cut and sold for not above six pence a-piece.'

The Brownes were of Elizabethan planter stock and were Catholic and Jacobite and evidently, though well-enough disposed to the Irish, not averse to a bit of profit when the going was good – a trait they shared with most people. Their estates were about 400,000 acres, but were confiscated in the lifetime of Sir Nicholas, Valentine's father. They were later partially restored. One of their tenants was the poet Aodhgán Ó Rathaille of the family which had been ollamhs to the Mac Carthy Mórs whom the Brownes dispossessed. When Valentine refused – or was unable – to accede to a request of Ó Rathaille's, the poet complained in a poem called 'Valentine Browne', that begins:

> Do leathnaigh an ciach diachrach fám sheana-chroí dúr
> As taisteal na ndiabhal n-iasachta i bhfearann Choinn chugaim...
> (Covered by a mist of pain is my dour old heart
> Since the foreign devils came into the land of Conn...)

The great wood of the Picts, named from the place where Robert the Bruce bivouacked near Tara, had vanished by the mid-sixteenth century.

APPENDIX 3

Some Irish Statutes before 1800

(*based on a paper by M. L. Anderson and other sources*)

Charles I. Cap. XXIII. 1635
This Act shows that the authorities were alive to the need for preserving woods and plantations. It was a copy of an English statute of Elizabeth of 1543. The title is 'An Act to avoyde and prevent misdemeanours in idle and lewd persons in barking of Trees, etc.'

'Forasmuch as unlawful cutting and taking away of corn and graine growing, robbing of orchards and gardens, digging up, or taking away of fruite trees, breaking of hedges, pales or other fences, cutting or spoyling of woods or underwoods standing and growing, barking of growing trees, and such like offences, are now more committed by lewd and meane persons than in former times, and that the said offences are great causes of the maintaining of idlenesse... be it enacted... That all and every such lewd person and persons, which from and after the first day of May now next following shall put up, or take up any fruite tree or trees in any orchard, garden or elsewhere... or shall barke any tree or trees that are growing, or shall cut or spoile any woods or underwoods, poles or trees standing, not being felony in the laws of this realme, and their procurer or procurers and receiver or receivers knowing the same, on conviction shall pay recompense or if not able to pay eftsoons be committed to the constable to be whipped.'

William III. Cap. XII. 1697
Section XII of this Act runs as follows:
'The mears of lands between propriety and propriety... shall, at equal charge of the proprietors thereof, or their tenants, be enclosed with good ditches, where earth sufficient may be had to make the same, and thereon one or two rows of quick sets shall be planted, and where earth shall be wanting, such other fences shall be made as the nature of the soil shall permit.'

William III. Cap. XII. 1698
This is the first important legislation of a true forestry character in the statutes. The coming into force of the Act was post-dated for five years to the 25th of March 1703, to give persons concerned time to grow the trees required, which had to be four years old.

'Forasmuch as by the late rebellion in this Kingdom, and the several ironworks formerly here, the timber is utterly destroyed, so as that at present there is not sufficient for the repairing the houses destroyed, much less a prospect of building and improving in after times, unless some means be used for the planting and increase of timber trees...'

The first section of this statute enacted, that from the 25th March 1703, all persons, being residents within Ireland, having estates of freehold or inheritance therein to the annual value of ten pounds or tenants for which eleven years had still to run, paying rent of ten pounds, should plant every year at seasonable times of the year for 31 years: 'ten plants of four years growth, or more, of oak, firr, elm, ash, walnut, poplar, abeal or elder, in some ditch or elsewhere, on the said lands'.

It was also provided that the planters should preserve the trees. In addition, said Anderson, 'every person or society having iron-works had to plant 500 yearly every year he or they had the said iron works'. Incidentally, Forbes is in error in saying that the amending Act of 1705 (the fourth, not the fifth year of Anne) extended the trees by adding walnut, poplar, abele and alder. The first three of these were already included in this Act and the last was meant to be, but by mistake 'elder' was substituted for 'alder'.

The second section provided that occupiers of 500 acres or more (plantation, ie., Irish measure), other than tenants in common, over and above the aforesaid 10 trees, should enclose and plant one plantation acre thereof in seven years from November, 1698, with a good, sufficient fence or stone wall, ditch, hedge, pales or rails and...

'...plant one plant, at least one foot over the ground when planted for every ten foot square contained in such area in such method as he, she or they shall think fit'.

The acre had to be preserved for twenty years. The third section provided that every person, body politick or corporate in possession of any lands or anyone in possession of lands on dower, jointure, by courtesie, or who had possession of land as mortgages or as creditors or by any other ways or means, could be liable to plant a '... proportion of 260,000 trees of oak, elm or firr of the age and size afore-mentioned yearly for 31 years from 25th April, 1703, in such a manner and proportion as hereinafter is expressed'.

Anne. Cap. II. 1703

This Act attributes the destruction of the forests to the iron industry; proposes to encourage the importation of iron from abroad and also that of bark, barrel staves and other forest produce, while at the same time ensuring revenue. The first section reads as follows:

'Whereas the great duties laid on foreign iron are a great discouragement to the importation thereof, and tend to the lessening of her Majesty's revenue, and to the destruction of the woods of this Kingdom; and whereas there is a great scarcity of all sorts of timber in this kingdom, and the great duties laid on hoops, bark, laths, and on staves imported for making barrels, pipes, or casks are a great hindrance to the importation thereof for the remedy thereof be it enacted... after 1st November, 1703, all duties on unwrought iron, bark, hoops, laths, staves and timber for casks shall cease...'

The new duties substituted were five shillings custom and five shillings excise per ton of iron imported and sixpence custom and sixpence excise per 1200 staves for casks; one penny per 1200 for hoops or laths and the same per barrel of bark.

The second section was designed to restrict export of timber by increasing the existing duties, as follows:

'And for the preventing the exportation of timber out of this kingdom to any ports beyond the sea, other than to the kingdom of England, be it enacted that there shall be paid over and above all duties now payable £2.10s. per ton for timber or planks and proportionally for any greater or less quantity; 5/- per 1,000 for laths; £3 per 1,000 for staves.'

Timber made up as part of a ship or vessel was exempted. 'The last section', claims Anderson, 'is an illustration of the fallibility of the statute-maker. It provides that the word "elder" in the 10th of William III., XII., should be amended to "alder".'

Anne. Cap. IX. 1705
This was an explanatory Act to put into execution items two and four of the foregoing. The first eight sections made important changes in 10th William III., XII.

In the first place the liability to plant ten trees annually was to be determined on an area basis in place of a valuation basis, namely: 'such person as hath or holdeth 30 acres of land or more in a manner aforesaid and any other person or persons whatsoever'.

The following areas were specially exempted from liability – the city of Dublin, the city and suburbs of Londonderry, either within them or within one mile of them, and 'any area within one mile of any city or town corporate'.

The omission of ash in Section III of the main Act was rectified and ash was included along with oak, elm and 'firr'.

Section IX is of especial importance in view of the sudden disappearance of hazel at this time. It reads:

'And whereas great quantities of young trees are daily destroyed by the making of gadds and withs, and that it will very much conduce to the incouragement of the iron and hempen manufactures, that gadds and withs be no more used in this kingdom, be it therefore enacted... that from the 1st November no person or persons shall make or use in plowing, drawing of timber, or other work whatsoever, or in wattling the walls of houses, or cabbins, or out-buildings, any kind of gadd or gadds, wyth or wyths, of oak, ash, birch, hazel or other tree whatsoever...'

The Act further provided that:

'... any person by warrant of a justice may search suspected houses and places for any wood, under-wood, poles, trees, clap-boards, barrel staves, poles, rails, stiles, posts, gates or for any gadds, wyths, willows, hedge-wood, bark, rind or coat of any tree, unlawfully barked as aforesaid...'

Anne. Cap. V. 1710
'An Act for the further explaining and putting into execution an Act for planting and preserving timber trees and woods.'

Section V also prohibited the cutting of young growth for 'gads or withs', as follows:

'and whereas the cutting and using gads or withs is found to be very destructive to all young plantations of woods; be it further enacted... that any person or persons who shall from and after the first of September 1711, cut or make use of any gads or withs on his or their plows, carrs, carts, harness, tackle or otherwise; or in whose custody or possession any gads or withs shall be had or found, either selling or using the same; shall for every gad or with so cut, sold or found, forfeit the sum of two pence, to be immediately paid to the informer by the said offender or offenders'.

The offender could not refuse to pay. If he did, by warrant of the nearest magistrate the sum could be secured from him by levy, and if there was no convenient magistrate the constables could claim payment, and, if that was refused, could hail the offender before the next justice to levy double the sum and if the offender had no means of paying the levy he could finally be whipped.

George I. Cap. XVI. 1715
An Act for reforming abuses in the making of butter casks '... to be made of sound, dry and well-seasoned timber in different sizes for firkin, half barrel, three-quarters barrel and barrel, and every such cask, hereafter to be made, shall be made of three hoops in each quarter, to be set in with twiggs or sufficiently notched, and have two heads to be put into riggles and made tight, so as to hold pickle...'

Section VIII complained that, notwithstanding Section V of Anne's Act of 1710: 'great quantities of gadds and withs are daily sold in markets and fairs to the great destruction and almost utter ruin of the young growth of wood in this kingdom; for the better preserving whereof... Any one may seise gadds or withs found in any fair, market, town or place to his own use and any in due possession found, to be brought before the justice and whipped'.

George I. Cap. XII. 1717
Also governed cask making.

George I. Cap. V. 1721
'An Act to oblige proprietors and tenants of neighbouring lands to make fences between their several lands and holdings with white-thorn, crab, or other quick sets, where the same will grow, and in ground where such quick-sets will not grow, with furzz, with sallows, alder or other aquatick tree.'

George I. Cap: VIII. 1721
Repealed much of the 10th of William XIII, III.

Section II enacted 'That where any tenant or tenants for life or lives or years, of any lands in this kingdom of Ireland shall during his, her or their term plant in or upon the same any trees of oak, ash, beech, firr, wallnut, alder, elm, poplar, abeal or birch and shall preserve the same, such tenant or tenants or his, her or their executors, administrators or assigns respectively, shall at the expiration of such term or estate be entitled to, and shall have liberty, and is and are hereby authorized and impowered to fell and carry away for his and their use and benefit one-third part of the several kinds of such trees so by him, her or them planted and which shall at that time be standing and preserved on the lands so held in lease as aforesaid.'

Note the addition of beech and birch to the list of approved trees.

George I. Cap. IX. 1723
This was another butter-cask statute and is interesting for its mention of sycamore.

George I. Cap. V. 1725
Again the subject was butter casks.

George II. Cap. IX. 1731
This statute was 'An Act to encourage the improvement of barren and waste land and boggs, and planting of timber trees and orchards'. Section IX, referring to Section II of the preceding Act, provided that: '... such (a) tenant or person instead of such a third party shall have an equal moiety of all such trees as he or she or they shall hereafter plant in pursuance of the said act'.

George II. Cap. VII. 1735
This statute gave the executors or administrators of a tenant for life or tail title to a moiety of any timber the tenants had planted, except in an avenue or garden walks. 'This', says Anderson, 'was evidently done to round off the ninth section of the preceding Act.'

George III Cap. XVII. 1765
'An Act for encouraging the planting of timber trees.' Its first section relieved tenants from being impeachable for waste: 'whereas the distress this kingdom must soon be in for want of timber is most obvious; and it is equal to its inheritors, whether tenants do not plant, or have a property in what they plant; be it enacted by the King's most excellent Majesty... that from and after the 1st day of September, 1766, tenants for lives renewable for paying the rents and performing the other covenants in their leases, shall not be impeachable of waste in timber trees or woods which they shall hereafter plant, any covenants in leases or settlements heretofore made, law or usage to the contrary notwithstanding.'

The second section adds the following species to the list of trees given – pine, chestnut, horse chestnut, quick or wild ash. The revisioner had the right one year before the term of the lease expired to make a claim to buy the trees, whose value had to be determined by a jury.

George III. Cap. XXIII. 1768
'An Act for the further preservation of Woods and timber trees.' Its chief interest lies in the method of valuation adopted for small trees 'as follows... every ash, elm beech or sycamore tree of half an inch diameter not less than sixpence halfpenny; of one inch, 1/s.; two inches, 2/s.; three inches, 2/6d.; four inches, not less than 3/s.; five inches, 4/s.; six inches, 4/6d.; seven inches, 6/s; eight inches, 8/s.... and every oak tree at double the value ...over that size to be valued by two appraizers and the diameter of every such tree shall be measured at the butt end...'

George III. Cap. XXVI. 1776-7
'An Act for encouraging the cultivation, and for the better preservation of trees, shrubs, plants and roots.

'Whereas the several acts of parliament... have not had the desired effect, and to avoid confusion which may arise from the multitude of the laws relative to the same subject, it is thought expedient to repeal the said several acts and to make one new act containing all such parts of the said acts as are proper to be continued, with such alterations and additions as are hereinafter contained...'

Five earlier Acts were repealed. That of 1765 – George III. Cap. XVIII – was not repealed.

Section II provided: '...that from and after the first day of May, 1776, every person, who shall wilfully cut down or break down, bark, burn, pluck up, lop, top or otherwise damage, spoil or destroy any timber tree, or fruit tree, or any young trees or shoots, or any part thereof, without the consent of the owner or owners thereof firsthand obtained, or who shall be aiding or assisting in so doing, or who shall have in his, her, or their possession any timber tree, or any kind of wood, underwood, poles, sticks of wood, shoots or young trees, shrubs, plants or roots, and shall not give a satisfactory account, that he, she or they came fairly and honestly by the same, or who shall fix up in any church or chapel the green branches of any tree or shrub, or any part of tree or shrub, having the leaves on it, except holly, bay, laurastina, yew or ivy, shall thereof be convicted... be fined an amount not exceeding £5 or imprisoned for six months'.

Section III purported to be a definition of the term 'timber trees' and includes larix and sycamore – which come fifth and sixth respectively – cherry, lime, holly timber, sallow, asp and cedar. Nineteen species are named:

'Be it further enacted... That all oak, beech, ash, elm, larix, sycamore, walnut, chestnut, cherry, lime, poplar, quicken or mountain ash, holly timber, sallow, asp, birch, cedar, pine or fir trees shall be deemed and taken to be timber trees, within the meaning and provision of this act and of any other acts in force in this kingdom relative to timber trees.'

Section IV is a repetition of previous statutes in respect of theft, the gist of it being that anyone may be fined forty shillings or get three months in jail who is convicted, if he:

'... wilfully cut down, or break down, pluck up, or spoil, harm or destroy, or take, carry or convey away any shrub, plant or root, shrubs, plants or roots, out of nurseries, gardens, woods, or fields of any other person... or aid in so doing... or who shall make use of any gads, withs, bows or backbands, made of wood, on his or their plows, harrows, cars, carts, harness or tackle or... found in his possession...or shall make use of any scollops of oak or ash, or any other tree for thatching of houses... or set up any bush... or keep bark or rind of trees, not being a tanner...'

Section V prescribes a penalty against clerks of the peace for failing to file planting records in accordance with the provisions of the previous statute. This neglect led to the provision of very detailed instructions for registration.

George III. Cap. VII. 1779
Section XXIV of this Act granted £7500 to the Dublin Society, half at least to be used for the encouragement of agriculture and planting.

George III. Cap. XVI. 1796
Directed the application of £5500 to the Dublin Society for similar purposes.

George III. Cap. XXXIX. 1784-5
This statute improved the position of tenants as tree planters. Bearing in mind the situation of tenants already outlined and the unavoidable religious ingredient and subsequent corruption, the Act, in fact, probably benefited Protestant tenants in the main. Tenants were entitled to dispose of trees they had planted 'or caused to be planted' or any part of the same, not only at the expiration of the term or when the trees had attained maturity, but at any time during the term – subject to certain provisions.

Section II dealt in great detail with the first of these provisions, namely, the registration of the planting, prescribing that: 'any tenant so planting or causing to be planted, should, within twelve months after such planting, lodge with the clerk of the peace of the county, or county of a city where such plantations shall be made, an affidavit sworn before some justice of the peace of the said county, reciting the number and kinds of the trees planted and the name of the lands in the form following:

'I, A. B., do swear, that I have planted or caused to be planted, within twelve calendar months last past, on the lands of .... in the parish of .... held by me from .... the following trees (here reciting the number and kinds of trees) and that I have given notice to the person or persons under whom I immediately derive, or his, her, or their agents, of my intention to register said trees, twenty days at the least previous to this day, and that I have given notice of my intention to register such trees, by publick advertisement in the DUBLIN GAZETTE, thirty days at the least previous to the date hereof, or else, and that I have also given notice of the same in writing to the head landlord, owner or owners of said ground or his, her or their agent, twenty days previous to the date hereof (as the case may be).'

The second section gave further detailed instructions as to how the county clerks of the peace were to keep the records of the registered plantings, and what fees they were to receive and what penalties they incurred by failure to comply. The records were to be open to consultation by anyone on payment of a fee of threepence.

Section III gave tenants the right to enclose any piece of ground containing coppice wood.

'And be it enacted, that if any tenant as aforesaid shall inclose any piece of ground containing coppice wood, which he is not bound by his lease to inclose or preserve and which has not been inclosed or preserved from cattle for five years preceding the said tenant shall have power to cut, sell and dispose of the trees, which shall grow from said coppice at any time during his term, leaving one timber tree on every square perch of such coppice where timber trees are growing.'

Section IV provides for the giving of 12 months' notice in writing of intention to enclose to the landlord or his agent. Section V provides that a map and a certificate should be lodged with the clerk of the peace within six months after enclosure. Anderson comments: 'Trees left standing to the number apparently of 160 to the statute acre, remained the property of the landlord.'

Later sections of this statute deal with the safeguarding of the landlord's position; for the tenant selling his right to the standing trees planted by him to his landlord; on surrender of a lease for renewal or the granting of a new lease, for the tenant's existing rights remaining in full force, etc.

Sections XIII and XVI relate to stolen timber and the penalties for same, and Section XVII is a reminder that it is a felony to cut down trees between sunset and sunrise.

The final clauses exempted trees planted under any special covenant of a lease and tenants who were evicted for non-payment of rent.

Effects of the Act of 1784-5

Clearly the Act had two main thrusts. Its first purpose was apparently to protect and encourage trees and planting. Its second, consequential or otherwise, was to further penalise the Catholic peasant population. They could not build with wattle; they could not use timber for thatching, plowing or husbandry generally, without fear of fine, imprisonment or whipping. Stronger tenant farmers could benefit, but only with the consent of their landlords.

The following advertisement from the *Dublin Gazette* of 9 January 1830 helps to illustrate the position:

NOTICE. Take notice that I have planted or caused to be planted within twelve months last past, on the land of Cross in the Parish of Ballyclay, Lower Half-Barony of Antrim, the County of Antrim, held by me from Henry Joe Tomb, of Belfast, Esq., the following trees, viz: 3000 Alder, 2000 Beech, 2000 Sycamore, 1000 Elm, 2000 Ash, 1000 Mountain ash, 1000 Larch and 100 Silver Fir; and that I intend to register said trees, pursuant to the Statutes in that case made and provided – dated this 29th day of December, 1829. David Kirk.

To Henry Joy Tomb, of Belfast, Esq., the Landlord of the Lands and Premises in the foregoing notice mentioned; and all others concerned.

The proviso requiring registration in the *Dublin Gazette* meant that there is a record of much of the tree planting carried out between 1785 and 1850, and Anderson selected four years – 1805, 1810, 1829 and 1844 – as years in which to make a sample survey. He comments; 'considerable areas must have been planted – very much greater than those under the Dublin Society's premium scheme.' (See Appendix 4, giving details of the Dublin Society's premium scheme under which it is claimed that, for the period 1765 to 1808 some 2800 acres were planted with 24,767,245 trees being raised and sold.)

Anderson gives the following table of registrations published in the *Dublin Gazette*:

Year	Registrations	Number of Trees	Approximate Acreage
1805	57	360,177	180
1810	215	1,633,125	817
1829	117	986,258	493
1844	55	459,452	230
Total	444	3,439,012	1720

APPENDIX 4

Note on the Measures Adopted by the (Royal) Dublin Society to encourage the planting of forest trees in the eighteenth century, and the results attained

(Appendix 71, Report of the Departmental Committee on Irish Forestry, 1908)

Planting

The Premium system was instituted in 1739 by the Rev. Samuel Madden, one of the founders of the Dublin Society (now the Royal Dublin Society).

In 1741 the system appears to have been applied for the first time to planting and the sum of £10 was offered to the person who raised the greatest number of timber trees in nurseries. At this time a premium was also offered in the following terms, *viz*: 'To the person who shall plant in groves, coppice, or hedgerows the greatest quantity of timber trees and secure the same by good fences or enclosures from this time to the first of December, 1742, the sum of fifty pounds in plate'.

The money for these and many other similar premiums was subscribed by members, and it is evident from the minutes of the period that a great deal of care was taken to ensure that the funds were judiciously applied.

In 1747 the Society received its first grant in the form of £500 per annum from the Privy Purse, and no doubt the premium system underwent a corresponding extension, but it is difficult to throw any light on the matter because the MS. Minute Book of this period was lost many years ago.

In 1761 the Irish Parliament voted the Society a sum of £2000, and in subsequent years the grant was largely increased. This money was expended in premiums and bounties, chiefly in aid of manufacturing industries. It is difficult to trace the development of the system as applied to planting until we reach the year 1765. From that year to the year 1808, when premiums for planting ceased, the progress of the system and its ultimate abandonment are easily traced. It is only necessary here to refer to some leading features.

In 1765 the Society offered medals for the planting of forest trees, including oak, ash, elm, Weymouth pines, Scotch firs. At the same time premiums of £20, £15, £10 were offered to the person who had the greatest number of oaks, not less than 160 per acre, in a thriving condition, seven years after he had claimed the medal.

A few years later it was provided that renters of land should receive the sum of £5, instead of a medal. This amount was afterwards increased to a sum varying from £10 to £20.

In 1780 the sum of £224 was offered in amounts varying from £5 to £20 for propagating forest trees.

Under Grattan's Parliament a large increase took place in the premiums offered, and they assumed a new form. In 1783 a premium of £40 was offered 'to each person enclosing any quantity of ground, not less than ten acres with a sufficient fence proof against cattle, and planting the same with a number of oak, ash, beech, elm, maple, sycamore, chestnut, larch, fir, or pine, not less than two thousand plants on each acre. The nature of the fence and preparation of the ground must be fully stated in the claim, and security will be required in double the amount of the premium that the said fence and trees shall be preserved, and such trees as shall die, be continually replaced so as to keep up the number of one thousand trees in each acre for ten years. For this premium the sum of £400 will be granted; but if there should be more than ten claimants, the sum of £400 will be rateably divided.'

In 1786 it was provided that 'Every person, whether a member of the Society or otherwise (except nurserymen), to whom any premium for the encouragement of agriculture and planting, amounting to more than £10 shall be adjudged by the Society, shall not receive the same in money; but shall be entitled to receive the full value of the premium in any utensils of agriculture or planting, which he shall think proper, at such prices as shall be settled by the Society; the same to be made at the Society's manufactory in Poolbeg street, and to have the seal of the Society marked or painted thereon, with such inscription as the Committee of Agriculture shall direct.'

In 1796 persons receiving premiums were required to give security by a bond of £100 that fences would 'be preserved and trees kept up to 500 oak and 500 other trees on each acre for ten years, an affidavit to be lodged with the Society in May or June of each and every of the ten years that fence and trees are preserved'.

In 1800 three premiums of £100 each were offered 'for planting by contract, the greatest quantity of ground not less than 40 acres, and three of £50 to the next greatest quantity of ground not less than 20 acres'. The recipient of the premium was required 'to enter into a written contract with some nurseryman who keeps a nursery for sale, that such nurseryman shall furnish and plant in each acre 8,000 trees at the least and replace all that die or may be injured for three years at the least, so as to leave at the end of the third year 8000 good growing trees in each acre. Sallow, poplar, birch, and horse chestnut are not included; and there must be at least 3000 oak, or 3000 larch among the number contracted for in each acre.

'A copy of the contract and of the security of its performance is to be delivered with the claim, which must be made before the 10th day of May, 1801, and will be taken into consideration on the last Thursday of the said month.

'The person receiving the premium must give security in the sum of £100 that a sufficient fence shall be preserved around such plantation for ten years; that the number of 6000 trees, at least, shall be preserved or replaced on each acre thereof, during the said ten years so as always to have 6000 growing trees on each acre, and that proof shall be made thereof to the Society by affidavit in every month of May or June. All contracts where it shall appear that not more than £8 per acre is to be paid for the entire expense, exclusive of fencing and draining, will have a preference in the claim.'

Nurseries

The difficulty of obtaining a sufficient supply of trees seems to have been recognized from the outset and in 1765 it was announced 'That improvers in all parts of the kingdom may be better and more conveniently supplied with trees, the Society will pay for every person who shall first keep a well inclosed nursery of forest trees, within three miles of any of the county towns of this kingdom (the trees in such nursery being of two years' growth) a yearly rent of 20s. per acre for three years for the ground so occupied in a nursery, the whole of the yearly rent promised for any one nursery not exceeding £3.

'*Condition.* – N.B. – Five of the Grand Jury of the Spring Assizes, where this encouragement shall be claimed, are to certify concerning the condition of the nursery and the quantity of ground occupied therein.'

These premiums for nurseries were continued for many years, with such variations as experience suggested. In 1780 they took the following form: 'A premium of four shillings will be given for every thousand of oak, beech, chestnut, walnut, plane, elm, larch, fir or pine, not less than two nor more than five years old, which shall be sold between the first day of September, 1780 and 3rd May, 1781, out of any nursery or nursery grounds wherein the same have been raised. The said premiums not to exceed in the whole £400, and if premiums be claimed of more than two millions of trees, a preference will be given to those which shall be sold at the lowest prices, pursuant to their respective kinds. The claims are to be put before the 1st of June, 1781; the number, age, and species of trees; where and when raised; to whom, when, and at what price sold, are to be proved by the affidavit of the person selling the same; as likewise that none of the said trees were sold to any man who keeps a nursery for sale; and certificates are also to be produced, signed by the persons purchasing, setting forth the number and species of trees; where, when, and from whom bought, and at what price, and that none of said trees were bought with intent to be sold again for the purpose of being planted.

'N.B. – No premium is to be given for trees sold to any person who keeps a nursery for sale.'

Results

Under these schemes it appears that the sum of £12,460 13s. 11d. was awarded in the forty years ending 1806. The money was distributed as follows:

	£	s.	d.
Ulster	1371	14	0
Munster	2866	5	8
Leinster	4023	6	7
Connacht	4199	7	8
	12,460	13	11

In addition to the above a sum of over £6000, not including gold and silver medals, was awarded by the Society as premiums for planting sallows, willows, poplars etc., which were used in basket-making and other minor trades.

All the details relating to these awards are summarized in the Transactions of the Society, Vol. V, pp.103-104, published in 1806. It is unfortunately impossible to ascertain the area planted under the premium system, as particulars are not given in every case. The total area of the plantations specified is about 2800 acres, but this is probably far short of the actual area for the planting of which premiums were awarded.

A similar difficulty applies to nurseries. It appears, however, that at least one hundred and eight nurseries received premiums, and of these, sixty-seven are mentioned as new. They were situated in every county in Ireland except Leitrim, Longford, Monaghan and Sligo. Galway heads the list with sixteen nurseries. The number of trees mentioned as raised and sold is 24,767,245. The County of Galway contributed nearly half this number, – *viz*; 11,395,226.

Fraudulent Practices

The premium system afforded many opportunities for fraud and imposition, and that such attempts were made is evident from the minutes of the Society. Long before attention was directed to planting, the Society had the experience of administering the bounties and premiums for encouraging various forms of industry, and had learned that attempts at fraud invariably arise under such conditions. This was the origin of the very unusual provision in the Society's Charter (1750) empowering the Chairman at a general meeting: 'To administer an oath to any person or persons for discovering the truth or value of anything offered or proposed to the said corporation'.

The following extract from the minutes will serve to show how the Society attempted to meet the difficulty in the case of premiums for planting: 1782 – 'It appearing that several frauds had been lately attempted to be made on the Society for premiums under forged affidavits, therefore – Resolved: On the motion of Mr Ford and Col. Madden that for the future all claims relating to premiums to be paid by the Society shall be certified to the Assizes by the Foreman of the Grand Jury in the town or county from where such claims are made.'

1784 – 'All persons intending to claim any of the following premiums must give notice to the Society of the quantity and situation of the ground planted or sown as soon as the several crops shall be put into the ground, otherwise they will not afterwards be permitted to claim the same, and the Society will, from time to time, publish in newspapers all doubtful claims that may be made to them for premiums, in order to receive objections and to discover such frauds as may have been practised with respect thereto.

'All false affidavits presented before the Society being punishable in like manner as perjury in other cases, the Society intend sending a proper person as surveyor of claims through the kingdom at such times as they think necessary, to examine into the authenticity of the several claims and are determined to prosecute all perjurers in the most exemplary manner.'

1785 – 'A candidate for a premium or a person applying for a bounty being detected in any disingenuous methods to impose upon the Society, shall forfeit all such premium or bounty and be incapable of obtaining any for the future, and if any person shall be detected in offering any forged instrument in evidence to the Society, or in committing wilful perjury in proof of any claim, a prosecution will be carried out against such offender with the utmost rigour of the law.

'The Society being desirous of avoiding as much as possible the multiplication of oaths in the disposal of their premiums, request that the nobility, magistrates, gentry, clergy, in their several districts, will give their attention when applied to for certificates of the merits of any candidate for a premium, to examine the pretensions of such person that the Society may not be under the necessity of tendering an affidavit to him, which they apprehend has sometimes occasioned the misapplication of their fund and the guilt of perjury.'

Later in (1787) an inspector was employed, and he was instructed 'To inspect into the merits of the several claimants agreeable to the Society's determination in their publication of premiums for the three years 1785, 1786 and 1787, and that the sum of three half-crowns per day should be paid to him whilst on the circuit in the service of the Society out of the funds appropriated by Parliament for the encouragement of agriculture and planting.'

In 1808 the Committee of Agriculture were instructed to 'Employ proper persons to go to the several counties where premiums had been granted for planting, and to make their report on oath of the state of such plantations, and that the persons so employed be furnished with the schedules thereof by the Assistant-Secretary.'

And in the same year the following minute appears: 'Resolved – That in consequence of information having been received that persons who have obtained premiums for planting trees have not preserved the trees according to their engagements, together with some suspicious circumstances which throw a doubt on the truth of certificates and which magistrates are requested to observe and examine before they subscribe, the planting premiums be discontinued until the Society receive reports of the persons to be employed in taking a view of the several plantations for which premiums have been granted.'

This apparently was the end of the premium system for planting, and it looks as if the prevalence of fraud was the chief reason for discontinuing the premiums.

It was not only in relation to planting that fraud proved a serious obstacle in carrying out the premium system. Mr Isaac Weld, who was for many years an Honorary Secretary of the Society, stated in his evidence before the Select Committee appointed by the House of Commons, in 1836, to inquire into the administration of the Royal Dublin Society, that the number of applications for premiums received from poor renters for the reclamation of land became so great, and the frauds practised were so numerous, that it was found expedient to abandon the system.

Apart from this consideration the premium system could not have long survived. The Act of Union guaranteed for a few years the funds necessary for the maintainance of the system, but at the earliest opportunity the Imperial Parliament reduced the Votes-in-Aid to the Society to a sum insufficient even for the maintainance of the educational and scientific institutions the Society had founded, and all attempts at fostering industries by monetary assistance of necessity ceased.

The Condition of Premium Plantations early in the Nineteenth Century

The following information had been gleaned from the Statistical Surveys of the different counties in Ireland, which were published by the Society early in the last century.

Sir Charles Coote, Bart., in his survey of the Queen's County (1801) mentions seven plantations for which premiums were awarded between 1787 and 1794, varying from 5 to 16 acres each. Two belonged to Lord Portarlington, two to the property of Sir John Parnell and one the property of Sir John Tydd are stated to be in 'in good heart and well preserved'. Mr Thomas Drought's plantation of 10 acres is reported to have been 'injured by cattle about five years ago, but the damaged parts were replanted carefully and the whole is now in great vigour and well enclosed'. Of Mr John Pigot's plantation of 16 acres the writer says – 'a small part damaged by fire which fell from a peasant's pipe, but the rest of the plantation is in good heart'.

The same author in his survey of King's County (1801) refers to nine plantations varying from 2 to 14 acres each and reports favourably on three belonging to Mr Thomas Drought, and two belonging to Mr C. W. Bury. Three plantations, the property of Mr George Synge, he mentions as being in a very unpromising condition, except in the valleys approaching the house.

Mr Robert Frazer, in the survey of the County Wicklow, (1801), says, 'Before the late unfortunate disturbances [presumably the 1798 Rising] planting was going on rapidly in this county, and a number of candidates applied for the premiums of the Society annually. Some destruction has inevitably taken place in the plantations so made on account of the residence of all those, who were near the mountains and who had received the premiums (as it is adjacent to them here, planting has chiefly taken place) being destroyed, the fences being also impaired, cattle had injured them. However, the plants all seemed thriving, particularly at the Marquis of Waterford's lodge, where there is a fine screen of planting, extremely well-disposed on the brow of the mountains adjacent to the ruins of his lodge.

'Many of the farmers complained to me that the premiums of the Society did not extend to smaller spaces than ten acres as they could not always spare ten acres in one place for planting without encroaching on their best meadow or pasture ground. If premiums were granted by the Society or the Farming Societies in the different districts to small tenants, and even cotters, for planting a rood or two round their cabins, as is doing by Lord Fitzwilliam, it might have the effect of giving the lower class a love for trees and begin a greater means of preventing their wanton depredations on them than any penalties.'

Lieut. Joseph Archer in his survey of the Co. Dublin says nothing about the premiums, but gives a list of registered trees and comments unfavourably on the state of the plantations in the county, making an exception in the case of Mr Alexander Hamilton, near Balbriggan, who, he says 'raises and plants ten or twelve thousand yearly, which are very judiciously attended to and in a very flourishing state'. Mr Hamilton received a premium for planting in 1794. Referring generally to the subject of timber, the author says, 'Timber rises in value every year on account of the scarcity and is likely to continue rising, as the small quantity remaining is, on an average, through the county, cutting down in the proportion of four to one that is planted. This calls loudly for redress, to reinstate us in the possession of timber, which we had formerly in such abundance.'

In his 'Observations on Mr Archer's Statistical Survey of the County of Dublin' (1802), Mr Hely Dutton points out that amongst some 70,000 of the registered trees referred to there are only 3904 larch, and no less than 4495 mountain ash. This, he says, needs no comment. He complains that 'great numbers of the trees generally planted in this county are the very refuse of nurseries, or picked up as bargains from hawkers or jobbing gardeners; but as we may expect from this, and from unskilful planting and bad management, that very few of these trees will ever arrive to the state of timber, they are scarcely worth noticing, except to put those who wish to be informed on their guard.'

The author expresses the opinion that plantations have suffered more from want of judicious thinning than from any other cause.

Sir Charles Coote, Bart, in his survey of the Co. Cavan, (1802) mentions nine plantations of about 106 acres, in respect of which premiums had been awarded between the years 1789 and 1794, and with the exception of a 'Danish Fort', which had been planted by Mr John Piers, they are all spoken of as in good preservation.

In the survey of the County of Mayo by James M'Parlan, MD (1802), eight plantations, representing an area of seventy-three acres, planted within the period 1789-1798, are mentioned, and it is stated that they are all of 'good growth and preservation, except those of Mr Bourke, of which not a trace remains'. This was a small plantation on a 'Danish Fort'. The author goes on to say, 'As to the effect produced by those grants, it is clear that they had not been very considerable in diffusing example and emulation, as many more premiums had not since those dates been demanded. But whatever may be the cause, one effect is visible, that planting young trees is the pride and fashion of the county.'

The same author in his survey of the County Donegal (1802) refers to only two cases of planting for which premiums had been awarded, that of Mr Henry Stuart, of Tycallan, who, he says, 'exceeds all his county, and perhaps any individual of Ireland for planting. He has lived here but a very few years, yet already has he planted upwards of one hundred acres of birch, oak, ash, etc. His nursery, containing sixteen acres, is now at full growth for a succession of planting out annually twenty acres.' This gentleman received premiums amounting to £244 for planting sixty-one acres. The other case was merely the planting of a 'Danish Fort'. Both plantations are said to be 'very well enclosed, preserved and in full growth'.

The same author in his survey of the County Sligo (1802) mentions three cases of planting under the premium system, of which he says 'Mr Wynne obtained a premium in 1787 for enclosing and planting ten acres. This plantation is in the best state of growth and preservation.

'Mr O'Hara obtained a premium for ten years in 1798; that, too, is in luxurious growth and excellent preservation.

'But Mr George Dodwell's plantations of ten acres, for which he obtained a premium in 1793, is destroyed. Some changes occurred in that family which caused their removal from that seat, and the consequent ruin of the plantations.'

Mr John M'Evoy in his survey of the County Tyrone (1802), refers in favourable terms to the condition of several plantations for which premiums had been awarded. He gives a table showing that from the year 1791 to 1801 Lord Mountjoy planted under the premium system 178 acres 2 roods and 5 perches with 483,970 trees, in addition to which 136,762 trees were planted for which premiums were not claimed. An appendix of ninety-five pages is devoted to Lord Mountjoy's improvements, and a large part of it deals with planting. The author speaks very highly of the way the planting was carried out, and the way it is preserved, and considers that the work is equal to any in the United Kingdom.

In the survey of the County Meath by Mr Robert Thompson (1802) the nursery belonging to Mr Reilly at Ballybeg, near Kells, and occupying about thirty-nine English acres, is very favourably mentioned. This was one of the nurseries to which a premium was awarded. A few plantations for which premiums were awarded are referred to as flourishing and well fenced.

In the survey of the County Kilkenny by Mr William Tighe (1802) mention is made of seven plantations for which premiums had been awarded, and they are all, with one exception, stated to be thriving and well preserved. The exception is a plantation of twenty-seven acres belonging to Mr J. Davis. This was preserved for ten years 'but the land is now grubbed up for tillage'.

Mr James M'Parlan, MD, in the survey of the County Leitrim (1802) states that only two persons in the county received premiums, and in both cases the plantations were 'Danish Forts'. One of these belonging to Mr Lowther, is reported as 'in the highest vigour, bloom and preservation'. The other belonged to Mr Nesbitt, who died, and his family left the place, which has 'totally fallen into ruin and decay and the plantation also'. This author refers to the highly-wooded state of the county in former times. He says 'Living persons who saw it told me that about 100 years ago almost the whole county was a continual undivided forest. From Drumshanbo, I used to hear them say, to Drumkerrin, a distance of nine or ten miles, one could travel the whole way from tree to tree by the branches.' The destruction of the forests is attributed to the use of the wood as fuel in iron smelting. The author states that about thirty years ago (say 1772) 'a spirit of reviving and renovating the face of the county fortunately succeeded to that of indolence, devastation and want of taste'. He gives a list of eleven persons who planted 145,384 trees from the year 1770 to 1798. Premiums were not awarded to any of the persons mentioned, and this fact is interesting, as it shows that a great deal of planting took place at that time quite independently of the premium system. This may have been due to the example set by the premium system, but it is more likely to be simply a phase of the remarkable agricultural and industrial development that characterised this particular period in Irish history.

In the survey of the County Kildare by Mr Thomas Rawson (1807) eight plantations to which premiums had been awarded are referred to. These embrace an area of 47 acres, and are described as well enclosed. Two plantations of Mr Begot's (4 acres) are said to have 'flourished exceedingly'. One of Mr Brownlow's (13 acres) 'being on bleak bog, required to be replanted'; and one belonging to Mr Keating (10 acres) is described as not having had sufficient attention paid to the thinnings. The author refers to nine cases of planting in which no premiums were awarded. Referring to a plantation of 40 acres formed about the year 1757 he says 'For the past 30 years the noise of the axe has resounded through these forty acres; they have yielded from time to time £10,000 to the occasional possessors, leaving behind an ample stock of growing wood.' Another plantation of 30 years standing is referred to as a 'source of wealth'. Of Mr John La Touche the author says 'He has formed within a few years most magnificent and extensive plantations.' Mr D. W. O'Reilly is said to have planted 'with the greatest success' and several other plantations are mentioned in scarcely less laudatory terms. The most extensive and successful planting carried on in the county seems to have had no direct connection with the premium system.

In his survey of the County of Clare (1808) Mr Hely Dutton deals rather fully with the subject of planting, his remarks are perhaps worth quoting at some length. He says – 'The following gentlemen received premiums from the Dublin Society in the years annexed to their name: – James Molony, Esq., in 1785, 1786, 1789, 1793 and 1794; his plantations have been well preserved. Sir Joseph Peacock for planting oak, now completely destroyed by cattle. The late Charles M'Donnel, Esq., 1789; well preserved and flourishing. Robert O'Hara, Esq., 1790 and 1791; well preserved and beautiful. Boyle Vandeleur Esq., 1795; well enclosed and very thriving. There are some trifling plantations mentioned in the list of premiums granted, that I did not see, particularly for raths, which I confess I never wish to see planted whilst they are permitted to retain their present round shape; the money granted for the above premiums amounts to £403. 7s. 5d., and seems to have been very justly expended, except that given to Sir Joseph Peacock in 1793 whose plantation has been quite ruined by cattle, if it was the one that was shown to me in the barony of Tullagh. I beg leave to suggest that, as the public mind is not sufficiently pointed to the subject, and the value of plantations so well ascertained, a discontinuance of these premiums and the converting of the funds to some other beneficial purpose would be eligible.

'I beg also to mention that giving a premium for oak without limiting, or at least advising the proper soil, is so much money thrown away; for some of the plantations I have seen are upon dry, rocky, shallow hills, where larch would have been infinitely more valuable.

'What a reproach to the county, that in twenty-five years, one of such extent, and where trees are so much wanting, has had only ninety-six acres planted! It may be said that this is only the quantity that were planted for premiums, but I am convinced there has been very little more planted to the year 1795; of what has been planted since I have no account; but except the plantations of Sir Edward O'Brien, Bindon Blood, Esq., and William Burton Esq., the number is very small indeed. Whilst a whole county in twenty-five years has had only ninety-six acres planted, an individual in Scotland has in fifteen years planted 3,005 and one half acres.'

He then enumerates the various kinds of trees planted by the Earl of Fife, numbering nearly five millions, chiefly Scotch fir, and concludes – 'These plantations were well enclosed with walls, measuring in length upwards of forty English miles. When I inform my readers that the Earl of Egremont, Marquis of Thomond, Lord Conyngham, Marquis of Buckingham, Lord Milton, Mr O'Brien, Mr Westby and a long etcetera of absentees have thousands of acres of waste land as capable of being planted as Lord Fife's estate, what will they think?'

The same author in his survey of the County of Galway (1824) gives a list of six nurserymen who received between the years 1768 and 1795 the sum of £1,680.13s. 2d. for nurseries. This expenditure, he thinks, 'was of infinite use, as several extensive nurseries were established in this county, that probably would not otherwise have been. Since that period the nursery business has been at a low ebb, as most gentlemen, as they should do, have their own nurseries'. The author gives a list of twenty-six persons who received premiums for planting a total area of 702 acres, which, he remarks, is 'a considerable quantity in 27 years, when it is considered that those were only such as had obtained a bounty, for we cannot suppose that there was not infinitely more than that number that did not obtain any, for nothing under 10 acres received the bounty. At the same time the names of many gentlemen occur here that should individually have planted more than the whole amount; such were, amongst some others: – Right. Hon. Denis Bowes Daly, Sir Thomas Burke, Earl of Clanrickard, Mr Marcus Blake Lynch, Lord Riverston, Mr Richard Martin. All proprietors of extensive mountains well calculated to produce timber trees. It would have been a desirable condition, and I believe was intended in giving these premiums, that they should be well fenced, and thinned at the proper age, both of which the last especially, have been grossly neglected. But the planting that calls loudly for adoption is that of the extensive mountains and bogs that occur so frequently in this county, with the exception of Dalyston, Roxborough, Portumna and perhaps some few others, little has been done in this way. If the same spirit actuated the landed proprietors of this county, and I may say of Ireland, as those of Scotland, what a country would ours be?'

Mr Dutton expresses the opinion that the bounty system gave rise to the spirit of planting, and he says that having attained this object and made men think for themselves, the system was very properly discontinued.

I have not referred to the Counties Antrim, Armagh, Cork, Down, Londonderry, Monaghan, Roscommon and Wexford. The premiums awarded to these eight counties amounted to only £868. I can find nothing in the surveys of these counties throwing any light upon the operation of the premium system.

The author of the survey of the County Armagh (1804) in pointing out that the county is not well-wooded states, 'The population is so great, the linen manufacture so universally engaged in, and land in such request, that every spot in all but the mountainous districts is employed in tillage and pasture for milch cows, so great a supply of food is necessarily raised for the inhabitants.' It is not unlikely that the other counties in the north-east corner of Ireland took but little interest in the planting movement for a similar reason.

In the case of the Counties Carlow, Fermanagh, Kerry, Limerick, Longford, Louth, Tipperary, Waterford and Westmeath there were no surveys published. These nine counties received premiums for planting amounting to £3602. Two-thirds of this amount went to Kerry, Tipperary and Louth. It is therefore unfortunate that there are no records showing the effect the expenditure produced.

In these notes I have sought only to present the facts concerning the premium system in its relation to planting, as far as can be ascertained from the records of the Royal Dublin Society. These facts may be summarized briefly: The premium system was carried on for about half a century. It extended to practically the whole of Ireland. The entire expenditure on planting and nurseries, including planting in connection with basket-making and similar industries, was under £20,000, or less than £400 a year on the average. There is evidence that in many cases the money was well spent, and satisfactory results obtained; in a few cases the money was unwisely spent. The system seems to have done a great deal indirectly, and by force of example. It did not, however, lead to the creation of a single plantation on a really large scale, and it can scarcely be claimed that it promoted forestry in the proper sense of the term.

Richard J. Moss

Leinster House,
Dublin, January 31st, 1908.

APPENDIX 5

Land Ownership

The validity of 'ownership' of land under the Brehon system was under challenge for a considerable time from three quarters:
(1) English planters and *force majeure*.
(2) The Norman tradition of possessory title by then 300 years old.
(3) The decline and corruption of the Gaelic Order itself from within and in the face of these two powerful factors.

'The political system of ancient Ireland was no more communal than that of the Roman Republic. It was aristocratic. The ancient Irish "clan system" or "tribal system" are very modern inventions' (MacNeill, *op. cit.*, p.144).

With compelling analysis MacNeill challenges the view held by Joyce (P. W. Joyce, *A Social History of Ancient Ireland*, etc.) that:

'In theory the land belonged not to individuals, but to the tribe...' MacNeill points out that the confusion arises to a considerable extent by a mistranslation into English of the words 'tuath' – (meaning both the population and the areas governed by a king) and 'fine' (a small family group from whom the king might be elected). The words were, to some extent, used synonymously and there could be 'no question of alienation of the land of a tuath to a stranger, on the part of an ordinary member of the tuath, without the consent of the tuath'.

The conflicting views are demonstrated as follows: 'A fine (pron. finneh) was not a tribe or a clan. It was a group of persons who knew and could reckon their degree of relationship to each other, and it was based on this explicit knowledge. In other words, it was a family not merely of one household, but of all the households within a definite degree of consanguinity. The land owned by such a group of persons was not individual property, but it certainly was private property' (MacNeill, *op. cit.*, p.151).

Within the group (or group collective) the lower orders – serfs, slaves and bondsmen and bondswomen [*daor*; *daor-fuidhear*, a bond-tenant; fugitive, criminal, deserter or captive] – had no legal status so long as they remained members of this order (though status might be acquired). Except to indicate their value, which was important, or punishments for certain offences, the laws are silent about them. The basic rate of exchange was a *cumal*; the value of one female slave. Like other rates of exchange values floated from time to time. Mitchell estimates one cumal at about thirty-five acres of land in the eighth century, possibly enough to feed one cow for a year.

This group, collectively known as *fudir*, included slaves, who might be freed on manumission, but also prisoners, outlaws, renegades from other septs and other landless and titleless individuals who were neither slaves nor freemen. All could change status by release, purchase, inheritance or for some other reason.

Together with the lowest order of freemen they constituted the bulk of the population in what was an hierarchical society with a form of elective leadership confined to blood relations of the dominant families.

Farmers, craftsmen, freemen of one degree or another, were members of the *tuath*, or district(s) ruled by a sept or clan in a relationship of mutual obligation. The chieftain or king gave legal decisions and organised defence and protection: the members of the *tuath* accepted his leadership and adjudication, served the *tuath* militarily and provided food and services as laid down by law. The king or chieftain, in turn, gave his support and allegiance to more powerful lords and so on until the five kings of the five provinces were reached, one of whom (usually) was elected, or became by superior force, *Rí Éireann*, King of Ireland (the term High King was introduced by the Normans; Meath became a notional province, presumably to accommodate the kingship of Tara, thus introducing a dimensional complication to the fifth fifth).

The root *fine* related to a family unit that was also a legal unit, whether it was *gelfhine* (of three generations), *derbfine* (of four), *iarfine* (of five) or *indfhine* (of six). The *derbfine* was the customary active legal group in terms of succession, inheritance and disposal of goods and property. It was vital to retain a continuum in order that the *fine* group retain not alone office, but position and status as well. If, for instance, an entitled member of a *derbfine* failed to be elected to the kingship within three generations, the right was lost to the whole line of succession. In other words if an eligible great-grandson failed election to office after it had been held by his great-grandfather and no closer kin succeeded, that line failed and was relegated to a lower order (MacNeill, p.125).

If they failed to secure (the kingship) within three generations for one of themselves they dropped to a lower social order, emphatically characterised by '*cúig gluine ó rígh go rámhainn*'.

While a king held power of grant over the territory under his jurisdiction, this power was limited.

Therefore, as we have seen in Chapters 2 and 3, Dermot MacMurrough's 'grants' of land to the Normans were either illegal under Irish law, or interpreted by the Normans in a way that was alien and irrelevant to Irish law.

There is confirmation of the authority to 'grant' (subject to the limitations already noted) in the *Book of Armagh*:[26] 'Patrick, coming into the territory of Calrige (Co. Sligo), baptised MacCairthin and Caichan, and after he baptised them, Mac Cairthin and Caichan granted a fifth part (of the lands) of Caichan to God and Patrick...' within the boundaries, 'four oakwoods are mentioned by name' (*Book of Armagh*, Folio 17).

However equally important to recognising the conditional right of grant in the Brehon Law is to recognize that neither 'ownership' nor the power of transference under grant was as absolute as under Norman or feudal law. In Irish law it was essentially 'the use of' the land that was granted or tranferred and, in many cases, it was liable to re-transferral if the status of the custodian altered.

Above all occupants of land, tenants, could not be arbitrarily handed over to a new owner or lord without their agreement. As we have noted Irish law did not allow them to be evicted and therefore their consent was necessary in a peaceful, legal transfer.

Woodlands, however, – especially those adjacent to a settlement – were often common land and, presumably deriving from this, there were rights in common in each *tuath* – 'the full property of every tuath, belonging in equal right to every condition (of person)' (MacNeill, p.167).

These rights included the right to supplies of kindling, firewood, nuts; timber frames for vehicles and corpses (the dead were buried in a timber, usually leafy, wrapping), ploughs; shafts and handles for spears, horse-rods (*echlachs*) and implements; timber for harness making and other domestic uses, besides the right to unappropriated timbers.

So that not only were (many) woodlands unappropriated and common property, specific freedom of usage as well as specific penalties were recognised.

MacNeill clarifies the position thus: 'The ancient law tracts indicate that in Irish law, the lordship of the land belonged to the political rather than to the economic order of things. It imparted authority rather than ownership.

'This Irish idea of dominion over land very much facilitated the Norman conquest. The Irish lords thought they were submitting under duress to a new political authority. The feudal invaders believed that they were acquiring a rigid, complete and perpetual ownership of the "land" from the zenith to the uttermost depths – an ownership more complete than that of any chattel – an ownership that they imagined to be self-existing even when the person in whom it should be "vested" was, for the time being unknown and unascertained. They called this sort of ownership an "estate", i.e. a status, something that stood of its own virtue; and they could think of it as existing without a shred of the political authority which was the outstanding character of the relation of the Irish *flaith* to the land and its inhabitants. To my mind, the Irish idea of ownership of land was derived from the natural law, and the feudal idea, which is embodied in English law, was the artificial product of a feudal mentality.'

The Irish system contributed to society a cohesive mix of expectation and proper responsibility, the result of considerably more than a thousand years of development. It was a system that did not display major inadequacies until the onslaught of, first the Norse with a centralized urban trading economy and then the even more disruptive Norman feudal system with its totally alien concepts of ownership. One wonders what natural development in response to external stimulus, rather than imposed change, might have brought about. It might even be argued that today a form of basic expectation has been encouraged to resurrect itself, powerfully, from the folk memory, but without any necessary accompaniment of proper responsibility.

APPENDIX 6

Irish Statutes since 1920 Relating to Forestry

(1) British Forestry Act, 1919 (which affected forestry in this country with decreasing influence until the mid-1920s): An Act to promote the economic development of the United Kingdom and the improvement of roads therein. (Section I deals with forestry research and teaching, planting and the acquisition of land. The Forestry Commission was established under this Act) – (IX & X) George V. Cap. LVIII, 1919.

(2) Acht Foraoiseachta, 1928 (Forestry Act, 1928): An Act to make Further and Better provision for promoting forestry and for that purpose to amend the Forestry Act, 1919, to restrict the Felling of trees, and to make other provisions connected therewith – No. 34 of 1928, S.O., Dublin.

(3) Acht Talmhaíochta, 1931 (Agriculture Act, 1931); An Act to Dissolve the Department of Agriculture and Technical Instruction for Ireland, to transfer the functions now exercised by the said Department as to part thereof to the Minister for Agriculture and as to the remainder thereof to the Minister for Industry and Commerce, to make provision in respect of persons now or lately employed by and remunerated out of funds of the said Department, to make provision for the formulating and carrying out of schemes in relation to Agriculture and other rural industries, and the preservation of existing woods, and for other purposes incidental to or arising in connection with the matter aforesaid. (Sections 35 and 36 relate directly to forestry). No. 8 of 1931, S.O. Dublin.

(4) An tAcht Foraoiseachta, 1946 (The Forestry Act, 1946): An Act to make further and better provision in relation to Forestry. No. 13 of 1946, S.O. Dublin.

(5) An Act to Amend Agriculture Acts, 1931 to 1948: No. 2 of 1955.

(6) An tAcht Foraoiseachta, 1956 (The Forestry Act, 1956): An Act to facilitate Acquisition of Land for the purposes of the Forestry Act, 1946. No. 6 of 1956, S.O. Dublin.

(7) An tAcht um Fiadhulra, 1976 (The Wildlife Act, 1976): An Act for the Conservation of Wildlife (including game) and for that purpose to protect certain wild creatures and flora, to enable a body to be known in the Irish Language as An Comhairle um Fhiadhulra and in the English Language as The Wildlife Advisory Council to be established and to define its functions, to enable certain other bodies to be established and to provide or administer certain services, to enable reserves and refuges for wildlife to be established and maintained, to enable dealing in and movement of wildlife to be regulated and controlled, to make certain provisions relating to land, inland waters and the territorial seas of the State, to amend certain enactments and to make other provisions connected with the foregoing. No. 39 of 1976.

(8) An tAcht Foraoiseachta, 1988 (The Forestry Act, 1988): An Act to make provision for the development of forestry and to provide for the establishment of a company for that purpose and for the assignment to the company of functions heretofore exercised by the Minister for Energy; and to provide for related matters. No. 26 of 1988.

APPENDIX 7

'A National Scheme of Afforestation'

The assumption that what the 1907 Committee proposed was a programme of National State afforestation such as we have today is – or at all events was until the point was raised during the course of this work – virtually universal in the Forest Service, in spite of the existence of the Committee's Report and reference to it from time to time by various authorities. The Report of the committee does not support that interpretation. It might be argued that there is some ambiguity in the phrasing of the Report, but this does not bear examination. What seems to have happened is that, *a posteriori*, the circumstances of State afforestation as they developed were unquestioningly attributed to the Committee, who were concerned with very different facts and circumstances, not least that the *only* forestry in the country at the time the Report was written was private forestry.

Taken in context the wording of the recommendation in the Report leaves no room for doubt.

It was neither envisaged nor recommended, as has been assumed, that a State programme of afforesation should be undertaken which would result in a State forest holding of 80 per cent, such as there is today. To have made such a proposal in 1908 would simply not have been realistic. The idea of wholly owned State enterprise of any kind was then, at best, a dubious concept; the idea of State forestry on a limited scale in the United Kingdom was still experimental and under critical scrutiny; the example of Knockboy was only nine years away. Above all, the forestry and forest landowners that existed were private (with the exception of Avondale, and even that was specifically intended for the encouragement of private, not public, forestry). The idea would not have been considered. Nor did the Committee attempt anything so foolishly radical.

What the Committee did do was urge the State to take action to encourage 'A National Scheme of Afforestation' of about 1,000,000 acres (initially), consisting of the 306,000 acres of existing (private) woodlands; of State forest of between 200,000 and 300,000 acres, with a further 500,000 acres through private owners and county councils – 'thus establishing within a period of about eighty years an ultimate forest area in Ireland of at least 1,000,000 acres', which would be 'managed' by the State.

The remarkable fact is that this target was achieved within the specified time.

The relevant paragraphs from the Report of the Committee are:

1 Part IV, par. 53, pp. 24/25.
2 Part V, par. 54, pp. 25/26.
3 Part V, par. 55, p. 26.
4 Part V, par. 71, pp. 33/34.
5 Part V, par. 72, pp. 34/35.
6 Part V, par. 73, p. 35.

The appropriate extracts are:

...this, then, is the problem to be worked out: The proper utilization, through private owners and County Councils, of about 300,000 acres of existing woodlands.

The creation of a State forest on between 200,000 and 300,000 acres of mountain or other rough land in large blocks to be specially acquired for the purpose and directly managed by the State.

> The planting of a further 500,000 acres, chiefly in comparatively small blocks, through private owners and County Councils.
>
> Thus establishing within a period of about eighty years an ultimate forest area in Ireland of at least 1,000,000 acres.
>
> (2) It is important to have it clearly established that, if this work is to be done, action must be taken primarily by the State. A national scheme of afforestation cannot be undertaken by private individuals... such help as can come from them in the preservation and development of the existing woodland area can be effective only if it is given in close cooperation with a scheme *managed* by the State.
>
> ...The 306,000 acres of existing woods must be regarded as an area part of which would come into the possession of the State and the County Councils, and part of which would remain in the possession of the present owners. But whether it be dealt with by the public authorities or by private owners, it must be considered, *as regards its management and its extension*, in a category apart.

The Report went on:

> Our national scheme of 1,000,000 acres contemplates the addition of a new forest area of 700,000 acres. Of this amount 200,000 to 300,000 acres would be State forests directly created on the larger tracts of mountain land by the Forestry Authority. The remaining 500,000 acres *must* come mainly in the form of extensions of the existing woodlands upon the holdings of tenant purchasers which are larger than 100 acres, upon demesnes, upon inferior pasture or mountain land in the hands of larger proprietors, and upon lands in connection with woods acquired by the County Councils and the Central Authority...
>
> Although, as we have pointed out, a considerable proportion of the existing woods must pass from their present proprietors, the probability is that the greater portion will remain in their hands ... In every country where attention is paid to State forestry, the work of the State is invariably accompanied by the development of private enterprise in forestry operations, and private enterprise is in various ways encouraged and facilitated by the State. As a matter of fact, the proportion of forests in the hands of private landowners greatly exceeds in most countries the proportion in the hands of the State...

In a memorandum to the Committee (Appendix 22 of the Report) J. L. Pigot, Chief of the Mapping and Survey Branch of the Land Commission, said: 'Once the fee-simple of the land is vested in a tenant-purchaser, the latter has full right to fell and sell timber growing on the land, unless, in so doing, he lowers unduly the value of the land as security for the purchase money advanced. Yet, even if the Land Commission has the right to intervene it lacks the power to prevent. Consequently, trees on purchased holdings are being felled at an increasingly rapid rate, and often in sheer ignorance of the real value of the timber...'

In its Report (Par. 55) The Committee made the following powerful and compelling observation under this heading: 'Exceptional Obligation on the State in Ireland ...in Ireland there are special and peculiar reasons which put the idea of any other method of dealing with the problem entirely out of consideration. The State has a responsibility which does not obtain elsewhere. At this moment the process of destruction of the woodlands which is going on... is due to the legislation of the State, and, as we have already pointed out, this grievous waste of the woods, with its menace to industries depending on them, must continue unless the effects of this legislation are checked by further State action. But in the past it may be broadly stated that the excessive reduction of the woodland area of this country is due either to what the State has done or to what it has neglected to do... Throughout the 17th and 18th centuries the grantees were allowed to do with timber as they pleased, and what they pleased was in the main to realise with reckless extravagance, with the result that the greater part of the country for which at least shelter might have been preserved, is in its present bare and wind-swept condition. Had provident and intelligent government action been applied to the subject in Ireland, undoubtedly the forest area and the general agricultural wealth of the country would be in a far better position than they now are. It is, moreover, an important factor in the case that the proceeds of the Quit and Crown Rents, which are entirely derivable from these lands and which have yielded a return of upwards of £60,000 a year, have never, since the union of the Irish and the British exchequers, been directly spent in Ireland or applied to Irish purposes, but have been, with the general crown revenues, invested in Great Britain, sometimes even in promoting forestry. There is finally the fact that the State, in abolishing the landlord through the Purchase Acts, is bound to provide and has not provided a machinery to discharge his functions in respect of several matters, including woods, which cannot be left to individual tenant purchasers, and in which the general community, as well as the tenant purchasers, has now a specific interest. Having regard to what is at present occurring in the country, we cannot hesitate to say that, not only does the responsibility lie on the State for taking action, but that if action be not taken at once it will mean a gross neglect comparable with the improvidence of the past and far less excusable.'

It is worth remembering that this statement was made in a context that conceived of a programme of national afforestation combining State and private woodlands under one national umbrella.

The Committee's conclusions and recommendations were summarized and the following is a summary of that Summary:

(1) That forestry has been deplorably neglected by the Government in Ireland with the result that the country, particularly well-suited for tree-growing, had the lowest percentage of any country in Europe; that this was too low for the welfare of the country and that, nevertheless, it was being wastefully diminished chiefly under the influence of the Land Purchase Acts 'with great loss to the country and imminent danger to existing Irish wood-working industries'. That with the transfer of land under the Acts to peasant owners the State had an obligation in respect to existing woods which was not effectively provided for in these Acts.

(2) That a comprehensive scheme of forestry (of at least 1,000,000 acres) was essential for the industrial and agricultural requirements of the country and could only be carried out by or under the direction of the State.

(3) That an exceptional opportunity, which cannot recur for acquiring land suitable for forestry, and not so suitable for any other use, presents itself now in Ireland, in connection with the Land Purchase Acts.

(4) That such a scheme, including the preservation and extension of existing woods, and the creation of a new forest area, is a sound investment for a nation, necessary for her agricultural and industrial development, and for the provision of an important capital asset which must otherwise be wasted.

The recommendations go on to outline how all this might be achieved, administered, costed and financed (Report of the Departmental Committee on Irish Forestry, Part X).

The Concluding Remark of the Report also contains some strong words: 'In presenting our Report we desire to express in the strongest way our sense of the obligation which lies upon the State to act immediately in this matter. Grievous mischief, loss, and waste, accompaniments of legislation and other State action, are going on, and ought to be checked without delay. Furthermore, as another accompaniment of legislation a great opportunity for husbanding and developing one of the resources of the country is now available, and if it be allowed to pass without being used, it means not only missing possible gain to the country but producing further actual loss. Legislation is not needed in order to carry out the major part of the reforms we suggest...'

APPENDIX 8

Some Senior Officials Responsible for Forestry since 1904

1904-06: – Department of Agriculture and Technical Instruction (DATI), Secretary, T. P. Gill; Assistant Secretary responsible for Forestry, J. R. Campbell; Senior Inspector of Technical Instruction, G. Fletcher; later W. Vickers Dixon.

1907-18: – (DATI), Secretary, Gill; Asst. Sec., Campbell; Forestry Expert (Chief Forestry Inspector from 1912), A. C. Forbes.

1919-22: – British Forestry Commission (DATI officers seconded therefrom) Gill, Campbell and Forbes (for that period Assistant Commissioner for Ireland).

1923-4: – DATI, Secretary, Gill; Asst. Sec., Campbell; Assistant Commissioner (later, 1925, Director), Forbes.

1925: – Department of Lands and Agriculture, Secretary, F. J. Meyrick; Asst. Sec. H. G. Smith; Director, Forbes.

1926-33: – Department of Agriculture, Secretary, F. J. Meyrick; Asst. Sec. H. G. Smith; Director, A. C. Forbes; Chief Inspector, John Crozier.

1933-4: – Department of Lands, Secretary, H. E. M. Bradley; Acting Director, Crozier.

1934-6: – Department of Lands, Secretary, Bradley (later M. Deegan); Acting Director M. L. Anderson.

1936-9: – Department of Lands, Secretary, Deegan; Director, Otto Reinhardt.

1940: – Department of Lands, Secretary, Deegan; Director M. L. Anderson; Chief Inspector, J. A. K. Meldrum.

1942-6: – Department of Lands, Secretary, Deegan; Director M. L. Anderson; Chief Inspector, J. A. K. Meldrum.

1947: – Department of Lands, Secretary, Deegan; Director, J. A. K. Meldrum.

1948: – Department of Lands, Secretary, Deegan; Director, Meldrum; Chief Inspector, P. Barry.

1949: – Department of Lands, Secretary, Deegan; Director, Meldrum; Chief Inspector, T. Manning.

1950-1: – Department of Lands, Secretary W. F. Nally; Director, Meldrum; Chief Inspector S. M. O'Sullivan.

1952: – Department of Lands, Secretary, Tim O'Brien; Director Meldrum; Chief Inspector, O'Sullivan.

1953-6: – Department of Lands, Post of Director abolished. Secretary O'Brien; Asst. Sec. (Forestry), S. Mac Piarais; Chief Inspector, O'Sullivan. H. J. Gray appointed principal officer in 1955.

1957: – Department of Lands, Secretary, O'Brien; Asst. Sec. (Forestry), Mac Piarais.

1958-63: – Department of Lands. Secretary, O'Brien; Two Inspectors General appointed, S. M. O'Sullivan and T. Manning.

1964-5: – Department of Lands, Secretary, O'Brien; Asst. Sec. (Forestry), Mac Piarais; Inspectors General, O'Sullivan and Manning.

1966: – Department of Lands, Secretary, O'Brien; Inspectors General, Manning and M. S. Ó Fiacháin.

1967	– Department of Lands, Secretary, O'Brien; Asst. Sec. (Forestry), H. J. Gray; Inspectors General, M. S. Ó Fiacháin and T. McEvoy.
1968-71:	– Department of Lands, Secretary, O'Brien; Asst. Sec. (Forestry), Gray; Inspectors General, McEvoy and Ó Fiacháin.
1972	– Department of Lands, Secretary, O'Brien; Asst. Sec., Gray; Inspectors General, McEvoy and Ó Fiacháin.
1973	– Department of Lands, Secretary, O'Brien; Inspectors General, McEvoy and Ó Fiacháin.
1974-7:	– Department of Lands, Secretary O'Brien; Posts of Inspectors General abolished. Chief Inspector, T. McEvoy.
1978	– Department of Fisheries, Secretary, A. W. Duggan (Forest and Wildlife Service); Chief Inspector, McEvoy; Asst. Sec. (Forestry), J. K. McGerty.
1979-80:	– Department of Fisheries and Forestry, Secretary, Duggan; Asst. Sec. (Forestry), H. J. Sullivan; Chief Inspector, McEvoy.
1981	– Department of Fisheries and Forestry, Secretary, Seamus de Paor; Asst. Sec. (Forestry), Patrick Callanan; Chief Inspector, McEvoy.
1982-3:	– Department of Fisheries and Forestry, Secretary, de Paor; Asst. Sec., Callanan; Chief Inspector, Niall Ó Muirgheasa.
1984-5:	– Department of Fisheries and Forestry, Secretary, Patrick Whooley; Asst. Sec., Callanan; Chief Inspector, Ó Muirgheasa.
1986	– Department of Tourism, Fisheries and Forestry, Secretary, Whooley; Asst. Sec. (Forestry), Thomas Rea; Chief Inspector, Ó Muirgheasa.
1987	– Department of Marine, Secretary, Whooley; Asst. Sec., Rea; Chief Inspector, Dr Niall O'Carroll.
1988	– Department of Energy (Forest Service), Secretary, J. C. Holloway; Asst. Secs. (Forestry), Rea and Paul Cassidy; Chief Inspector, O'Carroll. Announcement of Coillte and transfer of sundry staff preparatory to vesting day, January 1989.
1989	– Department of Energy, Secretary John Loughrey; Coillte, Chief Executive, Martin Lowery.

APPENDIX 9

Ministers Responsible for Forestry since 1904

1904-21: Department of Agriculture and Technical Instruction (DATI).
Vice Presidents
- Horace E. Plunkett (1900-07)
- Thomas W. Russell (1907-19)
- Hugh T. Barrie (1919-21)

1919-22: (First and Second Dáils)
Ministers for Agriculture
- Robert Barton (1919-21)
- Art O'Connor (1920- deputizing during Barton's imprisonment) (1921-2)

1922-32: Saorstát Éireann (Irish Free State).
Minister for Agriculture
- Patrick Hogan (1922-32)

1932-48: First Fianna Fáil Administration.
Minister for Agriculture
- James Ryan (1932-7)

(Forestry Transferred to Lands, 1 December 1933)
Ministers for Lands
- Joseph Connolly (1933-6)
- Frank Aiken (1936 five months)
- Gerald Boland (1936-9)
- Thomas Derrig (1939-43)
- Sean Moylan (1943-8)

1948-51: First Inter-Party Government.
Minister for Lands
- Joseph Blowick (1948-51)

1951-4: Fianna Fáil Administration.
Minister for Lands
- Thomas Derrig (1951-4)

1954-7: Second Inter-Party Government.
Minister for Lands
- Joseph Blowick (1954-7)

1957-73: Fianna Fáil Administration.
Ministers for Lands
- Erskine Childers (1957-9)
- Michael Moran (1959-68)
- Padraig Faulkener (1968-9)
- Sean Flanagan (1969-73)

1973-7: Coalition Government (Fine Gael/Labour).
Ministers for Fisheries and Forestry
Tom Fitzpatrick (1973-6)
Patrick S. Donegan (1976-7)

1977-81: Fianna Fáil Administration.
Ministers for Fisheries and Forestry
Brian Lenihan (1977-9)
Patrick Power (1979-81)

1981-2: Coalition Government (Fine Gael/Labour).
Minister for Fisheries and Forestry
Tom Fitzpatrick (1981-2)

1982-6: Fianna Fáil Administration.
Minister for Fisheries and Forestry
Brendan Daly (1982-6)

1986: Coalition Government (Fine Gael/Labour).
Minister for Fisheries and Forestry
Liam Kavanagh

1987-9: Fianna Fáil Administration.
Minister for Fisheries and Forestry
Patrick O'Toole (1987)
Minister for Marine (and Forestry)
Brendan Daly (1987)
Minister for Energy (and Forestry)
Ray Burke (1987-8)
Minister of State for Forestry
Michael Smith
Minister for Energy (and Forestry)
Michael Smith (1988-9)
Minister of State for Forestry
Liam Aylward (1989)

1989: Coalition Government (Fianna Fáil/Progressive Democrats).
Minister for Energy
Robert Molloy (1989-)

Notes
Index
Select Bibliography

ABBREVIATIONS

CAS Report - Centre for Agricultural Study, Strategy for the United Kingdom, Forest Industry, CAS Report 6, University of Reading

CSPI - *Calendar of State Papers relating to Ireland.*

DCIFR - Departmental Committee on Irish Forestry (1908), Report

ED - *Economic Development*

FAOR - Food and Agricultural Organization of the United Nations: Report (Cameron) on Forestry Mission to Ireland (15 February 1951)

FIPAP – Forestry in Ireland, Policy and Practice

IFP - Convery, Frank, Irish Forest Policy (1979), National Economic and Social Council (no. 46), S.O., Dublin

MRIA – Member of the Royal Irish Academy

NDP – National Development Plan

OFP – Operational Forestry Programme

PRIA - Proceedings of the Royal Irish Academy

RRGF - Report of the Review Group on Forestry

RSAI – Royal Society of Antiquaries of Ireland

TFI - *The Forests of Ireland*

TCFF - *The Case for Forestry*

NOTES

CHAPTER ONE

1. The first forest phase in Ireland, marked by the spread of pines and birchs, was known as the pre-temperate period. Improvement in soil conditions following the Ice Age enabled trees to survive into the early temperate period; species such as oak, elm, hazel, alder and ash proliferated. Temperature and soils again deteriorated during the late temperate period and mixed oak forests declined. Hornbeam, beech, silver firs and spruce increased. The post-temperate period saw a further decline in temperature and deterioration of soil quality and forests of pine, birch and spruce developed. The present interglacial period began about ten or twelve thousand years ago and by 8000 BC the country was free of ice. By 5000 BC the country was virtually covered in forest and the climate was 1 to 2 degrees warmer than it is now. Many species previously found in this country did not develop at this time, even though they are found in Britain and on the European mainland, e.g. beech, hornbeam and the limes, spruce and fir.
2. 'The Forests of Ireland', *The Society of Irish Foresters*, (ed.) Niall O'Carroll (Dublin 1984), p. 18.
3. Sonia Cole, *The Neolithic Revolution* (London 1959), pp. vii/viii.
4. Frank Mitchell, *The Irish Landscape* (London 1976), p. 103.
5. *Ibid.*, p. 104.
6. *Ibid.*, p. 106.
7. 'The Forests of Ireland', p. 31.
8. Mitchell, *op. cit.*, p. 116.
9. *Ibid.*, p. 119.
10. Case, 'Land Use in Goodland Townland, Co. Antrim from Neolithic times until today,' *Journal of the Royal Society of Antiquaries of Ireland* (1969), p. 99.
11. Mitchell, *op. cit.*, p. 120.
12. *Ibid.*, p. 117. He adds: 'In any case he was going to till the soil with a mattock and not a plough...'
13. Carl Sauer, American geographer.
14. Cole, *op. cit.*, p. 59.
15. Mitchell, *op. cit.*, p. 122.
16. The quotations in this passage are from *Celtic Heritage*, Alwyn and Brinley Rees (London 1961).
17. *Ibid.*
18. *Ibid.*
19. *Revue Celtique*, vol. XXII, no. 9.
20. P. W. Joyce, *A Social History of Ancient Ireland* vol. I, part II, chap. 9, sec. 5.
21. A. T. Lucas, 'The Sacred Trees of Ireland,' *Journal of the Cork Historical and Archaeological Society*, 1962-3, p. 34.
22. Giraldus Cambrensis, *Topographica Hibernica*.
23. *Tripartite Life*, p. 185.
24. Joyce, *op. cit.*
25. Lucas, *op. cit.*, has a valuable commentary on trees and sacred sites: 'The Book of Armagh (8th century)... tells us that St Patrick founded a church near the Bile Tortan. Under the year 995 both the Annals of the Four Masters and the Annals of Ulster have entries which state that the monastic town of Armagh was set on fire by lightning and that, along with the buildings, the sacred wood (fidnemedh) was burned... The neimed is "probably a small chapel or oratory (or else the spot on which such stood)". It is used for ecclesiastical land assigned to a church in 1148 and again in 1196 in an obituary notice of

the heir presumptive to the high kingship, who is described as the "founder of churches and fair sanctuaries (*caoimhneimheadh*)". In the Laws a *neimed* is declared to be an inviolable sanctuary. From these definitions it might appear as if this particular wood (*fid*) in Armagh was called a *fidnemedh* merely because it happened to be near or within the sanctuary (*neimed*). However, the compound *fidnemedh* is reasonably frequent in other contexts; *neimed* is well known from Celtic areas on the Continent where it was often applied to sacred groves. There appear, also, to have been sacred woods independent of ecclesiastical sites and one such is mentioned as attaching to a dun or fortified homestead. The term used for it is *defhid* or *deid*, which also contains the element *fid* and which is sometimes, apparently, synonymous with *fidnemedh*. Moreover *fidnemedh* is explained in one place in the laws as *fid cilli*, "wood of a church", but there is no rational reason why a wood should be annexed to a church as a matter of course. It seems to the writer that the only reasonable deduction from these data is that the wood at Armagh derived its name not from its accidental contiguity to a church sanctuary but from a *fidnemedh* or sacred grove which existed there from pre-Christian times... A passage in the canon of the Laws, *im lot do aibinne*, has been translated as "for injuring thy meeting hill". Another version of the commentary, however, runs: "Thy aimmine, i.e. seat, or mound or beautiful tree, &c (*do aimmine .i. suideach, no fert no bile cain agurl*)". Whatever the exact renderings which should be given to the various technical terms in the above contexts, it seems reasonably certain that they refer to places of assembly, some of which may also have been inauguration sites, each such place, in the mind of the ancient writer being associated with a noteworthy *bile* tree which evokes the appelation "beautiful (*cain*)". The Laws also specify "distress of ten days... for the appropriated tree which is in the forest (*im crand ngabala bis i mdithribh*)," the commentary on this containing the following extraordinary details which, at first sight, seem to refer to some sort of topiary but, more probably, have to do with the lopping of the branches before or after felling, couched in the *mathematical form* beloved by the ancient jurists: "i.e. the crossed tree, its stay is ten days, that of its first shaping five days, that of its first preparation three days, that of its full preparation one day (*.i. in crann crosta, ar dechmaidh, se cet cepta ar cuicti, cet urlum ar treisi, lan urlum ar aine*)"'.

26. *Ibid.*
27. *Ibid.*
28. Joyce, *op. cit.*

CHAPTER TWO

1. F. C. Osmaston, *The Management of Forests* (London 1968), p. 307.
2. Thomas Hinde, *Forests of Britain* (London 1985), p. 19. It is uncertain when the word 'forest' was introduced to England. Its origins are Latin and Old French. The Normans brought it to Ireland, but it was not in common use in this country until the English language became commonplace. In England related words were once in common use, but have changed their meaning; the word 'lawn', for instance, originally meant an open grass glade for the benefit of deer within a woodland. Lawns were tended by a forest official known as a 'lawnder'. The word came to mean, successively, any level area of mown grass and, finally, the common or – no pun intended – garden lawn.
3. *Ibid.*
4. A. C. Forbes, 'Some Legendary and Historical References to Irish Woods and their Significance', *Proceedings of the Royal Irish Academy*, 1933.

5. Mitchell, *op. cit.*, p. 135.
6. Eoin MacNeill, *Celtic Ireland*, p. 170.
7. *Ibid.*, p. 167.
8. Vol. v., p. 483. cited MacNeill, *op. cit.*, p. 167.
9. Fergus Kelly, 'The Old Irish Tree List', *Celtica* 11, 1976. This tree-list may have originated in the lost law-tract *Findbretha* (Tree-judgments) referred to in the 7th *Bechbretha* (Bee-judgments).
10. *Ibid.*
11. *Ibid.*
12. *Ibid.*
13. Kelly, as follows: 'Mr Oliver Mooney (now retired) ... of the Forestry and Wildlife Service observed the cutting of holly branches to feed cattle at Glencar, Co. Kerry, about 35 years ago' (1976).
14. Mitchell, *op. cit.*
15. *Ibid.*
16. Mitchell, *PRIA*, 57. B, p. 232.
17. Kelly, *op. cit.*, p. 111.
18. *Ibid.*, p. 111.
19. Gerard Murphy, *Early Irish Lyrics*, 8th-12th century (Oxford 1956).
20. Forbes provides the following commentary on the inclusion of the pine:

'Octach, a pine, is given in connection with *in crand giuis* ridge-pole, and another phrase used is *ochtae, a bi a tulcha*, which O'Donovan renders as "pine: its being in the puncheon". Pine has no definite connection with "ridge-pole", which latter may have been any material of either metal or wood, and one of O'Donovan's glosses is *fo octae findruinne*, which can be translated as "under a roof-tree of bronze". *Tulcha*, which O'Donovan translates as "puncheon", would rather suggest the word "cup", and a Scottish-Gaelic speaker interprets this as meaning "cup-shaped roofs"... While it is difficult to trace a logical connection between the words "roof-tree" and "puncheon", there is a plausible theory that the Irish phrase might well be translated as "roof-tree: that is found in the cup-shaped roof". This roof-tree was probably the centre-pole or one of the main support beams of the building. With the exception of the word "octach", no difficulty exists in making out the meaning of the other trees given in the Irish text, or the reason for their inclusion as "chieftain trees".

'Dr E. J. Gwynn, to whom the point was referred, wrote as follows:
"...There seems to be no doubt that *octach* is used in *Laws*, Vol. IV, pp. 146, 148, 150, as the name of a forest tree, being classed along with oak, hazel, holly, yew, ash and apple. It seems to occur also in an early poem in the form *octgach* as the name of a tree (Meyer, *King and Hermit*, st. 4)". Everywhere else that I have met the word it is used to denote part of the structure of a house. Instances:

1. *Cormac's Glossary* gives a fanciful explanation, saying that "the house is more perfect for being upon it" (i.e., it supported the house in some way).
2. *Irische Texte*, ed. Windisch, *vol. i*, p. 286, 1.12: "Conall throws the wheel up to the *ochtach* of the King's house" (the roof, or part of it).
3. *Irische Texte, vol.* 2, pt. 1, p. 179, first line: "Houses with *ochtacha* of white metal."
4. Story of 'Tain Bo Fraich' in RIA Irish MSS. Series, 1870, p. 140, 1.4: "Four *ochtga* (plural) upon (over) the couch of Ailell and Medb." Seemingly four poles.
5. *Anecdota from Irish MSS.*, vol. 1, p. 5, line 2: "though the full of the house

up to the *ochtach* were given, it (a precious bracelet) shall not be sold for it". Here the octach again is the roof, or part of it.

6. *Silva Gadelica*, I, p. 75, line 29 (Transl. 11, 80): "'tis the *octach* of the house – the ridge-pole – that shall fall on thy head". Here *octach* is equated with *feicc* (or *feice*), which O'Grady translates by "ridge-pole": it is elsewhere rendered by *mullach in tige*, "top of the house".

'Perhaps the word originally meant "a pine" (what sort?) and then "a beam of pine-wood", especially one used to support the roof.'

The precise meaning of the word '*octach*' is not in dispute. It is its common or colloquial use in particular circumstances that is in question. One wonders what, on the basis of similar evidence only, future analysts might make of such phrases as: 'He made the rafters ring', or 'He threw it up to the rafters' to say nothing of 'He raised the rafters', or the word 'roof-beam' used for concrete? Is it possible that the enduring (and abundant) bog-pine contributed to the confusion about '*octach*' as supplies of fresh pine became exhausted? Or that it derives from common vernacular semantics, the word moving from the specific to the general, with altered meaning as so often happens in the vernacular?

21. Mitchell, *op. cit.*, p. 177.
22. Osmaston, *op. cit.*, p. 307.

CHAPTER THREE

1. MacNeill, *Celtic Ireland*, p. 169.
2. *Ibid.*, p. 147.
3. Hinde, *Forests of Britain*, p. 111.
4. Osmaston, *The Management of Forests*, p. 309.
5. Hinde, *op. cit.*, p. 23.
6. FitzNigel, *English Historical Documents*, Vol. III, eds. David C. Douglas and George W. Greenway (1953); Manwood, 3rd edition, 1665; Turner, *Select Pleas of the Forest*, 1901.
7. Calendar of State Papers of Ireland, *Sweetman's Calendar*, 1171-1307, Patent Rolls of Henry III.
8. Forbes, 'Some Legendary and Historical References to Irish Woods, and their Significance', *Proceedings of the Royal Irish Academy*, 1932, p. 18.
9. CSPI.
10. *Ibid.*
11. CSPI, *Sweetman's Calendar*.
12. Gilbert's *Historical and Municipal Documents of Ireland*, pp. 539, 540: 'Grant to the Church of Dublin and Luke, Archbishop-Elect thereof, and his successors, that all woods in the following lands be disafforested, namely, Sauf-Kevyn, Ferchir and Coillac, all of the See of Glendalough and of Thomas, Abbot of Glendalough. These lands extend in the east towards the King's land of O'Brun and Othee; on the south to Wykingelow, land of the baron of Nas; thence to Arclow, land of Theobald Pincerna; thence to O Mail, land of Philip Fitz-Resus, of the demesne of Walter de Rydelsford; on the west to the land of Nas; thence to Rathmor, land of Maurice Fitzgerald; on the north to the King's land of Tach Saggart; thence to Belach, land of Walter de Rydelsford; thence to Cruack (Cruagh), land of Richard de St. Michael; and thence by the brow of the mountain to Senchal, land of the Archbishop of Dublin. All those who have woods within these metes, although formerly in the forest, may enclose them, make parks, and dispose of the woods as they please, have full right of way without interference of the foresters: the woods, when assarted, shall be

quit of waste, regard, and view of the foresters, warders and regarders: all men dwelling within the metes and their heirs shall be forever quit of suits of pleas and amercements of the forest, lawing of dogs, and of all summonses, pleas and presentments belonging to forests and foresters. Witness, &c...
"*Mandate accordingly to Richard de Burgh,
Justiciary of Ireland. December 4, 1229*'.
13. Scots: the name by which the Irish were known until the sixteenth century. Scotia Major, Ireland; Scotia Minor, Scotland.
14. Hovedon, Roger of, thirteenth-century English chronicler.
15. Joyce, *op. cit.*
16. *Ibid.*

CHAPTER FOUR

1. Osmaston, *The Management of Forests*, p. 313.
2. Herbert Francis Hore, 'The Woods and Fastnesses of Ancient Ireland', *PRIA*, 1931.
3. Mitchell, *The Irish Landscape*, p. 185.
4. *Ibid.*, p. 185.
5. Hore, *ibid.*, p. 148.
6. Mitchell, *op. cit.*, p. 190.
7. Forbes, 'Some Legendary and Historical References to Irish Woods and their Significance', *PRIA*, 1931.
8. N. D. G. James, *A History of English Forestry*, p. 35. While there is no specific reference extant concerning Roman policy on forestry in Britain, it may be assumed that individuals selectively applied established Roman principles of forestry there. The Romans distinguished between coppice and high forest, regulated amounts cut and introduced penalties for cutting and damaging trees. Pliny maintained records of rotation using coppice – eight years for sweet chestnut and eleven years for oak. In the Roman influence may lie the reason for the English readiness to accept subsequent alien forest legislation so readily. The Romans extended the cultivated land of Britain at the expense of its trees. During the Roman occupation, which lasted for 350 years, England became an important grain-producing country and large areas were cleared for this purpose. Elsewhere, especially on the Welsh and Scottish borders, woodlands were cleared for military reasons. The Romans also brought wood-consuming industries to England – iron-smelting and forging, and pottery-making. Hence the expansion of tree-cultivation, probably coppicing, in England at an early date. An analysis of Roman charcoal suggests that they were burning uniform-sized wood, in turn suggesting coppicing. There may be other explanations. In Saxon times certain areas of woodland were defined as 'perambulations', and were included in charters.
9. Venison offences were not simply straightforward poaching; receiving meat or giving refuge to offenders were included. Death or mutilation and trial by ordeal were at one time frequent for poaching offences, especially during the reigns of William I and William II. Penalties eventually became so light that a poacher was likely to be less severely punished by a forest court than by a Court of common law or a manor court.
10. Osmaston, *op. cit.*, p. 312.
11. Calendar of State Papers, Elizabeth, Vol. II, p. 196. See also Arthur D. Innes, *England Under the Tudors* (London 1905). It was not a new policy. 'At the close of the thirteenth century a law was enacted for cutting passages through the

forests, and the policy of felling the woods as a military measure was afterwards pursued by the English on a gigantic scale during the wars of Elizabeth and in the long peace that followed.' William Lecky, *History of England in the 18th Century* (London 1878-90).

12. James, *op. cit.*
13. Note in the State Papers Office, London, accompanying a map of Co. Down dated 1566.
14. Hore, *op. cit.*, p. 145.
15. An English or Scottish planter on forfeited lands in Ireland who 'undertook' certain 'developments', or one with a franchise to do so. The Phillip Cottingham mentioned was a London carpenter and his account for 'money by him disbursed for hewing and carriage of timber and planks in the woods of Kilbarro and Kilcoran in the County of Waterford' is recorded in the CSPI, James I, Vol. II (p. 255), as amounting to £71.3.4½d. Cottingham was sent back to Ireland in 1623 to inspect the woods of Connacht – Dr Lyons, *Journal of Forestry* Feb., 1883, p. 661, Hore, *op. cit.*
16. *Ibid.* The word 'coal' was originally used to describe charcoal (the carbon residue obtained by partially burning wood in the process known as charcoal-burning), and the name 'colliers' was applied to those engaged in charcoal-burning and not to those working in coalmines. Coal as we know it today was called either 'sea-coal' or 'pit-coal', depending on whether it had been transported by sea or was dug from a pit (not a mine).
17. In 1283 William le Devenies, Keeper of the King's Demesne Lands and Remembrancer in the Exchequer of Dublin (at 5p per day, 'equivalent to the wages of a good carpenter'), was permitted to have twelve oaks fit for timber as the king's gift from the king's wood in 'Glinery' to build a house at St Audeon's parish ('then and long after the best part of Dublin'). The house was 'fitted up with the latest sanitary improvements, obtaining a supply of water from the city conduit, at a rent of sixpence a year, by a pipe the thickness of a goose-quill', Le Fanu, *op. cit.*, p. 274.
18. An Act that Timber shall not be felled to make coals for burning of iron I Eliz. I, C.15. Almost one hundred years later, in 1655, following Cromwell's ruthless and pragmatic campaign, a similar instruction was issued in Ireland prohibiting the felling of trees 'within four miles of the sea, or one mile of the Shannon, or of any other navigable river, no oak timber fit to be used for shipping to be felled or cut down' – instructions issued to the Commissioners at Loughrea, 16 June 1655 – Dr Lyons, *op. cit.*, p. 660.
19. McCracken, *op. cit.*, p. 29.
20. Charles Smith, *Antient and present State of the County and City of Cork*, 1751.

CHAPTER FIVE

1. Forbes, see 'Some Legendary and Historical References to Irish Woods, and Their Significance', *Proceedings of the Royal Irish Academy*, 1932.
2. Robert Greenhalgh Albion, *Forests and Sea Power, the Timber Problems of the Royal Navy 1652-1862*, Introduction, p. x.
3. *Ibid.*, p. viii.
4. Mitchell, *op. cit.*, pp. 193-4. 'Thus began a system of devastation which we have rarely if ever seen equalled in the history of any country, the effects of which, though extensive, were not publicly cognisable in the turmoil and trouble which followed on the confiscations of Cromwell' – Dr Lyons, MP, *The Forestry Journal*, Feb. 1883, p. 656.

5. Spenser, *The Faerie Queene*.
6. Mitchell, p. 195. The anxiety behind much of the early legislation was caused not so much by the shortage of timber (though this later became a factor), but by the fact that timber which, in the opinion of the legislators, should have been employed for ship and house-building purposes was, in fact, being used for the more profitable stave trade.
7. In *The Irish Woods Since Tudor Times* McCracken has a comprehensive estimate of the extent of the woodland in Ireland between 1600 and 1800, which I summarize: – 'The largest and densest areas of woodland lay to the north-west of Lough Neagh, in the Erne basin, along the Shannon, in the river valleys of the west and south, and on the eastern slopes of the Wicklow and Wexford hills. Smaller but significant areas of wood existed...' (elsewhere). In Munster 'the extent of forest was almost legendary'. These areas of woodland, which at the beginning of the seventeenth century covered about an eighth of the whole country, were progressively diminished until by the beginning of the nineteenth century probably only one per cent of the land was forested. 'So remote (so far as Planters were concerned) was the Lough Erne district that anything that was irretrievably lost was said to be "got beyond the Arney", one of the rivers entering Lough Erne.' These were the woods eventually to be consumed by the iron works of Boyle, Arigna, Drumsna, Dromod and Drumshanbo, the country of Coill Conor, which gave McDermott Roe his title 'lord of the woods'. During Phillip O'Sullivan Beare's retreat from Kinsale he and his people took an entire night to pass through a wood near Ballinlough in Roscommon. A few years earlier Essex found it more convenient to build a fleet of ships to take his soldiers from the east side of Lough Neagh to the west than to march them round, because the lands were so wooded.
8. Dardis, *The Occupation of Land in Ireland* (London 1902).
9. Crawford, *Depopulation Not Necessary*, p. 12.
10. CSPI, Vol. III & Vol. v.
11. McCracken, *op. cit.*, pp. 30-1.
12. *Ibid*.
13. Arthur Young, *A Tour of Ireland* (London 1780).
14. Some of the last (if replanted) remnants of the Shillelagh Woods fell victim in the 1980s to Irish planning procedures, the framers of which never envisaged the kind of situation that arose. New owners of the Coolattin estate decided, in the 1970s, to develop it for building purposes. The owners sought planning permission to fell the woods, but this was refused on the grounds that the woods were a public amenity. The owners, accordingly, quite properly sought compensation for being forced to maintain a public amenity at cost – if notional – to themselves when to do so was the responsibility of the local authority. The local authority, unwilling or unable to pay compensation, then granted permission to clear-fell part of the ancient oak woods. A similar procedure was followed in respect of a second portion of the wood. When a third, and final, portion was threatened State intervention suspended the process. Interestingly in parallel cases local authorities faced with similar problems have been known to seek to thwart legitimate claims for compensation (where no such national or genuine amenity interest was involved) by manipulating regulations to their advantage in a manner that might be described as angarious. One method is to introduce a spurious, non-compensatory ground of refusal to the overall refusal, which, in law, has the effect of nullifying a claimant's statutory right of compensation to a refusal based on the grounds of public amenity.
15. Le Fanu, 'The Royal Forest of Glencree', *RSAI*, XIII, p. 3.268. In 1589 it was 'recommended that a high steward should be appointed over the royal

manors in Ireland, who should also be "wood-warden and chief forester" in this province' (Munster) – Hore, *op. cit.*, p. 158. Le Fanu writes: 'There was a regularly organized forest department for these counties (Wicklow/Wexford), consisting of a woodreeve, who received a salary of £100 per annum, four assistants at £26, and a clerk at £20. There was also a similar but smaller establishment for the counties of Carlow and Kildare.' These were clearly counties within the influence of the Pale if not always or actually in it. The evidence clearly indicates that it was in this area that forestry tradition in this country was strongest and survives. In 1655 'wood-reeves' were appointed 'to take charge of woods and forests in the King's and Queen's counties, when Richard Wright and John Bryan were designated, with full power and instructions for the care of the woods' – Lyons, *op. cit.*, p. 661.

16. The late John Tyrrell, Arklow.
17. State Papers, *Carew MSS*.
18. Elm is a dense timber which does not easily split. It might also be noted that the elms on the canal banks in Dublin, many of which were felled in the 1980s because of Dutch elm disease, were planted primarily in order to provide urban conduits and water-pipes for the city.
19. William Reid, *Weapons Through the Ages* (London 1984), p. 111.
20. The Chevalier de Latocnaye, *A Frenchman's Walk Through Ireland* (Paris 1798; reprint Belfast 1985).
21. McCracken, *op. cit.*, p. 105.
22. James, *op. cit.*, p. 137.
23. *Ibid.*, p. 135.
24. 'During this period of history corruption was commonplace.' James, *op. cit.*, p. 128.
25. McCracken, *op. cit.*, p. 64.
26. *Ibid.*, p. 28.
27. *Ibid.*
28. Osmaston, *op. cit.*, p. 316.
29. Rock Church, King's Surveyor, *Olde Thrift*, cited James, *op. cit.*, p. 137.
30. McCracken, *op. cit.*, p. 99.
31. *Ibid.*, p. 105 *et seq.* The woods of Glanconkeyne were granted wholly to the city of London in perpetuity in 1609 – CSPI, James I, Vol. II, pp. 88-9.
32. Le Fanu, 'The Royal Forest of Glencree', *RSAI*, p. 271.
33. *Ibid.*
34. McCracken, *op. cit.*, p. 99.
35. Carew MSS 1575-88.
36. T. W. Moody, *The Londonderry Plantation, 1609-41*.
37. McCracken, *op. cit.*, p. 110.
38. *Ibid.*
39. *Ibid.*
40. *Ibid.*, p. 75.
41. *Ibid.*, p. 78.
42. Moody, *op. cit.*, p. 136.
43. Frank Mitchell, 'Littleton Bog, Tipperary', published by the Geological Society of America, 1945.
44. Henry, *The Woods and Trees of Ireland*. The prohibition on the use of gads was reinforced seven years after the Act of 1703, with another in 1710. See Appendix 3 above.
45. McCracken, *op. cit.*
46. *Ibid.*, p. 82.
47. Hennessy, 'Raleigh in Ireland', cited McCracken.

48. McCracken, *op. cit.*, p. 88.
49. *Ibid.*
50. *Ibid.*
51. 'As veins, bedded and as bog deposit' – McCracken, *op. cit.*, p. 90.
52. Departmental Committee on Irish Forestry Report (Appendix 71, p. 442).
53. McCracken, *op. cit.*, p. 93.
54. *Ibid.*, p. 95.
55. Le Fanu, *op. cit.*
56. McCracken, p. 95.
57. *Ibid.*, p. 113.
58. Feehan, *The Landscape of Slieve Bloom*.
59. John Evelyn, *Sylva*.
60. Arthur Standish, *The Common Complaint*, 2nd ed. 1611.
61. Rock Church, *An Olde Thrift newly Revived Wherein is Declared the Manner of Planting, Preserving and Husbanding yong Trees of Divers Kindes for Timber and Fuell... Discoursed in a dialogue betweene a Surveyour, Woodward, Gentleman and a Farmer*, 1612.

CHAPTER SIX

1. Frans G. Bengtsson, researching for *The Long Ships* for instance.
2. Albion, *Forests and Sea Power* (London 1968), p. viii.
3. *Ibid.*
4. John Rushworth, *Historical Collections*.
5. William Marshall, *Planting and Rural Ornament*, Vol. 1 (2nd ed. 1796).
6. James, *A History of English Forestry*, p. 183.
7. Albion, *op. cit.*, pp. 9 and 45.
8. *Ibid.*, pp. 84-5.
9. *Ibid.*, p. 11.
10. James, *A History of English Forestry*, pp. 156-7.
11. But see Albion, pp. 45-6.
12. T. P. Le Fanu, 'The Royal Forest of Glencree', *RSAI*, XIII, p. 2/1.
13. CSPI. Jac. I., Vol. III.
14. *Ibid.*, Vol. IV.
15. Arthur D. Innes, *England Under the Tudors*, pp. 311 and 314.
16. It is said that interpreters were employed on the English fleets at Trafalgar and elsewhere in the eighteenth century to interpret, in Irish, the officer's commands to the (preponderantly Irish) crews.
17. A. T. Lucas, 'The Dugout Canoe in Ireland'.
18. John de Courcy Ireland, *Ireland's Sea Fisheries, A History*, pp. 20-1.
19. Faulkner's *Dublin Journal*, February 1778.
20. Richard J. Scott, *The Galway Hookers* (Dublin 1983), p. 52.
21. According to the late Jack Tyrrell, grandson of the founder. He said that as far as he could remember and as far as his records show there was very little timber in north Co. Wexford in either his lifetime or his father's lifetime suitable for shipbuilding. The unattributed quotations are from a private conversation, 1985.
22. *Ibid.*
23. James, *op. cit.*

CHAPTER SEVEN

1. Patrick G. Dardis, *The Occupation of Land in Ireland*, p. 53.
2. Richard J. Moss, Registrar, Royal Dublin Society, in his submission to the DCIFR, 31 January 1908. See Appendix 4.
3. McCracken, *op. cit.*, p. 124.
4. Jones & Simmons, *The Story of the Saw*.
5. McCracken, *op. cit.*, p. 126.
6. *Ibid.*, p. 135.
7. Not only a traditional heritage from times of acute land hunger. Few farmers will view with equanimity a woodland, the refuge of foxes, rabbits, rats and other pests, on the boundaries of their arable holdings.
8. McCracken, *op. cit.*
9. Forbes, *op. cit.*
10. Dardis, *op. cit.*, p. 2.
11. Parliamentary Papers, 1861.
12. Albion, *Forests and Seapower*, p. 407.
13. Parliamentary Debates, 1862, clxvi.
14. Albion, *op. cit.*, p. 409.
15. Jane Isbet, *The Forester*, Vol. 1 (1905).
16. Dardis, *The Occupation of Land in Ireland*.
17. Crawford, *Depopulation Not Necessary*.
18. Dardis, *op. cit.*
19. Commons papers, Vol. 7, 1825.
20. McCracken, *The Irish Woods Since Tudor Times*, p. 138.
21. *Ibid.*, p. 136.
22. Commons papers.
23. Dardis, *op. cit.*, p. 48.
24. *Ibid.*, p. 31.
25. O'Brien, *The Great Famine*, p. 27.
26. Dunraven, *Legacy of Past Years*, p. 234.
27. Dardis, *op. cit.*, p. 127.
28. Terms of Reference, Select Committee, 1885; Annual Report of Commissioners of Woods and Forests, Great Britain 1905; also *cf.* James, *op. cit.*, p. 192.
29. Annual Report of Commissioners of Woods and Forests, Great Britain 1905.
30. F. C. Osmaston, *The Management of Forests*, p. 319.
31. *Ibid.*, p. 322.
32. Osmaston, *op. cit.*, p. 338.
33. Albion, *Forests and Seapower*, p. 404.
34. The following figures are obtained from the Forestry Report for period 1 October 1923 to 31 March 1925; Report for Period 1 April 1925 to 31 March 1926 (reprinted from the 24th General Report of the DATI for Ireland, 1923-6), as follows: (1) Area of State Forests. – The total area of land acquired by the State for afforestation in Saorstát Éireann up to 31 March 1925 was 29,563 acres... (p. 1): (2) Privately owned Woods. – Woods still in the possession of private owners comprise about 84 per cent of the total woodland area of the country – (p. 11), e.g. 29,563 equals 16 per cent.
35. Forestry, while labour-intensive (though increasingly less so), is also seasonal. Because of the nature of the work it is not deemed to be a practical answer to acute unemployment. In this country, where on-stream forest output is relatively recent, the problems associated with marketing and processing further limit the practicality of considering forestry as a direct substantial alleviator of unemployment, while its potential as an indirect stimulator

NOTES TO PAGES 110-118

of downstream, smallholder and co-operative employment is held to be considerable. Increased afforestation and the maturing of considerable quantities of forest produce should provide considerable local employment.

36. James, *op. cit.*, p. 209.
37. The difference in day-wages for 'an ordinary labourer' and its significance as between England, Wales, Scotland and Ireland in 1883 might be noted:

England	Scotland	Wales	Ireland
2/6d to 4/-	2/6d to 3/4d	2/4d to 2/10d	1/6d to 1/10d

– *Journal of Forestry* 1883, Vol. 7, p. 126 (attached table). Forest produce prices varied comparably, *Journal of Forestry*, Vol. 5, p. 104 (attached table).

38. 'Howitz concluded that five million acres of Ireland's 21 million acres was more fitted for forestry than anything else, and that of these five million acres three million should be planted... Howitz's opinions were more or less in line with those expressed by Sir Arthur Griffith who, in his Valuation Report of 1843, suggested that half of the six million acres of waste and unproductive land could be planted,' McCracken, *op. cit.*, p. 143. Forbes's views on Howitz and his proposals are generally known; he expressed them with characteristic pithiness himself in 1947 in an article in *Irish Forestry* (Vol. IV, No. 1), as follows: 'No doubt Howitz was a respectable man in his own sphere of life, but to regard him as an authority on Irish conditions in general, and on forestry in particular seems bordering on the ludicrous.'
39. John Mackey, *Forestry in Ireland, A Study of Modern Forestry and of the Interdependence of Agriculture and 'Industry'* (Cork 1934), pp. 63-5.
40. In 1883 Dr Lyons's efforts bore first fruits when an experimental forestry scheme was begun at Glencolumcille, Co. Donegal. The land was provided by a Mrs McGlade of Belfast and a number of unemployed local men were engaged for the work of preparation and planting. Dr Lyons was appointed honorary secretary of the 'meeting'. The Bishop of Raphoe, who presided, said that 'the work in which Dr Lyons was engaged and of which he was that day about to make a beginning, appeared to be a work which would lead to very great practical results' *Journal of Forestry* (1883), Vol. 7, p. 69.
41. Forbes, 'Tree Planting in Ireland During Four Centuries', *PRIA*. XLI.
42. C. Litton Falkiner, 'The Forestry Question Considered Historically', paper to the Statistical and Social Inquiry Society of Ireland, 23 January 1903, pp. 16-17.
43. Forbes, 'Some Early Economic and Other Developments in Eire, and their Effects on Forestry Conditions', p. 15.
44. J. F. Durand, unpublished thesis citing File No. 2992/1898, Report 15 Jan. 1891. SPO.
45. Evidence of Henry Doran, Chief Land Inspector of the Congested Districts Board, before the 1907 Committee. Minutes of Evidence.
46. Professor Augustine Henry, Minutes of Evidence, Departmental Committee on Irish Forestry, 4046, p. 166.
47. 'Afforestation in the Budget', *Quarterly Journal of Forestry*, Vol. III, 1909.

CHAPTER EIGHT

1. Report of the Departmental Committee on Irish Forestry (DCIFR). The scope and objects of the 1907 Committee were almost certainly influenced by an earlier Departmental Committee on British Forestry, 1902, which reached broadly similar conclusions in an English context. The terms of reference

of this 1902 Departmental Committee were 'to enquire and report as to the present position and the future prospects of forestry, and the planting and management of woodlands in Great Britain, and to consider whether any measures might with advantage be taken either by the provision of further education facilities, or otherwise, for their promotion and encouragement'. The programme outlined in 1919 for the United Kingdom by the 'Forestry Sub-Committee of the Ministry of Reconstruction', known as the Acland Report, which defined a forest strategy for Great Britain and the means of employing it also diverged from the findings of the 1907 Committee. The principal findings were later included in the British Forestry Commissioner's report of 1943 entitled 'Post War Forest Policy' – a clear indication of the importance attached to them.

2. The Irish economy has been fashioned in great part by the circumstances of the later nineteenth century, a period in which the principles of international free trade were fully applied to this country. 'International free trade favoured those countries which, unlike Ireland, were rich in natural resources – particularly coal and iron, the very foundations of economic progress in the nineteenth century. Ireland also differed from countries which, through colonial conquest, had command of the sources of industrial raw materials on the one hand and of profitable markets abroad for their finished products on the other. Ireland formed part of an imperial system in which her special interests were not safeguarded. The price of agricultural produce and the amount of industrial employment were both determined by the play of economic forces in the world rather than by trends in Ireland. These circumstances influenced both the manner and the extent of economic development in the years between the Famine and the Anglo-Irish Treaty of 1921.' *Commission on Emigration report*, 1948-54 (Dublin 1955), p. 24.

3. Recognition is due to the work of landowners who developed 'Estate Forestry' in the late eighteenth and early nineteenth centuries. They did plant, and much of what they planted consisted of exotic species, which provided important information about performance and yield to later generations of foresters. Moreover they were planting at a time when interest in silviculture and neighbourly rivalry were jointly stimulated by continental techniques. Accordingly modern forestry outlook in this country was, from the outset, more influenced by continental methods than by English practice. Finally, as a third leg for this unusual forest tripod to stand on, was the benevolent climate of Ireland, perhaps the most ideally suited to forestry in Europe.

4. *A Register of trees for Co. Londonderry, 1768-1911*, Public Record Office of Northern Ireland, Introduction by E. and D. McCracken, p. 15.

5. See Report, and Professor Alfred O'Rahilly, 'Notes on' *Forestry in Ireland* by John Mackey, p. 166.

6. *Ibid.*, p. 168.

7. Henry Doran being queried, DCIFR, 2102, 2103, 2104.

8. Schlich estimated that there were 3,900,000 acres of bog, marsh and barren mountain land in Ireland. 'Assuming', he says, 'that one-half of this area is fit and available for planting, an area of 2,000,000 acres in round figures could be converted into forests.' Schlich, *Report on the Afforestation of Great Britain and Ireland*, Pamphlet, 1886, given as Appendix 35, DCIFR. Schlich was not unequivocally enthusiastic about the prospect of forestry in Ireland, stating with caution: 'The effects of the forests in Ireland have been decidedly overestimated' (par. 37). However, having pragmatically stated his case, he also pointed out that afforestation might prove of great value to Ireland in respect of labour (pars 46-52).

9. The absence of scientific findings on the performance of exotic (foreign) species in Irish conditions naturally confined the cautious Schlich's view. He accordingly limited his forecasts to native trees. Native species are today much subordinate to the exotic species that thrive so well and, in the light of the interim scientific experimentation and findings, Schlich's opinion that 'of the indigenous and well-established species, the Scotch fir will be the ruling tree in the dryer localities, and spruce in the moister places', sounds curiously erratic.
10. Durand, *op. cit.*, pp. 8-9.
11. It may be a surprise to note that Schlich, in his report, stated: 'As already considerable outlay has been incurred at Knockboy, I should not abandon the experiment, because, with a comparatively small additional outlay, final conclusions will be arrived at.'
12. Forbes, 'Tree Planting in Ireland During Four Centuries', PRIA, XLI, No. 6., p. 192.
13. Durand, *op. cit.*, p. 38.
14. Forbes, 'The Forestry Revival in Eire', *Irish Forestry*, Vol. IV, No. 1.
15. Undated typescript report of Assistant Commissioner Forbes, addressed to the Secretary of the DATI; Departmental files, p. 6.
16. Council of Agriculture, 5 March 1903, cited Durand, p. 39.
17. John Mackey, *op. cit.*, p. 19.
18. Forbes, 'Tree Planting in Ireland During Four Centuries', p. 174.
19. *Ibid.*, p. 193.
20. The source for this figure of 2,000,000 acres, something of a magic figure, is obscure. In 1883 Dr Lyons MP proposed, out of 'the ample margin of waste' (land) in the country, 'to afforest a C-shaped belt on the north, west and south of about one million acres' (*The Journal of Forestry and Estate Management*, Vol. VI, Feb. 1883). Thereafter it cropped up from time to time until that figure was afforded something like official recognition by the 1907 Committee. In 1900 Dr Robert Cooper gave a lecture in the Rotunda Rooms, Dublin, entitled 'Ireland's Real Grievance, the Deforestation of the Country' in which he proposed a Forestry Department with a programme of two million acres of afforestation as being the desired minimum acreage. The state of afforestation in the Republic today (1991) is approximately half that figure.
21. 'Report on Woods, Plantations and Waste Lands in the South Eastern Counties of Ireland, 1904', John Nisbet, Forestry Division Files – Ref. also Durand p. 79. This is a clear public utterance with regard to State responsibility for afforestation.
22. Report to DATI, 1903, Forestry Division Files.
23. Transactions of the Council of Agriculture, DATI, 28 November 1908 (cited Durand, p. 56).
24. Durand: 'under the 1889 Act DATI could spend money, but it was not empowered to borrow it, with the result that it was not disposed to enter on so large an undertaking as forest ownership'.
25. Arthur Griffith, *Sinn Féin National Programme*, 1905.
26. Durand, *op. cit.*, p. 65.
27. *Ibid.*, p. 68.
28. Gill asked Nisbet to suggest a suitable expert to be employed full-time in Ireland, and he suggested Arthur Charles Forbes who, prior to becoming lecturer at Armstrong College of Science, had been forester-in-charge of the Longleat estate of the Marquess of Bath. Forbes was interviewed on 8 October 1902 and given a four-year contract.
29. *Irish Forestry*, Vol. I, No 1, editorial comment, p. 3.

30. Forbes, *Tree Planting in Ireland*, p. 193.
31. 'Avondale Forestry Station, General Description and Progress of work, 1906-12', Dept. of Agriculture and Technical Instruction, Mr A. C. Forbes, FHAS, MRIA, Inspector of Forestry, p. 2.
32. Forbes, 'Tree Planting in Ireland during Four Centuries', p. 193.
33. 'General Description and Progress of Work, 1906-12'. The eight recruits were to be trained as foresters for private estates, the question of State forestry not having been seriously considered at that time.
34. H. J. Gray, 'The Economics of Irish Forestry', *Statistic and Social Enquiry Society of Ireland*, paper, 20 December 1963, Appendix 1, p. i. Plunkett had lost his seat in Co. Meath with the defeat of the Conservative government in the great Liberal landslide of 1906 and was no longer eligible for re-appointment as vice president of DATI.
35. DCIFR, pp. 3-4.
36. 'In October, 1908 the Vice President of the Department, in answer to one of a series of Parliamentary Questions as to the Government's attitude, said it was true that the Report "had the approval of the whole public". Official response to the Report... was expressed in the provision of an annual Vote of £6000 in each year from 1909/10 onwards to enable more land to be purchased and planted. Meanwhile in Britain the idea of direct State effort in the field of afforestation was gaining some ground. A Commission on Coast Erosion and Afforestation in 1909 put forward the first proposals in that country for State afforestation extending beyond the establishment of demonstration forests... their Report led directly to the power given to the Development Commissioners set up under the Development and Road Improvement Fund Act, 1909, to aid forestry, *inter alia*, by purchasing and planting land. The Development Commissioners, though reluctant to promote large-scale afforestation in Britain, decided to sponsor State Forestry in Ireland, having regard to the very low percentage of woodland in Ireland and the poor prospects of private forest development due to the Irish land-tenure position. The Commissioners recommended to the Treasury advances of up to £25,000 to finance the purchase and planting of land by the State in Ireland and indicated that they would be prepared to recommend advances to cover certain ancillary matters. These proposals were accepted by the Treasury and the Report of the Department of Agriculture and Technical Instruction for 1912-13 records the first land purchases financed by the Development Fund' (H. J. Gray). The only comment necessary is the reminder that in 1909 Ireland was still *de jure* part of the United Kingdom.
37. DCIFR, 32.
38. DCIFR, 34-5. The effects of the 1903 Act on timberlands were greeted with considerable alarm. A double premium was being set on the destruction of trees, 'first by the landlord, who naturally does not wish to convey to the tenant (the purchaser) more than is paid for; and secondly by the tenant purchaser, who seeks to partly recoup himself by the sale of any timber he may happen to have acquired, at the same time to extend and enhance, according to his likes, the value of his arable land – not much else can be looked for unless remedial measures are adopted', *The Quarterly Journal of Forestry* (transactions of the newly formed Irish Forestry Society), Vol. 1. no. 1. p. 46, 1907. 'In addition to the arable land, very much wasteland unfit for agriculture and economically considered unfit for grazing – in fact one might almost say hypothecated by the Creator for planting – is passing into the hands of the tenants, and incidentally is lost probably for ever for afforestation purposes.' The next volume of the *Quarterly Journal* pursued the theme even more trenchantly: 'The Land Purchase Act of 1903 is responsible

for a great deal of denudation and it is really this Act that is to be blamed, and not either the landlord or the purchasing tenant, on whom usually the odium is laid... The landowner must either make what he can by selling the trees to a timber merchant before disposing of his property, or else give them to the tenants, in which case experience had unfortunately shown that they are seldom spared for 12 months... It is unfair to abuse the purchasing tenant, who has never owned a tree in his life, and whose only knowledge of trees is an hereditary grudge against them in that, though exclusively the property of the landlord, they grow fat on his land, if in the first flush of ownership he indulges in the delight of a bargain with a local timber merchant and makes a clean sweep, not only of the hedgerow trees, but of immature or ornamental plantations and of shelter belts and clumps of trees, of which, too late he realises the use, and sincerely regrets', *Quarterly Journal of Forestry*, Vol. II. No. 1, 1908. Forbes, writing of the proposal to acquire 300,000 acres and plant 200,000 of these within forty years, said: 'As regards (this) ... there is no doubt that this was the estimate of the chairman (Gill) and the writer had nothing to do with it. It was far too rapid in both directions ...(and) would have brought in its train all manner of complications, as everyone acquainted with Irish land occupation can understand,' *Irish Forestry*, Vol. IV, No. 1, p. 19. In fact, by 1950, there were 242,343 acres of timberland in the country, of which 142,425 were State stocked and 94,871 were under private ownership. An additional 5000 acres were held by other public bodies. Forbes was also to write: 'Matters of small importance in the abstract often assumed large dimensions when land was touched on...', a point as valid today as it was from the outset, and often overlooked.

39. Durand, p. 94, see Appendix 10, 1908 Report.
40. 'Decision of the Court of the Irish Land Commission on a question of Law as to power of a Vendor to reserve to himself Timber on a holding being sold to the occupying Tenant under the Land Purchase Acts' before the Hon. Mr Justice Meredith, Tuesday, 9 May 1905 – cited DCIFR.
41. J. L. Pigot, Evidence before DCIFR, pp. 34 and 26.
42. *Ibid.*
43. Letter to Denis McGrath, Secretary, Departmental Committee on Irish Forestry, 26 October 1907, DCIFR, Appendix 35, p. 319.
44. DCIFR, 10, p. 4.
45. The origin of the misconception may lie with Forbes. In his undated, typewritten, report to the Secretary of the DATI for the year ended 1922 he states: 'The (1907) Committee recommended a programme of State afforestation, the chief features of which were the planting of 200,000 acres of poor grazing land, the acquisition by the State of demesne woods which were passing into the hands of the Estates Commissioners and a scheme for acquiring and maintaining small woods which could be worked by County Councils.' He makes no reference to the role envisaged by the committee for private forestry in the *national* scheme they proposed, but went on: 'At present the wooded area of the Free State does not exceed 248,500 acres, practically all of which is in the hands of private owners. The owners are under no obligation to maintain their woods or to replant land from which timber has been cleared, nor are they under any obligation to offer the cleared land to the State. The net result is that while the State may be acquiring and planting land wherever possible, a still larger area is being destroyed through the neglect of the individual landowner to repair his waste. Under such conditions it is merely a question of time for the country to become practically treeless, except for a few widely distributed woods which have been formed at great expense

from National Funds.' In Vol. 1 No. 1 of *The Quarterly Journal of Forestry*, of which Forbes was editor, the following appeared in 1907: 'When it is explained that this whole transaction (Land Purchase under the 1903 Act) is taking place at the rate of about a half million acres per annum, and there are a little short of 21 million acres in the whole of Ireland, it will readily be recognised how pressingly urgent is the need of securing at least a portion of the wasteland – preferably by the State – in order to earmark it for timber growing...'

46. DCIFR, Appendix 22 – 'Memoranda on the working of the Irish Land Act, 1903, as bearing upon the question of the destruction of timber trees and plantations, and the reservation to the State of surplus waste lands for forestry and other purposes of general public utility', J. L. Pigot (1) First Memorandum (d), p. 294. Pigot was a former forester, a graduate of the Nancy Forestry School, who had been head of the Ceylon Forest Service.
47. DCIFR, p. 26.
48. DCIFR, Part X, Summary of Principal Conclusions and Recommendations, I. (1), (2) and (3), pp. 57-9.
49. *Ibid.* Concluding Remark, 115. pp. 59/60. In the same year, 1908, three Scotsmen were recruited to the Forestry staff of the DATI, J. D. Crozier, A. McRae and David Stewart.
50. Gray, 'Economics of Irish Forestry', p. 2.
51. A. C. Forbes, 'Tree Planting in Ireland During Four Centuries', p. 196.
52. It is true that because of the decline during the previous forty years or so there was something of a similar appearance there, but, as became quickly evident when the war focused attention on the question, it was a decline in practice only, not in fundamental attitudes towards forestry, whereas in Ireland it was the fundamental attitude that was the problem.
53. In view of the cost, the requirement for long-term management, and the value to the community, the question that this work should be undertaken by the State rather than by private interests was raised, but it was acknowledged to be an extremely radical approach and for this reason the committee offered more acceptable alternatives. In this context the following from 'Commercial Forestry', published by the British Ministry of Reconstruction in 1919 (pp. 7-8), is interesting: 'It is evidently no part of a Government policy to create a State monopoly in forest produce, but rather, while encouraging private enterprise by assistance during the expensive period of creation, to maintain on State property such an area of forests that by efficient and economical management the price of produce can be regulated to some extent in favour of those industries which it is hoped may become more dependent on the home market.

'An alternative to purchase lies in leasing land suitable for afforestation. The advantages of this system are obvious. The owner retains possession of the soil, the land comes under control as it is utilized, so that there is no sudden change in management, and thus initial expenditure is reduced. The upkeep of accurate statistics becomes easy while resident labour is to some extent existent and more may be attracted. The part of the State is confined to afforestation on suitable portions of a given estate.

'Another system provides even more fully for the retention of the owner's interest in the soil. It is that of proceeds sharing, in which afforestation is carried out by the State and the owner contributes a share of the expenditure, in foregoing his rental, in sharing the cost of maintenance or in such other ways as may appeal to him, the proceeds of the undertaking being shared in proportion to the capital invested by either party. For the encouragement of afforestation by the individual by the method of sharing

NOTES TO PAGES 134-136

his risks and providing technical skill this system has undoubtedly great advantages.

'In all of the varying systems above briefly alluded to the State supplies the whole or the major part of the capital to be invested in the scheme. In the first it undertakes the responsibilities of full ownership and must consider the disadvantages that a newcomer may have to encounter. In the other two the owner remains undisturbed, and gives his assistance to the introduction of a new industry. Moreover, he ultimately regains full control of his property, on terms to be agreed upon, and would then find himself in possession of an organized forest area with the necessary experience to continue its management.'

54. It was broken down to the rehabilitation of an estimated 300,000 acres of existing (privately owned) woodland and some 700,000 acres of plantable land, 200,000 acres of which should be acquired by the proposed Forest Authority. Alternatively they proposed that the sum of £80,000 (plus an annual income of £30,824) from unredeemed quit and Crown rents should be used to finance forestry with an additional land annuity of £13,000 a year from the same source, which would enable an acquisition of 50,000 acres. £7000 was estimated as the amount necessary to cover the cost of administration and technical and expert advice on training. Summarised by Durand: 'the Committee hoped to effectively deal with the problem as they saw it of establishing an ultimate forestry area in Ireland of at least one million acres. This they hoped to do by maintaining the 300,000 acres of private woodlands, by the creation of State forests of 200,000 or 300,000 acres and by the planting by private owners of a further 500,000 acres'.

55. This point, that private ownership would be responsible for a major share of the total forestry involved, is of considerable interest. It may be seen how the disparity between it and what later came about, namely State afforestation on a scale approaching 80 per cent, led to something quite different from what the Committee envisaged. This divergence began to be redressed during the 1980s, with the developing interest by commercial concerns in forestry as an investment, and the decision by the government to establish a Forest Authority.

56. Gray, *op. cit.*, p. 1, citing the Report.
57. Durand, *op. cit.*, p. 121.
58. H. M. Fitzpatrick, 'State Action with the Brake On', *Irish Times*, 10 August 1955. This practice has been, and continues to be, roundly condemned by foresters.
59. Resolution of the Association of Municipal Councils to the Secretary of the Local Government Board, June 1907.
60. Gray, 'Economics of Irish Forestry', p. 1.
61. *Northern Standard*, 26 Feb. 1986.
62. Ballyhoura, Co. Cork; Glendalough, Co. Wicklow and Mountrath (then Bunreagh), Co. Leix.
63. Second Report, Royal Commission on Coast Erosion and Afforestation.
64. Augustine Henry, 'Co-operation of State and Citizen in Irish Forestry', *Studies* (December 1924), p. 261.
65. 'Afforestation in the Budget', *Quarterly Journal of Forestry*, Vol. III, 1909, cited James, *op. cit.*, p. 203.
66. 1912 Committee appointed by the British Board of Agriculture and Fisheries to advise on matters relating to the development of forestry.

CHAPTER NINE

1. 'In the meantime (the 1890s) steps were being taken to deal with the Crown Woods and Forests which were now feeling the full effects of the use of iron and steel in place of timber for naval construction... During this period there were large areas of oak, much of which had been planted at approximately the same time and which subsequently had been so heavily thinned that the density of the remaining stock had been reduced to a minimum. The system of thinning that had been followed in the past was designed to produce timber that was suitable for shipbuilding but the demand for oak for the Navy no longer existed and this open-grown timber was of restricted value for other purposes. The following steps were proposed... No further thinning was to be undertaken and the poorer stands of oak were to be clear-felled and replanted with conifers...' – James, p. 196.
2. *Ibid.*, p. 203.
3. Hinde, *op. cit.*, p. 15.
4. James, *op. cit.*
5. Forbes, 'Tree Planting in Ireland During Four Centuries', p. 196.
6. Henry, p. 620.
7. *Ibid.*, p. 642.
8. Report of Royal Commission on Coast Erosion and Afforestation, 1907.
9. James, p. 212. 'Before 1914 it was known that the importation of foreign timber amounted in value to over 45 million sterling annually, and that its price was steadily on the upgrade,' *Commercial Forestry*, pamphlet (British) Ministry of Reconstruction, 1919.
10. An Act for Establishing a Forestry Commission for the United Kingdom, and promoting afforestation and the production and supply of timber therein, and for the purposes in connection therewith, 191, 9 and 10, George V. Cap LVIII. This Act provided for the appointment of assistant commissioners for Ireland, Scotland and England and Wales and also for the establishment of a representative consultative committee for each 'for giving Commissioners in accordance with the provisions of the Order advice and assistance with respect to the exercise and performance by the Commission of their powers and duties under this Act'.
11. James, p. 225.
12. F. C. Osmaston, *op. cit.*, p. 333.
13. Cited Durand, *op. cit.*, p. 143.
14. Introduction of the Forestry Sub-Committee of the British Ministry of Reconstruction (The Acland Committee), 1916. The main recommendation of the Acland Report was: 'In order to render the United Kingdom independent of imported timber for three years in an emergency it is necessary, while making due allowance for an improved yield from existing woods, to afforest 1,770,000 acres. Taking eighty years as the average rotation, we advise that two-thirds of the whole should be planted in the first forty years.'
15. British Forestry Commission, 'Post-War Forest Policy'.
16. Forbes, 'Tree Planting in Ireland During Four Centuries', p. 196.
17. James, p. 217. Under the Forestry Act the first Commissioners were appointed on 29 November 1919, with a statutory provision of £3.5 million for the first decade and with authority over afforestation throughout the United Kingdom, including Ireland.
18. The staff transferred and seconded consisted of Forbes, who became Assistant Commissioner for Ireland from being Chief Inspector, Crozier, Kilmartin and five district officers. In addition there were two foresters, four foremen,

four caretakers, 138 labourers and five boys, besides an administrative staff consisting of one chief clerk, a junior staff officer, an executive officer, a mapping clerk, eight clerks and three messengers. The total area of forest land acquired was 20,000 acres approximately.

19. Sir Henry Beresford Peirse, *The Role of Forestry in the Economy of Britain*, Summary, p. 7. The same point was made in *Commercial Forestry*, Ministry of Reconstruction, 1919: 'In this country it is the creation of forests which is chiefly in question.' The journal (p. 9) went on to emphasize the dominance of the strategic aspect in English forest policy. 'Not ideals, but necessity must be our guide. We require an area from which we can fell in wartime, lasting three years, five times the normal yield, without seriously endangering the future welfare of the forest.'
20. Henry, p. 625.
21. Professor Clear believes the Matthew J. O'Byrne mentioned (source: FWS) to be Michael O'Beirne, ARCSI.
22. Forbes, *op. cit.*, p. 9.
23. Undated typescript report of Assistant Commissioner Forbes addressed to the Secretary of the DATI; Departmental files.
24. Forbes, *op. cit.*
25. The third annual report of the British Forestry Commission for the year ending 30 September 1922 (published 1923) contains the following: 'Ireland. Following the Government of Ireland Act 1920 the care of Irish forestry passed on 1st April 1922 to the Provisional Government of what is now the Irish Free State and the Government of Northern Ireland.' The policy to which the Commissioners sought to give effect is outlined: 'Under Section 3(2) of the Forestry Act, 1919, which came into operation on 1 September 1919 the powers and duties of the Department of Agriculture and Technical Instruction for Ireland in relation to forestry were transferred to the Commissioners who were appointed on 29 November 1919. Following the report of the Departmental Committee on Irish forestry the Department of Agriculture and Technical Instruction began in 1908 to purchase and preserve the woods on estates which were being sold to tenant purchasers under the Land Act. For that purpose a sum of £6,000 was annually voted by Parliament.'

CHAPTER TEN

1. John Mackey, *A Study of Modern Forestry and of the Interdependence of Agriculture and 'Industry'* (Cork 1934), p. 89.
2. Personal interview, October 1985.
3. 24th General Report of the Department of Agriculture and Technical Instruction for Ireland, 1923-6; forestry section, p. 10.
4. Cited Professor Alfred O'Rahilly, 'Notes on' *Forestry in Ireland* by John Mackey, pp. 137-8.
5. The official British pamphlet *Commercial Forestry* (Ministry of Reconstruction, 1919) puts it thus: 'What is meant by the term commercial forestry? It is intended to emphasize the distinction between arboriculture, or the establishment of single or groups of trees with no financial purpose, and the systematic growth of masses of trees for profit', p. 4.
6. Mackey, p. 36.
7. Professor Ostenfeld in 1924, cited by Mackey, p. 40.
8. W. E. Hiley, *Economics of Forestry* (London 1930).
9. Mackey, *op. cit.*, p. 41.

10. *Ibid.*, p. 42.
11. *The Bulletin of the Forest Service of Washington*, 1933. In the Forest Products Handbook of the Svenska Teknolog Foreningens Forlag of Stockholm for the same year the direct products of wood distillation alone, including pitch, creosote, turpentine oil, covered fourteen closely printed octavo pages, and the *Literary Digest* for 19 February 1927 stated that '36,000 cords of wood are required to produce this magazine for a year'. A cord = 4.74 cubic yards; a spruce forest at forty years is taken as yielding 4660 cubic feet of pulpwood to the acre. 692 acres will yield 36,000 cords.
12. Mackey, p. 34. An interesting cautionary note sounded in Britain in *Commercial Forestry* (Ministry of Reconstruction, 1919): 'In the peculiar conditions of land ownership in these islands it is difficult to feel assured that popular interest would be aroused and maintained in forestry pure and simple.' This seems applicable to the Irish situation both then, when the private forests dominated, and later on when State acquisition was the criterion.
13. *Ibid.*, p. 55. The following comment from *Commercial Forestry* is illuminating: 'In all newly established industries there must be indeed an unproductive period ... and ... in forestry even an assurance of success is so long deferred that no reasonable estimate of the ultimate value of the output is possible,' p. 2.
14. DCIFR, p. 20.
15. *Journal of the Department of Agriculture*, 1924.
16. O'Rahilly, *op. cit.*, pp. 175-8.
17. James, *A History of English Forestry*, p. 218.
18. 24th General Report of the Department of Agriculture and Technical Instruction for Ireland, 1923-6; forestry section, p. 11.
19. James, p. 221.
20. Report on the Census of Woodlands and Census of Production of Home-Grown Timber, 1924, Forestry Commission, 1928.
21. Forbes, 'Tree Planting in Ireland During Four Centuries', PRIA, Vol. IV, Section C, No. 6, pp. 197-8.
22. Mackey, pp. 55-63.
23. A. C. Forbes, 'The Forestry Revival in Eire', *Irish Forestry*, Vol. IV, No. 1, p. 21.
24. Durand, *op. cit.* But that this could properly be called a meeting of the Dáil is doubtful. The Dáil had been repeatedly prorogued since the civil war began the previous June, while the Provisional government continued to meet. Saorstát Éireann did not come into existence until 6 December. This was almost certainly a meeting of the Provisional government.
25. Durand, p. 191.
26. Gray, 'The Economics of Irish Forestry', Statistical and Social Enquiry Society of Ireland, paper, 20 December 1963, p. 3.
27. McGlynn, private conversation, 1985. In a note to the author (1990) O'Carroll claims to be unaware of this survey and casts doubt on its outcome.
28. *Ibid.*
29. *Ibid.*
30. Mackey, *op. cit.*, p. 16.
31. Acht Foraiseachta, 1928 (Forestry Act, 1928), An Act to make further and better provision for promoting afforestation and for that purpose to amend the Forestry Act, 1919, to restrict the felling of trees, and to make other provisions connected therewith (10 December 1928). Its main provisions were to transfer to the Irish Land Commission the powers of decision on compulsory State purchase of lands vested in the Forestry Commission under the 1919 Act,

with powers of appeal and to introduce felling restrictions, penalties and preservation orders.
32. McGlynn, private conversation.
33. *Ibid.*
34. Mackey, p. 20.
35. MacBride, private conversation, 1985.
36. McGlynn.
37. 24th General Report of the Department of Agriculture and Technical Instruction for Ireland, 1923-6; forestry section. The State holding grew from 59,554.5 acres in 1933 to 116,603.5 acres in 1938 – Report of the Minister for Lands on Forestry for the period from 1 April 1933 to 31 March 1938.
38. McGlynn.
39. Vol. XXVII, 1927.
40. DATI Report 1925-6.
41. *Journal of the DATI*, XXVII, 1927.
42. 'On 1 December 1933 under the Forestry (Re-Distribution of Public Services) Order, 1933 (No. 158) all matters relating to forestry were transferred from the Minister for Agriculture to the Minister for Lands and Fisheries – Report of the Minister for Lands on Forestry for the period from 1st April, 1933 to 31st March, 1938. The Ministers and Secretaries Act, 1924, assigned "The Forestry Commission" to the Department of Lands and Agriculture which functioned as the State Forestry Authority until 12th August, 1927, when the Saorstát Éireann Forestry Commission (Transfer of Functions) Order, 1927 (No. 64), was made, definitely transferring all the powers, duties and functions of the Forestry Commission to the Minister for Lands and Agriculture, subsequently styled the Minister for Agriculture. This remained the position until the transfer, as from 1st December, 1933' – Report of the Minister for Lands on Forestry for the period from 1 April 1933 to 31 March 1938.
43. Forbes, 'Forestry Revival in Eire'.
44. Durand, *op. cit.*, p. 209.
45. *Ibid.*, p. 208.
46. 'The heart of forestry is silviculture, the science of tending a crop of trees to produce a desired end-product and to ensure perpetuity. This science was developed and preserved in Europe throughout and perhaps in spite of the industrial revolution and the principles are now adapted to suit various conditions and needs in different countries,' *The Forests of Ireland, History, Distribution and Silviculture*, p. 55. Silviculture is described by Dr Fernow, sometime doyen of American forestry, as 'the art of producing and tending timber crops. Silviculture, the production of wood crops, is the pivot of the whole forestry business'. Silvics is defined as 'the science which treats of the structure, life, habits, and behaviour of trees in the forest. Forestry, with a wider significance than silviculture, embraces several distinct branches of knowledge: (a) silviculture, (b) forest management and protection, (c) forest policy and economics. It implies a scientific acquaintance with silvics.' Mackey, *op. cit.*, pp. 18-19. Connolly had proposed a target of 500,000 acres.
47. Gray, *op. cit.*, p. 2, citing the Annual Report of the Minister for Agriculture for 1931-2. It should be borne in mind that Gray was a civil servant at the time he made this observation, and that it was he who assumed the mantle of 'Mr Forestry', and did so with energy and imagination. 'No man did more for forestry than Henry Gray. I had more rows with him than probably any other man in the Department. But I'll give him credit for that. He was the greatest loss ever to forestry. I said to him once, "The social aspect of forestry, particularly in the early days, lies as much in giving the farmer who sells the

land a decent price for his land, as anything else. They should be given a decent price for their land. And that would be a far better policy all round than trying to buy cheap land which is of no use to anyone"' (McGlynn).
48. Bulmer Hobson, *A National Forest Policy*, privately printed 1931, p. 15.
49. Durand, pp. 213-4.
50. Hobson, *op. cit.*, pp. 21-2.
51. Cited Durand, p. 229.
52. Durand, p. 229. But in fact this is a comment by Crozier in response to Bulmer Hobson's *A National Forest Policy*.

CHAPTER ELEVEN

1. The Report of the Minister for Lands on Forestry for the period from 1 April 1933 to 31 March 1938, summarizes some of the progress made: 'First policy remains ... to create a home supply of new timber sufficient to meet home requirements, so far as possible to grow in this country the types of timber required ... the area of lands acquired for forestry purposes, together with the area planted annually, have increased progressively. These increases were particularly marked in the years 1934-35 and 1935-36... technical staff has been markedly increased, a Director appointed (Reinhardt) and several new Inspector posts established. Greatly improved facilities for technical training have been provided. In 1934-5 an Arbor Day was instituted for schools – Substantial organizational changes also took place.'
2. Seán MacBride, personal interview, 1985.
3. O'Rahilly, *op. cit.*, pp. 182 and 184.
4. Theodore Woolsey on French forest mentality, *Studies in French Forestry* (New York 1920), cited by Mackey, *op. cit.*, pp. 53-4.
5. Mackey, *op. cit.*, p. 84.
6. P. J. O'Loghlen, Commission of Enquiry into Banking, Currency and Credit, 1938, Minority Report No. 111, paragraph 26, pp. 34-35, privately printed.
7. O'Rahilly, 'Notes on' *Forestry in Ireland* by John Mackey, pp. 167-8.
8. *Irish Forestry*, November 1943.
9. N. D. G. James, *A History of English Forestry*, p. 195.
10. *Ibid.*, p. 196.
11. *Ibid.*, citing an anonymous report titled 'Sir William Schlich's Work in Britain' in the *Quarterly Journal of Forestry*, Vol. xx, 1926.
12. Files of the Forestry Division cited Durand (twice), pp. 221 and 223.
13. Report of the Minister for Lands on Forestry, 1933-8.
14. Actual figures 6919 and 7321 respectively. Connolly Papers, by kind permission Joseph Connolly executors.
15. Bulmer Hobson, *A National Forest Policy*, pp. 11-12.
16. Forbes, 'Tree Planting in Ireland During Four Centuries', PRIA, Vol. XLI, No. 6, p. 197.
17. Hobson, p. 13.
18. When State control is 'almost non-existent, as in Great Britain, many woodland estates present a picture of neglect which is in striking contrast to the productive condition of those to be seen in countries where efficient forest management is enforced by law,' Professor R. S. Troup (Director, Imperial Forestry Institute, Oxford), *Forestry and State Control* (Oxford 1938).
19. James, p. 220.
20. *Ibid.*, pp. 221-2.
21. *Quarterly Journal of Forestry*. The closest Irish equivalent was in 1934-5 when the Forestry Division 'was represented at a conference of Timber Merchants

called by the Department of Industry and Commerce to deal with certain proposals for control of the home timber supplies to the sawmilling industry', Report of Minister, 1933-8, p. 36.
22. Mackey offers an interesting and accurate description of the word (and function of) propaganda: 'Propaganda has so frequently been put to base uses that its importance – the importance of systematic direction of effort to the attainment of a definite goal – is sometimes obscured', p. 33.
23. James, p. 223.
24. *Ibid.*, p. 224.
25. James, *op. cit.*, p. 224.
26. McGlynn. There was considerable industrial unrest in State forest activities in the pre-war years, especially in 1937-8, when several strikes for higher wages took place at forests throughout the country (Report of Minister, 1933-8, p. 29).
27. The hiatus in the second case arose because, following the death of Augustine Henry, Professor of Forestry, in 1930, the chair had lapsed. This gap was temporarily filled by the appointment of Forbes and H. M. FitzPatrick as part-time lecturers which, although it was not a satisfactory solution, resulted in five graduates coming on stream in 1935, recruited as temporary Assistant Junior Forestry Inspectors. Durand comments (pp. 232-3): 'One of the five graduates was Thomas Clear, who was shortly afterwards awarded a post-graduate scholarship. It arose from criticisms made by Connolly.... Connolly spoke of the poor facilities afforded forestry students – e.g. they were receiving only part-time forestry lectures at this time – and challenged the University to provide more funds and facilities. Clear won the subsequent award and Reinhardt arranged a visit to Eberswalde Prussian Forest College where he remained several months and followed an itinerary and programme arranged in consultation with Reinhardt, whose cousin, Professor Wittich, held a chair in soil science at Eberswalde. Wittich later visited Ireland to study the soils associated with old red sandstone at Ballyhoura Forest, Co. Cork.'
28. Personal communication from Anderson to H. M. FitzPatrick. F. J. Meyrick, Secretary of the Department from 1922-34, was also a Scot. He invited Anderson back to Ireland. This may have been seen in some quarters as a move to preserve the Scottish/British influence in the top echelon of State forestry that had continued uninterruptedly since inception. Anderson introduced another Scot named Petrie. It is suggested that this was to head off the advancement of dynamic young native forestry graduates like S. M. O'Sullivan and H. M. FitzPatrick, who was Divisional Inspector and also part-time lecturer in forestry at University College Dublin. This may have been a cause of some of the friction between Connolly and Anderson. Certainly there was a generally increased awareness of the need for a forest inventory and a rudimentary attempt was undertaken at the time, with Petrie, for instance, alleged to have counted trees from his car.
29. The author recalls using the description 'tree-farming', which seems an accurate and realistic one (and which he believed, erroneously, was original at the time), at a meeting of officials of the Forest and Wildlife Service of the Department of Lands in 1973-4. The reaction was one of barely concealed offence at the use of such an unorthodox and 'vulgar' phrase to describe the activities of the service, as was later indicated in no uncertain manner by the patronizing route generally known as 'a quiet word in your ear'. It is an expression commonly used by published authorities. But – 'I do not favour the use of the term "tree-farming" even to distinguish small-scale from large-scale forestry', O'Carroll, 1990, note to the author.

30. Mackey, *op. cit.*, p. 21.
31. No special legislation was enacted for forestry purposes, but some amendments were made to the 1928 Forestry Act. Amongst the Emergency legislation affecting forestry was some taken by other departments:

    The Holiday (Employee) Act, 1939, affected conditions of labour.
    The Agricultural Wages Act, 1936 (Minimum Rates) Order, 1942, affected forestry workers' wages.
    Government Order No. 143, Emergency Powers (No. 169) Order, prohibited the felling of any trees within 30 feet of the centre of any road, or of any tree in a militarily declared prohibited area, for military reasons.
    The Department of Supplies issued a number of Emergency Powers Orders affecting forestry, mainly Control of Timber Orders (Nos. 294, 1941; 378, 1941; 439, 1942) and Control of Prices Orders (Nos. 89, 1941; 438, 1941; 439, 1941; 277, 1942; 278, 1942; 5, 1943) etc.

32. MacBride, private conversation, 1985.
33. *Irish Forestry*, official publication of the Society of Irish Foresters, Vol. I, No. 1. He referred to Orders Nos. 121 and 148. By 1943, according to Forbes (*The Forestry Revival in Eire*, Irish Forestry, Vol. IV, No. 1, p. 32), the total area of woodland in the country was 282,888 acres. That compared with 250,000 acres in the State in 1921 (mostly in private hands), indicating that some 33,000 acres had been acquired in between. There was to be a further decline. By 1950 there were 242,343 acres of timber-land in the State, of which 142,425 were State-stocked and 94,871 were in private ownership. The Report of the Minister for Lands on Forestry for 1933-8 (p. 7) succinctly states one of the main problems facing the State forest developers: 'Owing to the absence in Eire of large areas of plantable land in the hands of individual owners the process of building up forest units of sufficient extent and sufficiently compact to be economically workable is a slow, complicated and difficult matter which involves initial acquisition of small areas and the gradual enlargement of these by subsequent repeated small additions. This process is unique as far as State afforestation is concerned.' See also Reports of the Minister for Lands for the period from 1 April 1938 to 31 March 1943.
34. Mackey, pp. 17-18.
35. Dáil Reports, May 1940.
36. FAO Report (FAOR), p. 10.
37. Connolly Papers, p. 22.

CHAPTER TWELVE

1. James, *A History of English Forestry*, p. 200, and bear in mind that Schlich's strongest advocacy of forestry for Ireland was on the grounds of the employment it would provide.
2. *Ibid.*, quoting an anonymous source in the *Quarterly Journal of Forestry*, Vol. I, 1907.
3. Royal Commission on Coast Erosion and Afforestation, Vol. II, Second Report (on Afforestation) of the Royal Commission Appointed to inquire into and report on certain questions affecting Coast Erosion, the Reclamation of Tidal Lands and afforestation in the United Kingdom, Cd. 4460, 1909.
4. Gray, *op. cit.*, p. 2.
5. Select Committee on Industries (Ireland), 1885, p. 266.
6. O'Rahilly, *op. cit.*

7. Bulmer Hobson, *A National Forest Policy*, p. 6.
8. Henry, 'Cooperation of State and Citizen in Irish Forestry', *Studies* (1924), p. 614.
9. *Ibid.*, p. 616.
10. Mackey, *op. cit*. Total employment in State forestry, 1989, with an acreage of close to about half this, was 2903. No doubt mechanization accounted for most of the difference.
11. Mackey, pp. 73-4. Today the nation is self-sufficient in softwoods for the building industry.
12. Seán MacBride.
13. Frank Aiken, Minister for Finance, Dáil Éireann, 14 May 1947.
14. Report of the Minister for Lands on Forestry for the period 1 April 1943 – 31 March 1950.
15. 'The Economics of Irish Forestry', *Statistical and Social Enquiry Society of Ireland*, 20 December 1963, p. 2.
16. Durand, p. 256.
17. Dáil Report, 27 June 1934.
18. James, *op. cit.*, p. 224.
19. 'Post War Forest Policy', HM Forestry Commissioners. Cmd. 6447, 1943.
20. James, *op. cit.*, p. 234.
21. Post-War Forest Policy, Private Woodlands, Supplementary Report. Cmd 6500, 1944.
22. An Act to provide for the dedication of land to forestry purposes for the deduction from compensation of grants made by the Forestry Commission in the event of compulsory purchase of the land in respect of which grants were made; and for the execution on behalf of the Secretary of State of instruments relating to land placed at disposal of the Forestry Commissioners, 1947, 10 & 11, Geo. 6. c.21.
23. James, p. 236.
24. An Act to make provision for the reconstitution and as to the exercise of the functions of the Forestry Commissioners, the acquisition of land for forestry purposes and as to exercise of the functions of the Forestry Commissioners, the acquisition of land for forestry purposes and the management, use and disposal of land so acquired; and in connection with the matters aforesaid to amend the Forestry Acts, 1919 to 1927 and certain other enactments relating to the Forestry Commissioners, 8 & 9, Geo. 6 c.35, 1945.
25. Forestry Commission, Census of Woodlands 1947-9, Woodlands of five acres and over, Crown Report No. 1., 1952.
26. An essential ingredient of successful policy was a Forestry Commission, a gap that Connolly tried to plug with the introduction of Reinhardt. A new Forestry Act (1946), 'An Act to make further and better provision in relation to Forestry' was introduced in November of that year. It provided nothing radical, simply a consolidation of the powers of the Acts of 1928 and 1919, which it repealed.
27. Michael Viney, 'Down in the Forest', *Irish Times*, 10, 11, 12, 13 and 14 October 1966.
28. Indeed this question remains inflammatory and is indicative of the imprudent rivalry between administration and various technical arms that was at times evident in the Forest and Wildlife Service from then on.
29. John A. Murphy, *Ireland in the Twentieth Century*, pp. 123-4.
30. T. K. Whitaker, past Secretary of the Department of Finance, past Governor of the Central Bank, Chancellor of the National University of Ireland, private conversation, 1987.
31. Ronan Fanning, *Irish Department of Finance*, 1922-58, p. 409.

32. MacBride, private conversation, 1985.
33. Murphy, *ibid.*, p. 16.
34. Whitaker, in a written comment to the author.
35. Fanning, *op. cit.*, p. 408.
36. MacBride, Lecture to the Forestry Society, University College Dublin, 20 March 1984.
37. Thomas Rea, *Irish Forestry*, Vol. 42, No. 1., 1985.
38. Fanning, p. 434.
39. *Ibid.*, ECA – Economic Cooperation Act and Agreement (Washington).
40. Fanning, *op. cit.*, p. 435. External Affairs minute to the Government, 16 December 1948, Finance (Files) 121/10/48.
41. *Ibid.*
42. Finance (Files) 121/10/48, cited Fanning, pp. 437-8.
43. Finance 121/1/50. (Finance objections were overruled.)
44. Minute 30 June 1949, cited Fanning, p. 446.
45. Fanning, p. 454.
46. Seán MacBride, Lecture to Forestry Society, University College Dublin, 20 March 1984.
47. Viney, article 2.
48. Irish State Forestry, Government Policy, 1948-59.
49. Rea, *op. cit.* O'Carroll points out that in 1948 the rotation period would have been much longer than 40 years. 'Forbes recommended a felling age of 70-100 years for pine and spruce. The calculations quoted in the Cameron Report are based on a rotation of 50 years, but *only because the tables of yields available at that time did not extend beyond fifty years*. In practice foresters in the 1940s would have been thinking in terms of rotations of 60-70 years on average.' Written comment to the author, 1990.
50. MacBride, private conversation.
51. *Ibid.*
52. Written comment to the author, 1987.
53. This figure varies enormously, as we have seen. Now that the 'magic million' afforested acres has been achieved it is currently calculated that there are some 1.5 million additional hectares which could be profitably forested. The total target acreage is 2 million hectares, 20 per cent of the total land area of the country. Convery's report states (p. 56), 'In reviewing this legislation one is struck by the preoccupation throughout with the facilitation of land acquisition. Sixty-six pages – out of a total of 165 – of the 1946 Act deal with this issue, while this topic was the central theme of the 1956 Act.'
54. A situation that still obtained 34 years on in 1984, although the area covered in forest increased in the interval from 1 per cent of the total land area to 6 per cent. See FAO (Cameron) Report, par. 3, Part 1, and Newsletter on the Common Agricultural Policy, reprinted in the Report of the Review Group on Forestry, p. 202.
55. Viney, *ibid.*
56. *Ibid.*
57. Written comment to the author, 1987.

CHAPTER THIRTEEN

1. Michael Viney, *Irish Times*, October 1966.
2. Dr Cornelius Lucey, Bishop of Cork, Commission for Emigration Report, 1955, Minority report.

3. Henry, *op. cit.*, pp. 621-2.
4. Review and Advisory Memorandum submitted to the government, 1944, pp. 5,6,7,8 – Connolly Papers. Proposals for co-operative 'tree-farming' with State involvement had been advanced in *Commercial Forestry* by the British Ministry of Reconstruction, 1919, in Chapter 11, fn. 79.
5. Report to the Commission on Emigration, substitute paragraph for paragraph 366, p. 349, of that Report. The other important and far-sighted comments of Dr Cornelius Lucey are equally relevant.
6. Henry, *op. cit.*, p. 626.
7. *Report on Emigration*, par. 360, p. 151.
8. Gray, 'The Economics of Irish Forestry', pp. 2 *et seq*.
9. FAO Report, par. 8, p. 10.
10. McGlynn, private conversation.
11. Hobson, *A National Forest Policy*, pp. 14-15.
12. FAO Report, Sections 72 and 78. This recommendation was not acted on seriously for 35 years, until 1986.
13. *Ibid.*, Section 69, p. 26. This is a point made throughout this work. A positive and favourable 'forest morale', environmental, economic and strategic, needs to be developed. The groundwork now seems well laid and the attitude of the public is well-disposed in general.
14. Report on Emigration, Summary of Cameron Report, par. 366.
15. Viney, article 1.
16. Gray, p. 3.
17. Report on Emigration, par. 361, p. 151.
18. Gray, p. 4.
19. Review Group on Forestry, 'Report to the Minister for Fisheries and Forestry' (Dublin 1985), p. 16.
20. White Paper, November 1958 – pars. 81, 85-6.
21. Dáil Report, 27 April 1949.
22. Durand, *op. cit.*, p. 265.
23. 1946 Act; 1956 Act. See Appendix 6.
24. Viney, *Irish Times*, 9 August 1955.
25. Viney, *op. cit.*
26. Dáil Reports, July 1956. The point was emphasized in *Economic Development* (Chapter 14, par. 10) in 1958: 'Because of the substantial employment it provides in areas which are largely bereft of other economic activity, a continuance of the present planting programme would be warranted, notwithstanding the heavy financial burden on the Exchequer for some years to come, *if* there were a reasonable expectation that the financial outcome envisaged... would be realized in practice' (author's emphasis). (A Department of Lands calculation estimated that break-even point on expenditure/revenue would be reached in 1982-3.)
27. Professor Tom Clear, *op. cit.*, 1951.
28. Because of the new enthusiasm and new techniques this was further increased. The Second Programme for Economic Expansion (1963 and 1964) stated that 44 per cent of new planting was to be in western counties and that by 1970 the forest area was expected to be about 550,000 acres (actual reached 661,467).
29. Gray, p. 5.
30. Viney, *op. cit.*
31. Viney, *op. cit.*, article 3. These statistics, as we have seen, did not survive the impact of modern technology and mechanisation.
32. Dáil Éireann, Forestry vote, 25-6 June 1958. Childers was emphatic on the question of mountain land: 'there can be no conflict with the sheep economy

which is so important... if we were to go beyond the present maximum price we pay (for land), we would be starting to invade on sheep ground; and I do not believe in doing that'. He had, however, evidently come to grips with the question of private planting. With the Emergency experience behind him and an enormous planting commitment before him, Childers could benefit from hindsight no less than from the advice of his experts. His proposals were modest and still fell between the two stools of cautiously encouraging traditional large landowners and the farming community generally, at whom the main thrust of his exhortations was directed. His attitude to private planting was summarized: 'The emphasis on State planting has been far too great and the whole activities of the Department have been concentrated on State forestry... The Government has agreed that we make a tremendous effort to interest our people in private forestry...' This was a radical departure from past official attitudes. But Childers's somewhat unfortunate reference to planting an acre of trees being akin to laying down a bin of port drew from the inimitable James Dillon the following magnificent retort: 'I wish the Minister to know what port means in the west of Ireland. It is the beverage which temperance ladies drink by the glass on holidays in the belief that it does not break the pledge... I want the Minister to bear in mind that references to bins of port are not likely to awaken any favourable reverberations in the minds of the small farmers whom the Minister wants to plant an acre of Sitka spruce.' Childers also announced that commercial standards were to be set for home-grown timber.

33. *Economic Development* (Dublin 1958).
34. Review Group on Forestry, p. 25.
35. *Economic Development*, p. 147, par. 11.
36. *Ibid.*, par. 12.
37. *The Second Programme for Economic Expansion*, 1964.

CHAPTER FOURTEEN

1. The Case for Forestry, Forest and Wildlife Service, 1980 (revised 1983), par. 2.2, pp. 3-4. In fact as we have seen the 'policy' mentioned in this extract materialized rather unexpectedly. As expressed in the European Recovery Programme, *Ireland's Long Term Programme 1949-1953*, 1948, it was to 'plant a million acres by annual increments of 25,000 acres', pars. 81, 85 and 86, Government White Paper.
2. Fanning, *Irish Department of Finance, 1922-58*, p. 519. On 28 October Sean Lemass told the Fianna Fáil Ard Fheis that, within the next few days, a 'programme of national Economic Development' would be published as a white paper; the government had decided, he said, that it was 'essential to redefine the objectives of national economic policy'. The white paper was duly published on 12 November. Ten days later, *Economic Development* was published.
3. TFI, p. 62.
4. ED, Chap. 14, par. 4, p. 144.
5. *Ibid.*, pars 5-8, pp. 144-5.
6. Patrick Whooley, Secretary, Department of Fisheries and Forestry, 1985-7, in a written communication to the author, 1987.
7. ED, p. 8.
8. Report on Emigration, par. 362, p. 151.
9. Mackey, *op. cit.*, p. 94, and citing the British Empire Forestry Conference, 1920.

10. TCFF, Forest and Wildlife Service, 1980 (revised 1983) pp. 15-16.
11. FAOR, pp. 20-22.
12. McGlynn.
13. The European Recovery Programme: *Ireland's Long Term Programme 1949-1953*.
14. FAOR, par. 21 pp. 14-15.
15. Durand, p. 256.
16. FAOR, pars. 22, 23, 24, 25 pp. 15-16. Up to 1950 out of 970 acquisitions, only twelve exceeded 2000 acres, and none was over 5000. The largest category was those less than fifty acres (326), with the next largest those less than one hundred acres (195) – FAO Report, Table III.
17. Viney, *op. cit.*
18. *Second Programme of Economic Expansion*, Part II, Chap. 3, par. 2, p. 132, 1964.
19. *Ibid.*, par. 9, p. 135.
20. *Economic Development* (Dublin 1958).
21. H. J. Gray, 'The Economics of Irish Forestry', Paper to the Statistical and Social Inquiry Society of Ireland, 20 December 1963, pp. 5-6.
22. Gray, *op. cit.*, p. 9. MacBride adopted a more caustic view and was critical of what he saw as want of adequate planning 'for the utilization of the timber becoming available in fairly substantial quantities', comments held by the FWS to be overstated and unrealistic.
23. *Economic Development*, Chap. 3, par. 10, p. 146.
24. Connolly Papers.
25. *Ibid.*
26. Report of Review Body, 1985.
27. Interim Report on Afforestation Programme; Analytic Study of State Forest Undertaking (Confidential), Departmental document, 1974.
28. 'While output has been maintained close to pre-recession levels, losses have been sustained...', Convery, 'Irish Forest Policy 1979', National Economic and Social Council (no. 46), S.O., Dublin, p. 52.
29. ED, 1958, Chap. 14, par. 9, p. 146.

CHAPTER FIFTEEN

1. In 1984 Tourism was added to the Department. Following a change of administration in 1987, forestry as a division was transferred to the Department of Energy and Communications, acquiring for the first time a Minister for State with sole responsibility, Michael Smith, TD. That situation changed after the election of 1989, when no Minister of State for forestry was appointed.
2. Frank Convery, *Irish Forest Policy*, NESC No. 46 (1979), p. 38. Convery presents a framework in which to view forest development and provides some detailed explanations which are useful to both the specialist and the non-technical reader.
3. Each reappraisal confirmed the policy of planting what had now become 10,000 hectares a year. The 1967 review, *Appraisal of the Public Capital Programme, State Forestry*, was the work of an inter-departmental committee that recommended maintaining the status quo and undertaking a detailed cost-benefit study of State forestry. That study, marked confidential and interim, was completed in 1974 and did little more than reflect the differences in outlook on forestry between the Forest and Wildlife Service and the Department of

Finance, which had both carried out the analysis. It concluded that 'the rates of return (from forestry) ... are sufficient to justify continuance of present afforestation policy in the main'. The TCFF report emphasizes that 'pulpwood industries in Ireland (and in Europe generally) were severely affected by the 1974 recession. In recent years plants in Athy and Waterford closed and the production of pulp at Clondalkin has ceased'. The report also referred to the establishment of a fibre-board plant at Clonmel and to the closure of the chipboard plant at Scarriff, Co. Clare, subsequently reopened.

4. *Developing the Irish Timber Industry for the 80's*, p. 16, S.2.4.
5. TCFF, Introduction, p. 1.
6. TCFF, p. 9.
7. TCFF, p. 29.
8. CAS Report, p. 6.
9. TCFF, p. 12.
10. *Ibid.*
11. TCFF, p. 13.
12. Convery, *op. cit.*
13. CAS Report, p. 13.
14. Convery laid stress on what we already know, namely that: 'From the very beginning land judged to be suitable for agricultural use was prevented from consideration for State afforestation purposes...' by fixing the price of agricultural land at a figure that made its sale to the Department for afforestation purposes uneconomic.
15. TCFF, p. 17.
16. Joyce, *The Irish Press*, 19 September 1985.
17. McGlynn, private conversation.
18. *The Case for Forestry*, FWS, p. 13.
19. *Connacht Telegraph*, 26 January 1986.
20. *Ibid.*
21. *IFA Journal* w/e 28 April, 86.
22. Jack Whaley, *Irish Times* 28 May 1986.
23. Under Article 15/3 of EEC Regulation 797/85 for precisely that period in disadvantaged areas. Improved techniques and selection had reduced the period to first thinnings from 20 to 15 years.
24. Donal McDaid, *Longford News*, 24 July 1986.
25. *Leitrim Observer*, 3 January 1986.
26. This idea was revived three years later in 1987 when the new Fianna Fáil administration modified the proposal to one of selling off some of the uneconomic, small or inaccessible forest plantations amounting to 5 per cent of the State forest holding in order to raise £3 m. to fund the new semi-state board. O'Carroll, in a note to the author (1990), states that the purpose was to fund further forest expansion, not to fund the new semi-state board.
27. *Irish Times*, 4 December 1986.
28. *The Irish Press*, 25 September 1985.
29. RRGF, Introduction, p. 1.
30. 'In so deciding we were mindful of the fact that the forestry enterprise will not be in a position, for many years to come, to finance, from its own resources, its investment in new forests or even all of its current expenditure... We did not consider it appropriate to establish a State corporation or company which, although it would have commercial objectives, would require a considerable level of direct State financing on an annual basis from the beginning and for a considerable period into the future.'
31. RRGF, 5.4. to 5.9, pp. 35-7.

32. Among those industries which suffered were Clondalkin Paper Mills, Chipboard Ltd., Munster Chipboard Ltd., Irish Board Mills Ltd.
33. RRGF, p. 11.
34. *Ibid.*, pp. 16-18.
35. Brief supplied by the pre-Bord Coillte preliminary organization: Paul Cassidy, Assistant Secretary, later general manager, Corporate Planning and Development, Coillte.
36. RRGF, p. 28; references are cited in support of this contention.
37. *Ibid.*, p. 425 Appendix 4, footnote, p. 71. The system is similar to that used by the British Forestry Commission, estimating a future expenditure forecast and net income from sales on a given unit, with a resulting cash-flow (one entry for each remaining year in the rotation), that would be discounted to give a current estimated value. Thus the prediction of income would depend mainly on the forecast of volumes, available as forest management tables, and of corresponding prices.
38. *Ibid.*, 10.9, p. 58.
39. Siobhan O'Neill, *The Irish Press*, 19 March 1986.
40. Martin Fitzpatrick, *Sunday Independent*, 29 September 1985.
41. *Irish Times*, 8 July 1985. But in 1991 there are 100 sawmills operating using about 0.75 million cubic metres of timber (Roundwood) a year.
42. *The Theft of the Hills: Afforestation in Scotland*; Steven Tomkins, September 1986, Ramblers' Association, 1-5 Wandsworth Road, London, S.W.**8.** The points made by Tomkins were later raised in respect of large-scale conifer planting here, particularly by An Taisce, the National Trust for Ireland, in 1990. Their report and the State Operational Forestry Programme are considered in Chapter 18.
43. Viney, 'When Trees Get in the Way', *Irish Times*, 10 September 1988.
44. Frank McGuckian, spokesman for the Concerned Farmers' Organization of Leitrim, *Leitrim Observer*, 9 August 1986.
45. Peter Honeyman, Leitrim spokesman for the IFA, *Leitrim Observer*, 9 August 1986.
46. Padraic Joyce, *The Irish Press*, 19 September 1985.
47. *Ibid.*
48. Convery, *op. cit.*, pp. 70-1.
49. OFP, p. 15. But why since 1987? It would be more accurate to say since 1988.

CHAPTER SIXTEEN

1. Convery, *Irish Forest Policy*, p. 39.
2. TCFF, p. 4.
3. Review Group on Forestry Report, p. 15, and Peter Dodd/Liam Quinn, Coillte. See also table in Chapter 15. The tables do not necessarily indicate a reduction in *area* of some of these species, given that in 1958 the State forest lands amounted to one-sixth of a million hectares and in 1987 was almost half a million hectares. Norway spruce, for instance, 16 per cent of the 1958 total estate = 53,333 acres; 8 per cent of the 1987 total estate = 80,000 acres.
4. See Appendix 3.1, TCFF.
5. Mackey, p. 37.
6. *Ibid.*, pp. 30 et seq.
7. And were among the reasons for the initial experiments in State forestry being here rather than in England. Mackey, p. 45. Rotation periods are even less today.

8. Howard P. Weiss, Milwaukee conference on Forestry, 1932.
9. Professor Ostenfeld addressing a meeting of the British Forestry Association, 1924, cited Mackey, p. 40.
10. *Literary Digest*, 19 February 1927.
11. The Bulletin of the Forest Service of Washington.
12. Mackey, pp. 75 and 81.
13. 'The decision to plant a specific site should be made for economic reasons except where a policy decision is made to provide employment, or amenities. If afforestation is required to fulfil a social need then it should be approached in a different manner. A confusion of social and economic aims only leads to inefficiency' – (*Ireland's Forestry, A Review*, Union of Professional and Technical Civil Servants [1984], p. 7 4.1.). But this is yet another axiom that has been found to have a less than solid base. In OFP not only is social forestry (sometimes in a new guise) given a prominence never hitherto acknowledged, its projected development, especially in environmental, tree-farming and recreational/tourism terms are outlined. The impact of EC forest policy and funding, apparently, also concentrates minds wonderfully.
14. 'Rotation lengths may also be adjusted one way or another to achieve the requirements of the foresters' basic principle, which is the *principle of sustained yield*. This requires that the yield or output from a forest should never be raised above a level which can be maintained indefinitely, and its main purpose is to prevent short-term exploitation at the expense of a longer-term yield. This is one of the main reasons why forestry, throughout the world, is either carried out or closely controlled by government agencies on behalf of the people and for posterity.' TFI, p. 63.
15. TCFF, p. 34.
16. 'A normal forest is rarely achieved in practice – prevented by unpredictable factors such as fire, disease or storm damage – but forest management approximates to this ideal through successive rotations.' TCFF, p. 34.
17. T. McEvoy, 'Forestry in Ireland', a paper to the British Association, September 1957, published: The Advancement of Science, No. 56, March 1958.
18. *Ibid.*, p. 311.
19. TFI, p. 44.
20. O. V. Mooney, '*Pinus Contorta*', in *Irish Forestry*, No. 2, 1957.
21. McEvoy, *op. cit.*
22. *Ibid.*
23. RGF, p. 25.
24. McEvoy, 'A Review of Afforestation Expertise in Ireland', paper delivered at a FAO conference in 1966.
25. O'Carroll, 'The Progress of Peatland Afforestation in the Republic of Ireland', *Irish Forestry*, Vol. XIX, No. 1, Spring 1962, p. 93.
26. *Ibid.*, p. 95.
27. TCFF, p. 13. The term 'yield class' is a measure of potential production expressed in cubic metres per hectare per annum. The average yield class of the existing forest estate is fourteen, which is about 3 cu. m. higher than that in Britain and considerably higher than the estimated European average yields, which range from two to five cu.m. per hectare per annum (p. 13).
28. TCFF, p. 13.
29. TCFF, p. 21.
30. *Ibid.*, p. 23.
31. Developments in Forecasting and Regulating Forest Yield, Agricultural Record, 1949.
32. TCFF, pp. 23-4.

33. *Ibid.*, p. 31. Table 8.2.
34. Convery, *op. cit.*
35. TCFF, Chap. 10, p. 38.
36. *Ibid.* Interestingly, though, the second most conservative planting policy, policy number 5, although low in terms of projected self-sufficiency and employment compared with the first four, is kindest in terms of Exchequer finances over the period 1990-2020. (The respective cumulative figures in £m. for the six tables are 72.9; 76.7; 76.5; 80.5; 81.9 and 72.2.)
37. RGF., p. 17.
38. RGF., Diagram 3, p. 21.
39. *Ibid.*, p. 24.
40. *Ibid.*, p. 25 (footnote).
41. *Ibid.* (Appendix 4, Table 3).
42. *Ibid.*, table 7, p. 27.
43. *Ibid.*, 4.14, p. 28.
44. *Ibid.*, p. 51.
45. *Ibid* 10.10, p. 58. Banks, foreign and Irish insurance companies, the Irish Sugar Company and Swedish associates, building groups and timber companies all invested increasingly in forestry in Ireland in the 1980s.
46. 'Irish exports will compete without tariff or other protection with forest products from North America, the USSR, Scandinavia, etc,' Convery, *op. cit.*, p. 111.
47. *Ibid.*, Convery states: 'We should plan immediately for a horizontally and vertically integrated forest industry structure,' p. 112.
48. *Ireland's Forestry, A Review*, Union of Professional and Technical Civil Servants (1984), p. 11, 4.6.1 and 4.6.3. But this is contradicted by FWS sources who claim that because of the shortage of timber marketing was not, in fact, a priority, that the question was being adequately monitored by the IIRS and that a viable alternative was provided by way of grants to sawmillers.
49. Convery, *op. cit.*, p. 52.
50. OFP, pp. 9 and 57.

CHAPTER SEVENTEEN

1. 'In 1920 the first area for State afforestation in Ireland was acquired for the Forestry Commission at Baronscourt in County Tyrone,' *The Forests of Ireland* (ed. O'Carroll), p. 111.
2. Convery, *Irish Forest Policy*, p. 121. When the State came into existence in 1922 the situation was almost exactly the reverse: 'woods still in the possession of private owners comprise about 84 per cent of the total woodland area of the country' – 24th General Report of the Department of Agriculture and Technical Instruction for Ireland, 1923-1926, Forestry, report for period 1 October 1923, to 31 March 1925, p. 11. Convery pointed out that 'some planting, albeit on a modest scale, has continued since the 1930s', *op. cit.*, p. 148. 'In the period 1930 to 1934 the average annual rate of private planting was about 130 acres. In the period 1943 to 1945 the average was about 250 acres', according to *Economic Development*, Chap. 111, p. 147. Between 1958 and 1964, due to a campaign to encourage private planting, the average annual planting rate rose from 500 acres to 1000 and continued for about a decade, only to fall to 307 acres in 1977. That situation changed dramatically during the next four years. A table in the OFP shows that in 1989 overall private planting for that year was about 46 per cent of the total, and in 1990 it was about 50 per cent.

By 1993 private planting is expected to exceed State planting by almost 50 per cent according to a projected afforestation level table in OFP.
3. DCIFR, Part 1, 9 and 10, p. 4.
4. John Evelyn, *Sylva*, and subsequently.
5. Hore, citing Cottingham.
6. 10 Wm. III, Cap., XII, 1698.
7. McCracken, *Quarterly Journal of Forestry*, Vol. LVII, No. 2, 1963.
8. See Appendix 3. For example, George III Cap. XXXI, which deprived tenants of the common of estovers – the right of taking necessary wood for household use, making implements, repairs, etc.
9. Forbes, 'Some Early Economic and other Developments in Eire, and their Effect on Forestry Conditions', *Irish Forestry*, Vol. 1, No. 1, November 1943.
10. W. F. Baily, 'Forestry in Ireland', *Social and Statistical Enquiry Society of Ireland*, 2 July 1889.
11. Augustine Henry, *op. cit.*, p. 630.
12. Lord Powerscourt in evidence before the Select Committee on forestry, 1887, cited Durand, pp. 316-17.
13. In the Land Act, 1909.
14. Durand, *op. cit.*, p. 317. This statement is an oversimplification. Encouragement for private planters would have occurred anyway, as in England. The decline here was more specific, more serious and it required much greater and more enlightened action.
15. *Report on Emigration*, par. 361, p. 151. The *total* area under woods and plantations was 305,000 acres.
16. *Report of the Inter-departmental Committee on Irish Forestry*, 1908, pars. 71/72 (Recommendations), pp. 33-4.
17. Convery, *op. cit.*, p. 123.
18. FAO (Cameron) Report, pars. 5-6, pp. 9-10.
19. *Ibid.*, par. 21, p. 14.
20. Henry, 'Cooperation of State and Citizen in Irish Forestry', *Studies*, Vol. XIII, No. 52, 1924.
21. *Ibid.*, p. 626.
22. *Ibid.*
23. 24th General Report of the Department of Agriculture and Technical Instruction for Ireland, 1923-6, Forestry, report for period 1 October 1923 to 31 March 1925.
24. Hobson, *op. cit.*, p. 13.
25. FAO Report, par. 7, p. 10.
26. An inventory of private woodlands carried out in 1973 by Thomas J. Purcell, the first since 1944, gives the total area of private woodland in the country at that time as being 81,963 hectares (202,534 acres), of which about 42 per cent was scrub (33,102 ha.), unstocked (804 ha.) or undeveloped (156 ha.). Of the remainder, 15,054 hectares were under conifer and 32,847 hectares under broadleaves. The purpose of the survey was to show the area of private woodland in the country, assess volume and potential production and facilitate a private woodland policy.
27. *Economic Development*, 1958, Chap. 14, par. 11, p. 147. Planting grants were adjusted at intervals as follows: 1929, £4; 1943, £10; 1958, £20; 1972, £35.
28. *Ibid*
29. *Economic Development*, 1959, par. 13.
30. *Ibid.*
31. Forbes, 'Tree Planting in Ireland During Four Centuries', *PRIA*, Vol. XLI, Section C, No. 6, 1933, p. 199.

32. *Ibid.*, p. 197.
33. McGlynn, private conversation, 1985.
34. *Irish Forestry*, the Official Publication of the Society of Irish Foresters, Vol. 1, No. 1, November 1943.
35. *Private Forestry in the Irish Republic*, par. 1, 1956.
36. *Ibid.*
37. *Ibid.*, par. 23.
38. *Ibid.*, par. 28. p. 11.
39. Erskine Childers, Dáil Reports, 25 June 1958, p. 702. It was in this speech that Childers made the comparison of an investment in forestry with laying down a bin of port, drawing on his head Dillon's inimitable sarcasm for his pains.
40. *Ibid.*, p. 705.
41. *Ibid.*
42. *Second Programme of Economic Development*, pp. 136-7, 13 and 14.
43. Durand, *op. cit.*, p. 325.
44. Address to a public meeting, Castlebar, 15 October 1960.
45. *Ibid.*
46. P. M. Joyce, *The Irish Press*, 19 September 1985.
47. In 1958 private planting was 764 ha.; in 1986 it rose to 2560 ha., and to 3200 in 1987, according to FWS statistics.
48. *Investing in Forestry – A Practical Guide*, Touche Ross, in association with Celtic Forestry, Ltd, 1988.
49. FAO Report, pp. 15-16.
50. Maurice Harvey, Irish Creamery Milk Suppliers' Association, cited in the *Northern Standard*, 26 February 1986.
51. As noted there was frequent uncertainty concerning substantive policy, especially up to the 1970s. It may even have been in doubt as late as this. The statement referred to here was made in the Forest Estimate speech in Dáil Eireann, 1988, by the Minister for Forestry, Michael Smith, TD, when Niall O'Carroll was chief inspector of the Forest Service. But – 'I am not aware of this 1988 policy statement' – O'Carroll in a written comment, 1989.
52. McGlynn, private conversation.

CHAPTER EIGHTEEN

1. 9 & 10 Geo. 5. c.58.
2. Sutherland subsequently made several inspections of forest areas in Northern Ireland and submitted various reports and recommendations to the Northern Ireland Department of Agriculture on progress and development; Kilpatrick, *Northern Ireland Forest Service, A History*, p. 24.
3. 59,305 out of an insured register of 267,000.
4. Forestry in Northern Ireland, Ministry of Agriculture Forest Policy – unpaginated.
5. *Ibid.*
6. *Op. cit.*, p. 28.
7. Kilpatrick, *op. cit.*, p. 28.
8. *Ibid.*, citing the report, p. 38.
9. *Ibid.*, p. 52.
10. *Ibid.*, p. 54.
11. *Ibid.*, p. 76.
12. *Ibid.*
13. *Ibid.*, pp. 82 *et. seq.*

14. Hans Jurg Steinlin, *Unasylva*, Journal of Forestry and Forest Industries, No. 137, 1982, p. 2.
15. Peter Coxon, Department of Geography, Trinity College, Dublin, *Irish Times*, 20 March 1985.
16. *Endeavour*, Mason, p. 35.
17. *Unasylva*. The figure for Ireland was supplied by Niall O'Carroll.
18. *One World*, the journal of Trócaire, December 1985
19. *Ibid.*
20. *Ibid.*
21. *Ibid.*
22. The Prince of Wales, 'Earth in the Balance', BBC, 23 May 1990.
23. George Green, EC Commission Agriculture Information Service, No. 204. See also FIPAP, p. 7.
24. *Ibid.* As we have already noted that position is beginning to change dramatically with private planting in 1989 and 1990 running at about 50 per cent of the total.
25. Report by David Hickie, *Forestry in Ireland, Policy and Practice*, published by An Taisce in 1990, which makes the case for greater environmental controls in forestry – planned and active.
26. *Ibid.*, p. 4.
27. *Ibid.*, p. 28.
28. *Ibid.*, p. 30.
29. *Ibid.*, Foreword.
30. Operational Forestry Programme 1989-1990.
31. *Ibid.*, pp. 23-7.
32. O'Carroll (written communication 1988) states that 'in 1987 the government decided that Bord na Móna cutaway bogs should be used for State afforestation'.
33. FIPAP, p. 28.
34. *Ibid.*, p. 26.
35. Afforestation and Nature Conservation in Great Britain. Nature Conservancy Council, 1986, cited FIPAP, p. 14.
36. 'The accusation that conifers cause acidification and high aluminium in water has been contradicted on scientific grounds' – O'Carroll, written comment 1988.
37. Community Strategy and Action Programme for the Forestry Sector, 1988.
38. OFP, p. 62.
39. FIPAP, p. 27.
40. *Our Natural Heritage,* Union of Professional and Technical Civil Servants, 1987, and cited FIPAP, p.3.

SELECT BIBLIOGRAPHY
DOCUMENTARY SOURCES

Anderson, M. L., *The Natural Woodlands of Britain and Ireland*, London 1932.
A Register of Trees for Co. Londonderry, 1768-1911, Public Record Office of Northern Ireland.
Acts of the Privy Council (1615-26)
Agriculture and Technical Instruction Act, 1899.
An Inventory of Private Woodlands, Thomas J. Purcell, 1973.
Annual Report of Commissioners of Woods and Forests, Great Britain, 1905.
Appraisal of the Public Capital Programme, State Forestry, 1969.
Baily, W. F., 'Forestry in Ireland', *Statistical and Social Inquiry Society of Ireland*, 1889.
Bulfin, Michael, Gardiner and Radford, Agricultural Institute, 1981.
    Timber and Timber Products Trade, 1966/1973, Irish Journal of Economic Rural Sociology, vol. 5.
    Farm, Food and Research, 1986.
Bulletin of the Forest Service of Washington, 1933.
British Forestry Commission, Post War Policy.
Calendar of Carew Papers (1515-1624)
Calendar of State Papers Domestic (1667, 1670, 1671).
Calendar of State Papers Ireland (1588-1670).
'Case for Forestry, The', Forest and Wildlife Service, 1980 (revised 1983).
Celtica, 11. 1976.
Census of Woodlands, 1947-9, British Crown Report, 1952.
Commercial Forestry, British Ministry of Reconstruction, 1919.
Connolly, Senator Joseph, unpublished papers.
Coxon, Peter, Trinity College, *Irish Times*, 20 March 1985.
'Developing Timber for the 80's', Industrial Development Authority, 1981.
Durand, 'The Evolution of State Forestry in Ireland', unpublished thesis.
Edlin, H. L., Review of the Forests of Ireland. *Quarterly Journal of Forestry*, LXI, No. 2, 1967, 176.
European Recovery Programme, The: Ireland's Long Term Programme (1949-53).
Falkiner, C. Litton, *The Forestry Question Considered Historically*, Statistical and Social Inquiry Society of Ireland, 23 January, 1903, Sealy Bryers and Walker, Dublin.
FAO (Cameron), Report on Forestry Mission to Ireland, 1951.
Faulkner's *Dublin Journal*.
Fitzpatrick, H. M., 'State Action with the Brake on', *Irish Times*, 10 August 1955.
Forest Research Review, Forestry Division Department of Lands, 1965.
Forestry in Ireland, Policy and Practice, An Taisce, 1990.
Forestry in Northern Ireland, Ministry of Agriculture Forest Policy (Northern Ireland), undated, unpaginated.
Forbes, A. C., 'Some Legendary and Historical References to Irish Woods', *PRIA*, XLI, 1932.
—, 'Forestry Revival in Eire', *Irish Forestry*, Vol. IV, No. 1.
—, 'Tree Planting in Ireland During Four Centuries', *PRIA*, LI.
—, 'Some Early Economic and Other Developments in Eire, and their effects on Forestry Conditions', *Irish Forestry*, Vol. 1, 1943.
—, 'Undated typescript report addressed to the Secretary DATI for the year ending 1922'.
—, 'Avondale Forestry Station', 1906-11.
Gilbert's Historical and Municipal Documents of Ireland.

Government White Paper, *Economic Development*, 1958.
—, *The Second Programme for Economic Expansion*, 1964.
Gray, H. J., 'The Economics of Irish Forestry', *Statistical and Social Inquiry Society of Ireland*, December 1963.
Green, George, EC Commission Agriculture Information Service, No. 204.
Henry, Augustine, 'Cooperation of State and Citizen in Irish Forestry', *Studies*, December 1924.
Hore, Herbert Francis, *Woods and Fastnesses of Ancient Ireland*, UJA, 1st ser. VI.
Hobson, Bulmer, *A National Forest Policy*, privately printed, 1931.
Howitz, Daniel, A Preliminary Report on the Reafforestation of Waste Lands in Ireland, 1884.
Interim Report on Afforestation Programme; Analytic Study of State Forest Undertaking (Confidential), Departmental Document, 1974.
Inventory of Woodlands of the Forest and Wildlife Service, Stationery Office, 1973.
Investing in Forestry in Ireland, Touche Ross, 1988.
Ireland's Forestry, A Review, Union of Professional and Technical Civil Servants, 1984.
Our National Heritage – A Policy for Conservation in Ireland, UPTCS, 1987.
Isbet, Jane, *The Forester*, Vol. 1, 1905.
Irish Forestry, official publication of the Society of Irish Foresters.
Journal of the Cork Historical and Archaeological Society.
Journal of the Royal Society of Antiquaries of Ireland.
Journal of Forestry and Estate Management, The, Vol. VI, 1880-88.
Joyce, Prof. Padraic, *The Irish Press*, 19 August 1985.
Kelly, Fergus, 'Old Irish Tree List', *Celtica*, Vol. XI 1976, pp. 101/124.
Land Purchase Act (Wyndham Act), 1903.
Le Fanu, T. P., 'The Royal Forest of Glencree' *JRSAI*, XIII.
Lucas, A. T., 'The Dugout Canoe in Ireland'.
—, 'The Sacred Trees of Ireland', *Journal of the Cork Historical and Archaeological Society*, 1962-3, pp. 67-8.
Lucey, Dr Cornelius, Minority Report, Commission for Emigration Report, 1955.
MacBride, Seán, Unpublished lecture to the Forestry Society of University College, Dublin, 1984.
McEvoy, T., 'Forestry in Ireland', *The Advancement of Science*, no. 56, 1958.
—, 'A Review of Afforestation Expertise in Ireland', a paper, 1957.
—, 'Irish Native Woodlands and their Present Condition', *Irish Forestry*, 1943.
Mitchell, Frank, 'Littleton Bog', a paper published by the Geological Society of America, 1965.
National Development Plan 1985-1993, Dublin 1989.
O'Loghlen, P. J., Commission of Enquiry into Banking, Currency and Credit, 1938.
Operational Programme for Forestry, 1989-1993, Dublin 1990.
O'Rahilly, Alfred, Notes on Mackey's *Forestry in Ireland*.
Origin and Development of the Irish Forestry Service 1900-1966, Forestry Division, Department of Lands, 1967.
Pierse, Sir Henry Beresford, *The Role of Forestry in the Economy of Britain*.
Private Forestry in the Irish Republic, Irish Landowners Convention, 1956.
Proceedings of the Royal Irish Academy, 1933.
Promise and Performance – Irish Environmental policies Analysed, UCD, 1983.
Post-War Forest Policy, British Forestry Commission, 1943.
Quarterly Journal of Forestry, 1907-29.

SELECT BIBLIOGRAPHY

Report of the Departmental Committee on British Forestry, 1902.
Report of the Departmental Committee on Irish Forestry, 1908.
Report of the Department of Agriculture and Technical Instruction for Ireland, 1923-6.
Report on the Woods, Plantations and Waste Lands in the South Eastern counties of Ireland, 1904, John Nisbet.
Report of the Review Group on Forestry, 1985.
Report (Second) British Royal Commission on Coast Erosion and Forestry, 1909.
Report of the British Forestry Commissioners, 1943.
Report of the Commission on Emigration, 1948-54, 1955.
Report of the Forestry Sub-Committee (The Acland Committee) of the Ministry of Reconstruction, 1919.
Report on the Census of Woodlands and Census of Production of Home-Grown Timber, 1924, British Forestry Commission, 1928.
Report of the Minister for Agriculture, 1931-2.
Report of the Minister for Lands on Forestry, 1933-8.
Report of the Minister for Lands and Forestry, 1938-77.
Report of the Minister for Fisheries and Forestry, 1977-85.
Report of the Minister for Tourism, Fisheries and Forestry, 1986.
Report of the Minister for Marine, 1987.
Reports of the Department of Energy 1988-9.
Review of State Expenditure on Forest and Wildlife Service, Dáil Committee on Public Expenditure, 1986.
Revue Celtique.
Schlich, William, Report on the Afforestation of Great Britain and Ireland, 1886.
—, Report on the Plantations at Knockboy, 1895.
Sinn Féin National Programme, 1905.
Spenser, Edmund, *The Faerie Queene.*
Stanvel, Charles A., 'Irish Forestry and the Land Purchase Acts'.
Statistical and Social Inquiry Society of Ireland, 1907.
Strategy for the United Kingdom, Forest Industry, CAS Report 6, University of Reading.
Terms of Reference, Select (Recess) Committee, 1905.
The Theft of the Hills: Afforestation in Scotland, Steven Tomkins, Ramblers' Association, London 1986.

PAMPHLETS AND NEWSPAPERS

Bulletin of the Forest Service of Washington, 1933.
Bray People.
Connacht Telegraph.
Forest Products Handbook, Svenska Teknolog Foreningens Forlag, Stockholm.
The Farmers' Journal.
Irish Independent.
The Irish Press.
Irish Times.
Leitrim Observer.
Longford News.
Literary Digest, 1927.
Northern Standard.
One World, Trócaire, 1985.
Rea, Thomas, Extract from *Irish Forestry,* vol. 42. 1985.
Sunday Independent.

Speeches of:
    Erskine Childers, 1958.
    Eamon de Valera on the Forestry Estimate, 1950-1.
    James Dillon, 1958.
    Michael Moran, Castlebar 1960.
    Robert Molloy, 1989.
Unasylva, Journal of Forestry and Forest Industries, 1982.
Viney, Michael, *Irish Times*, 10, 11, 12, 13, 14 October 1966.

SECONDARY SOURCES

Albion, R. G., *Forests and Sea Power*, Harvard 1926.
Butcher, W. *Treatise on Forest Trees*, Dublin 1784.
Cambrensis, Giraldus, *Topographica Hibernica*; Gerald of Wales, *The History and Topography of Ireland*, transl. John O'Meara, Dundalk 1951; Dublin 1982.
Church, Rock, *An Old Thrift*, London 1612.
Cole, Sonia, *The Neolithic Revolution*, London 1959.
Convery, Frank, *Irish Forest Policy*, National Economic and Social Council, Dublin 1979.
Cook, Moses, *The Manner of Raising Ordering and Improving Forest Trees 1717* (2nd ed.), London 1717.
Cooper, R. T., *Forestry in Ireland*, privately printed 1902.
Crawford, *Depopulation Not Necessary*.
Dardis, Patrick G., *The Occupation of Land in Ireland in the first half of the 19th Century*, London 1920.
De Courcy Ireland, John, *Ireland's Sea Fisheries, A History*, Dublin 1981.
de Latocnaye, The Chevalier, *A Frenchman's Walk through Ireland*, Paris 1798; reprint Belfast 1985.
Douglas, D. C. and Greenway, G. W. (eds), *English Historical Documents*, London 1969-77.
Dunlop, R., *Ireland under the Commonwealth*, 2 vols, Manchester 1913.
Dunraven, *Legacy of Past Years*, London 1911.
Evelyn, John, *Sylva, or a Discourse on Forest Trees*, London 1664.
Fanning, Ronan, *The Irish Department of Finance*, Dublin 1978.
Feehan, John, *The Landscape of Slieve Bloom*, Dublin 1979.
Forbes, A. C., *The Development of British Forestry*, London 1910.
Halley, *Atlas Baratimos et Commercialis*, London 1728.
Hayes, Samuel, *A Practical Treatise on Planting and the Management of Woods and Coppices* (2nd ed.), Dublin 1822.
Henry, Augustine, *The Woods and Trees of Ireland*, Dublin 1914.
Hiley, W. E., *Woodland Management*, London 1967.
—, *Economics of Forestry*, London 1930.
Hinde, Thomas, *Forests of Britain*, London 1985.
Hoctor, E. *The Department's Story – A History of the Department of Agriculture*, Dublin 1871.
Innes, Arthur D., *England Under the Tudors*, London 1929.
James, N. D. G., *A History of English Forestry*, London 1981.
Joyce, P. W., *A Social History of Ancient Ireland*, 2 vols, Dublin 1903.
Jones and Simmons, *The Story of the Saw*, London 1961.
Kilpatrick, C. S., *Northern Ireland Forest Service, A History*, Belfast 1987.
Lecky, Wm, *A History of England in the Eighteenth Century*, London 1878-90.
Lucas, R., *Cork Directory* (1787).

SELECT BIBLIOGRAPHY

Mackey, John, *Forestry in Ireland, A Study of Modern Forestry and of the Interdependence of Agriculture and 'Industry'*, Cork 1934.
MacNeill, Eoin, *Celtic Ireland*, Dublin 1921; reissue Dublin 1981.
McCracken, Eileen, *The Irish Woods Since Tudor Times*, Newton Abbot 1971.
Marshall, Wm, *Planting and Rural Ornament*, London 1785.
Miles, Roger, *Forestry in the English Landscape*, London 1967.
Mitchell, Frank, *The Irish Landscape*, London 1976.
Moody, T. W., *The Londonderry Plantation, 1609-41*, Belfast 1939.
Murphy, Gerard, *Early Irish Lyrics*, Oxford 1956.
Murphy, John A., *Ireland in the Twentieth Century*, Dublin 1975.
O'Carroll, Niall (ed.), *The Forests of Ireland*, Dublin 1987.
Osmaston, F. C., *The Management of Forests*, London 1968.
Pigott, Stuart, *The Druids*, London 1968.
Pym, Sheila, *The Wood and the Trees*, London 1966.
Rees, Alwyn and Brinley, *Celtic Heritage*, London 1961.
Reid, William, *Weapons Through the Ages*, London 1984.
Rushworth, John, *Historical Collections*, London 1700.
Schlich, W., *Schlich's Manual of Forestry* (3rd ed.), London 1904.
Scott, Richard J., *The Galway Hookers*, Dublin 1983.
Smith, C., *Antient and Present State of Cork*, 1751, 2 vols, Cork 1815.
Standish, Arthur, *The Commons Complaint*, London 1611.
Troup, R. S., *Forestry and State Control*, Oxford 1938.
Woolsey, Theodore, *Studies in French Forestry*, New York 1920.
Young, Arthur, *A Tour of Ireland*, London 1780.

INDEX

abeal, 76
Abercorn, Duke of, 123, 279
Abercorn Committee, 283
Acland, F. D., 140-1
Acland Committee, 141, 142-5, 145, 165, 176, 203, 267, 350
Admiralty, British, 81
afforestation
    grants, 159, 163, 178, 220-1, 258, 275
    incentives, 92, 233-5
    opposition to, 98, 99, 105, 118, 161-2, 187, 200-3, 220-1, 263, 267-9, 276
    Co. Leitrim, 242-3
    in NI, 281
    sheep v. forests, 212, 218-20, 267-9
    report on, 135-6
Afforestation Conference, London, 134-5
Ager, W. E., 146
Agricultural Institute, 234
Agricultural Wages Act, 1936 (Minimum Rates) Order, 1942, 356
agriculture, 51
    cattle exports banned, 70-1, 74
    changes in, 272
    deforestation, 15-17, 27, 30, 76, 159-61
    see also afforestation, opposition to
Agriculture, Board of, 95, 110, 135, 140, 144
Agriculture, Council of, 122, 124, 134, 142
Agriculture, Department of, 119, 157, 165, 166, 171, 179, 182, 198, 220, 228, 270, 327
    annual report, 159
Agriculture, NI Ministry of, 280
Agriculture Act, 1931, 322
Agriculture and Food Research Institute; see Foras Talúntais, An
Agriculture and Technical Instruction, Department of (DATI), 119, 121, 123-4, 133, 146, 147, 158, 327
    and Acland Report, 144-5
    funding, 124, 125, 135, 142
    land acquisition, 124, 279
    report, 1923-6, 151-2
    Vice Presidents, 329
    woodlands trustee, 134
    see also Irish Forestry, Committee Report on, 1908
Aiken, Frank, 182, 329
alder, 28, 30, 31
Allied Irish Banks (AIB), 236

Amenity Land Act (NI), 1965, 283
Ancient Laws of Ireland, 28
Anderson, Jan, 243
Anderson, Dr Mark Loudon, 165-9, 178, 181, 182, 327
Anglo-Irish Treaty, 1921, 146, 147
Annals of Ulster, 45
Anne, Queen, 73
    forest legislation, 302-4
Antrim, Co., 14, 281
apple, 28, 29
arbutus, 28, 32, 76
Archer, Lt Joseph, 314
Arklow, Co. Wicklow, 85-90
Armagh, 22, 45
Armagh, Co.
    sacred wood, 333-4
    tree plantations, 317-18
Arts and Manufactures, Society for the Encouragement of, 92, 98
Asgard (vessel), 89
ash, 16, 17, 20, 63, 103
    Celtic classification, 28, 29, 30
    decline, 26, 30
    and glassworks, 75
    promotion of, 76, 93, 95
    propagation, 24
    sacred tree, 21
aspen, 28, 31
Augustine Henry Memorial Grove, Portglenone, 284
Australia, 10, 14
Austria, 106-7
Austrian Cameral Valuation method, 107
Avondale, Co. Wicklow, 88, 122-4, 134, 135, 138, 145, 259, 262
Aylward, Liam, 330

Babington Committee, 282
Bacon, Sir Francis, 76
badgers, 55
Bagneris, G., 173
Bailey, W. F., 126
Balbriggan, Co. Dublin, 44
Balfour, Sir Arthur, 111-12, 113, 119
Ballaghmoon, Co. Carlow, 17
Ballot Act, 103
Ballyarthur, Co. Wicklow, 87, 88
Ballyfad, 135
Ballykelly Wood, Co. Derry, 279
Ballynagilly, Co. Tyrone, 13

INDEX

Bandon Bridge, 56
Banking Commission, 192
bark, 23, 74
Baronscourt, Co. Tyrone, 279, 280, 281
barrel-stave industry, 68-70, 71
Barrie, Hugh T., 329
Barrington, Sir Jonah, 297
Barry, P., 327
Barton, Robert, 147, 329
basket-weaving, 94
beech, 80, 88, 89, 92, 95
Begot, Mr, 316
Betham, 47
Binchy, D. A., 31, 99
birch, 23, 28, 31, 63
Black, John, 125
Black Death, 48, 49
Blackhouse, Jonathan, 92
blackthorn, 28, 31, 32
Blood, Bindon, 317
Blowick, Joseph, 191, 198, 210-11, 270, 329
boar, wild, 55
boat-building, 12, 14
Boate, 76, 299
bog-myrtle, 28
boglands, 16
    afforestation of, 113, 253, 290, 368
    growth of, 10, 13, 26
Boland, Gerard, 182, 329
Bollem (forester), 107
Bord na Móna, 16, 181, 253, 290, 368
bore-tree, 28
Bourke, Mr, 315
Bowater Ltd, 283
Boyle, Sir Richard, Earl of Cork, 52, 53, 56, 61, 66, 70
    glassworks, 75
Boyle family, 63-4
bracken, 28, 32
Bradley, Diarmaid, 236
Bradley, H. E. M., 327
brambles, 28
Brandis, Dietrich, 108
Brazil, 15
Brehon Laws, 17, 33-4, 320
    carpentry, 45
    forest ownership, 27-8
    house construction, 43
Bretha Comaithchesa, 27
Brewster family, 76
Britain
    civil war, 65
    disafforestation, 100
    Irish timber exports to, 70-1
    outlaws, 49
    as sea-power, 49-50, 53, 59-60, 79-80
    timber imports, 137, 152-3
    timber reserves, 50, 138
    see also forests, British
British Forestry, Departmental Committee on, 122, 128
Bronze Age, 13
broom, 28, 32
Brown, Stephen, 126
Browne, Dr Noël, 199
Browne, Sir Valentine, 300
Brownlow, Arthur, 63
Brownlow, Mr, 316
Bruce wars, 48
Bryan, John, 340
Bulfin, Michael, 234-5
Burgess, Col. F. G., 280
Burghley, Lord, 51, 52, 64, 74
Burke, Ray, 330
Burke, Sir Thomas, 317
Burton, William, 317
Bury, C. W., 314
Byrne, Miley, 88

Caernarvon, Earl of, 111
Caesar, Julius, 18
Callanan, Patrick, 328
Cameron Report, 196-200, 205, 206, 211, 220-2, 264, 266, 277
Campbell, J. R., 327
Campbell, Professor Ritch, 126
Canada, 59
canoes, 12, 14, 84-5
Cardiff family, 85
Carew, Sir George, 51, 64, 297-8
Carlow, Co., 64, 253
Carrickfergus, Co. Antrim, 72
Carrington, Lord, 135, 185
Carton, Co. Kildare, 103
CAS Report, 232
Case, H. J., 14
Case for Forestry, The, 230, 231, 253-4, 254-5
    on marketing, 256-7
Cassidy, Paul, 328, 363
Castlecaldwell, Co. Fermanagh, 279
Castledermot, Co. Kildare, 48
Castletown, Lord, 123, 124, 126, 132
Catholic Emancipation, 102-3, 104, 262
Catholic Relief Act, 1793, 102
Cavan, Co., 253
    tree plantations, 314
Celts, 12, 13, 17-23
    forest ownership, 27-8
    land ownership, 24-32, 33-4
    social system, 319-20
charcoal, 16, 54, 75
Charlemont Report, 280, 281
Charles I, 71
    forest legislation, 301
Charles V, King of France, 67

INDEX

cherry, wild, 28
Chichester, Lord, 261
Chichester, Lord Deputy, 53, 70
Chichester, Sir Arthur, 83, 84
Childers, Erskine, 212-13, 266, 268, 270-1, 272, 274, 329
    Dáil speech, 359-60
chipboard, 283, 284, 362
Chipboard Ltd, 363
Church, Rock, 68
civil war, 159, 170, 264
Clancarty, Earl of, 300
Clann na Poblachta, 190
Clann na Talmhan, 190
Clanrickard, Earl of, 317
Clare, Co., 55, 206, 222, 253
    tree plantations, 316-17
Clear, Professor Thomas, 178, 211, 355
Clew Bay, Co. Mayo, 85
climate, 9-10, 13, 344
Clondalkin Paper Mills, 363
Clonmacnoise, Co. Offaly, 20
Cnut, King, 24-5
co-operative forestry, 161-2, 201, 205, 208, 221, 239, 259
    increase in, 136-7
    proposals, 233, 265
coal, 51, 54, 64, 73, 75, 107
Coillte Teoranta, 244-5, 254, 276, 290, 328
Coleraine, Co. Derry, 69, 283
Colmcille, St, 44
Commercial Forestry (Ministry of Reconstruction), 348-9, 351, 352
Commercial Policy of the Irish Parliament, The (Binchy), 99
Common Agricultural Policy (CAP), 278, 291
Compleat Measurer, The (Hodgson), 97
Condon, Lord, 52
Connemara hooker, 85-6
Congested Districts Board, 85, 112, 119, 145
conifers, 95, 97, 109, 119, 137
    complaints of, 241-2, 288, 368
    distribution of, 247-8
    introduction of, 78, 100, 114
    Irish percentage, 289
    planted in boglands, 290-1
    and shipbuilding, 81, 89
Connolly, Joseph, 174-5, 182, 190, 196, 201, 225, 274, 329, 355
    afforestation plans, 165, 166, 183
    and social forestry, 156, 168, 177, 185, 188
conservation, 66-7, 215, 224, 262-3
    in Northern Ireland, 283
Construction Corps, 180

Convery, Frank, 229-30, 232, 255, 257
Cooke, Moses, 68, 78
Coolattin, Co. Wicklow, 87, 89, 202, 339
Cooper, Dr Richard, 121, 123
Cooper, Dr Robert, 345
Coote, Sir Charles, 313, 314
coppicing, 29, 46, 54, 55, 72, 75, 76
    decline in, 107
Cork, 33, 85
Cork, Co., 69, 206
Corsican pine, 78
Cosby, Philip, 103
Costello, John A., 194
Cotta, Heinrich von, 106, 107
Cottingham, Philip, 52, 83-4, 261, 338
County Forest Schemes, 135
Coventry, Lord, 80
Coxon, Dr, 287
Crann, 243
Crimthan, 35
Crom Cruach, 19
Cromwell, Oliver, 53, 65
Crozier, John D., 145, 165, 168, 174, 327, 350
currachs, 14

Dáil Éireann, 146
    Forestry Committee, 147, 158
Dal gCais, 17, 21
Daly, Brendan, 330
Daly, Denis Bowes, 317
DATI; *see* Agriculture and Technical Instruction, Dept of
Davies, Sir John, 298, 299
Davis, J., 315
de Davenport, Sir John, 49
de Marisco, Geoffrey, 41
de Neville, Adam, 40, 41
de Neville, Hugh, 36
de Paor, Seamus, 328
de Valera, Eamon, 162, 183, 190, 225
de Vesci, Lord, 123, 143
deal, 42
Dean, Forest of, 106, 110
Dedication Scheme, 189
Deegan, M., 173, 327
deer, 42, 55
deforestation, 49-56, 60-4, 66-7, 103, 118
    fines for, 38
    17th c., 297-300
    worldwide, 287-8
Denmark, 16, 229
Departmental Endowment Fund, 123
Deputy's Pass, Co. Wicklow, 64
Derrig, Thomas, 182, 208, 329
Derry, 69
Derry, Co., 279, 281, 282
Desmond, Earl of, 52, 61

377

Development and Road Improvement Funds Act, 1909, 132
Development Commission, British, 132-3
Devonshire, Duke of, 92
Dialogus de Scaccario (FitzNigel), 37
Dillon, James, 271, 360
Dillon, John, 367
Dinely, Thomas, 298
disafforestation, 39
    Glendalough, 295-6
Dixon, W. Vickers, 327
Dobbs, A., 89
Dodwell, George, 315
Domesday Book, 36
Donegal, Co., 206
    tree plantations, 315
Donegan, Patrick S., 330
Donelan, Dermot O'C., 112
Doran, Henry, 121
Douglas fir, 89, 109, 114, 248, 252
Down, Co., 14, 280, 281, 283
Down Survey, 51
Drogheda, Co. Louth, 42
Drought, Thomas, 313, 314
druidism, 18-20, 22
Drummond, R. O., 283
Drummond, Thomas, 104
Drummond Committee, 283
Dublin, 33, 42, 44, 72, 79, 222
    decline of, 98-9
    shipbuilding, 85
    timber merchants, 96-7
    Wood Quay, 74
Dublin, Co.
    tree plantations, 314
Dublin Gazette
    advertisement, 308
Dublin Society; *see* Royal Dublin Society (RDS)
Duggan, A. W., 328
Dunboy, siege of, 56
Durand, J. F., 121, 124, 166, 196, 208
Dutton, Hely, 314, 316
Dymmock, John, 57

East India Company, 63, 70, 80, 83, 84
Easter Rising, 1916, 142
Eberswalde Prussian Forest College, 355
Economic Development (White Paper), 216, 244
Economic Expansion, Second Programme for, 222-3
Economic War, 161, 190, 267, 292
Économie Forestière (Huffel), 106-7
Edlin, G. L., 103
Edric the Wild, 49
education; *see* forestry education
Edward I, 47

elder, 28, 32, 76
elections, 99, 102-3
Elements of Silviculture (Bagneris), 173
Elizabeth I, 50-1
Ellison, 126
Ellison, Edwin, 125
elm, 71, 76, 340
    Celtic classification, 28, 30
    decline of, 15, 26, 31, 64
    dominant, 78
    promotion of, 93, 95
    and shipbuilding, 80
Emergency, The; *see* Second World War
Emergency Powers Order (No. 169), 356
Emigration, Commission on, 200, 205
employment; *see* social forestry
Energy, Department of, 290, 328
Energy and Communications, Department of, 361
English Arboricultural Society, 106
Enniscorthy, Co. Wexford, 51
environment
    and forestry, 202, 215, 229
    worldwide concern, 285-8, 289
Environmental Protection Agency, 290
Erris (Co. Mayo) Co-operative, 234
Essex, Earls of, 62
estate forestry, 91-5, 97-8, 98, 109-10, 133, 176, 266, 344
    acreage, 175
    decline in, 139-40, 176-7, 182, 187, 204
    effect of Land Acts, 118-19
    and Saorstát Éireann, 148-9, 155-6
    science of, 106-9
    and Second World War, 179-80
    between the wars, 174-5
Estates Commissioners, 125, 134, 135
European Communities, Commission of the, 255-6, 291
European Community (EC), 221, 245, 277, 278, 284
    grants, 260, 274-5
    private forestry, 218, 235
    timber demand, 229, 232, 235, 289
European Conservation Year, 229
European Economic Recovery Programme (Marshall Plan), 192-5
Evelyn, John, 200, 247
External Affairs, Department of, 191, 192, 193, 194, 197

Faerie Queene, The (Spenser), 61
Falkiner, C. Litton, 47, 76, 112, 298-9
fallow deer, 42, 55
Fanning, Ronan, 192
farmers; *see* afforestation, opposition to

INDEX

Faulkner, Padraig, 329
Faustmann Formula, 107
Fenians, 262
Fenwick, R., 92
Fermanagh, Co., 279
Fernow, Dr, 353
Feudal and Parliamentary Dignities (Betham), 47
Fianna Fáil, 164, 170, 187, 190, 208, 216
    forestry policy, 225
fibre-board plant, 362
fidnemed, 19, 20, 22
Fife, Earl of, 317
Finance, Department of, 162, 168, 216-17, 237, 244, 257
    and MacBride proposals, 191-9
Fine Gael, 190
Finglas, Co. Dublin, 29
Fintan of Munster, 18, 21
Fionn Mac Cumhail, 23
fir, 76
First World War, 132, 135, 137-43, 146, 152, 156, 249, 263
    timber shortages, 113, 138
Fisher, W. R., 126
Fisheries, Department of, 328
Fisheries and Forestry, Department of, 228, 237, 328
fitz Adam, Thomas, 40-2
FitzMaurice, Sir James, 61
FitzNigel, Richard, 37
FitzPatrick, H. M., 209, 355
Fitzpatrick, Tom, 330
Fitzwilliam, Lord, 314
Five Fifths, 18
Five Year Programme for Economic Expansion (White Paper), 193, 216
Flanagan, Sean, 329
Flannery, Thomas J., 113
Fletcher, G., 327
Flinn, Hugo, 168
flint tools, 11
Fodla Tíre (Divisions of Land), 30
Foir Teoranta, 257
Food and Agricultural Organization (FAO), 196-9, 232, 287
Foras Talúntais, An, 214
Forbes, A. C., 26, 40, 42, 48, 121, 126, 139, 159, 166, 172, 279, 295, 327
    Acland Committee, 143
    Afforestation Conference, 134-5
    Assistant Commissioner, 145-6
    and Avondale, 123, 124-5
    Chief Forestry Inspector, 155
    criticisms by, 156-8
    on early Irish forests, 57-9, 91
    on estate forestry, 174-5
    and First World War, 140-1
    on Irish Forestry Society, 122
    on Knockboy Experiment, 111-12
    on land acquisition, 164
    and NI forestry, 280
    planting scheme, 164-5
    on private forestry, 148, 176, 177, 211, 267, 274
    retirement, 165
    and Saorstát Éireann, 150-1
    on timber needs, 154
Forest, Charter of the, 49-50
Forest Advisory Service, 201
Forest and Wildlife Service, 215, 229, 232, 276
    examination of, 237-40
    Trees for Ireland, 241
Forest Commission, 144-5
forest law, 25, 27, 40-2, 337
    Eyre court, 39-40
    Norman, 24-6, 36-8
forest management, 46-50, 142, 231, 250
    after Second World War, 188-9
    benefits, 171
    continental developments, 35-6, 46, 106-9
    corruption, 48, 50, 68, 90
    cost of, 291
    environmental considerations, 202
    experimentation, 252-3
    forestry officials since 1904, 327-8
    funding, 170-1
    lack of, 39, 46-7
    legislation, 49, 53-5, 66, 67, 71, 73, 76-7, 144-5
    and MacBride, 190-8
    Ministers responsible for, since 1904, 329-30
    nationalist support for, 146-7
    need for conservation, 66-7
    Norman, 46
    policy, 153-5, 183, 203-6, 210
    rates of planting, 77-8, 123-4, 164-5, 182-3, 195-7, 205-6, 208, 210, 254-5
    Roman, 35, 337
    scientific forestry, 106
    Tudor, 53-4
    between the wars, 172-5
Forest Trees (Cooke), 68
Foresters of Great Britain; Society of, 175
forestry; *see* estate forestry; private forestry; social forestry; state forestry
Forestry, Advisory Committee on, 136
Forestry, Board of, 106
Forestry, Review Group on, 207, 213, 276
Forestry Act, 1919, 141, 144-5, 174, 176, 322
Forestry Act, 1928, 161, 163-4, 166, 176-7, 263, 266, 267, 322

379

INDEX

Forestry Act, 1946, 277, 322
Forestry Act, 1956, 322
Forestry Act, 1988, 322
Forestry Act (UK), 1947, 189
Forestry and Environment Division, EC, 255-6
Forestry Bill, 1988, 276
Forestry Commission, British, 117, 143, 147, 149, 151, 155, 165, 174, 238, 255, 327
    established, 140-1
    expenditure forecasts, 363
    and Northern Ireland, 279
    Planting Grant Scheme, 280
    Post-War Forest Policy Report, 167
    proposals, 161, 175, 186
    Second World War, 188-9
Forestry Commission Act, 146
Forestry Committee, Dáil Éireann, 147, 158
Forestry Division, Department of Lands, 163, 165, 200, 209, 244, 265
    annual reports, 211
    and co-operative forestry, 233
    control of, 191
    lack of policy, 187-8
    new department, 228
    rates of planting, 195-7, 212
    Reinhardt, director, 169
    Research Branch, 213-14
    and social forestry, 199
forestry education, 110, 134, 136, 157, 280, 355
    Avondale, 122-3, 124-5, 259, 262
    continental, 106-8, 178
    graduate intake, 177
    Northern Ireland, 282
Forestry Fund, 185
Forestry in Ireland (Mackey), 154, 170, 174
Forestry Service, 165-6
    personnel, 177-8
Forestry Sub-Committee of Ministry of Reconstruction, 141
Forestry (Transfer of Woods) Act, 1923, 140
forests, British, 66, 77-8, 90, 91, 132, 255
    Acland Committee, 142-3
    acreage, 100, 155-6, 222
    conifers, 251
    forest law, 36-8
    new forests, 256
    report, 1902, 122
    and shipbuilding, 81-2, 101, 261
    species, 248, 249-50
forests, Irish, 25-7, 37, 38, 83-4, 100, 108-9, 114
    acreage, 100, 110, 129, 133-4, 149, 151, 202-3, 222, 262, 265, 339, 345

assarted, 38
Celtic ownership law, 27-8
changing species, 247-9
clearance of, 15-17
common land, 27
conifers, 289
destruction of, 49-56, 60-4, 66-7, 103, 118
distribution of, 251, 252
effects of Land Act, 109-10
experimentation, 111-13, 252-3
during First World War, 138-9, 146
long-term investment, 162-3
neolithic period, 13, 15-17
Norman law, 58
pre-Tudor, 57-9
profitability, 202-3, 206, 209, 273-4
reafforestation, 101, 119-21, 261-2
rebel strongholds, 39, 45, 47, 48, 50, 56, 62-3, 74, 76
scrub, 57-8
17th-c. deforestation, 297-300
Tudor survey of, 52-3
see also Irish Forestry, Committee Report on, 1908
Forstbenutzung, die (Gayer), 108
Forstschutz, Der (Hess), 108
foxes, 55
France, 59, 69, 79, 102, 222, 289
    forest management, 35-6, 46, 90, 137, 152, 171-2, 173, 228
    nurseries, 95
    Ordinance de Melun, 67-8
    scientific forestry, 106-7
    timber market, 70
Frazer, Robert, 314
furze, 22, 28, 32, 44-5

Gaelic League, 132
Gaelic Revival, 106, 122
    14th c., 48
Gaeltacht, afforestation of, 166, 168, 188, 211, 224
Gallagher, Ray, 236-7
Galvin, John, 126
Galway, Co., 206
    tree plantations, 317
game; *see* sport
Gardner, Jack, 234
Garrett, George, 193-4
Gayer, Karl, 107-8, 108
George I
    forest legislation, 304
George II
    forest legislation, 305
George III
    forest legislation, 305-8, 366
Germany, 137, 141, 178, 222, 228, 289

INDEX

forestry, 95, 172-3
   scientific forestry, 106-7, 136
Gill, T. P., 119, 126, 147, 327
Giraldus Cambrensis, 20, 29, 30, 47, 57
glaciation, 9-10
Gladstone, W. E., 111, 120
glass manufacture, 73, 74-5
Glen Mama, battle of, 20
Glenariff Forest Park, 283
Glencolumcille, Co. Donegal, 343
Glenconkyne, Co. Derry, 71
Glendalough, Co. Wicklow, 40-2, 58
   disafforestation, 295-6
gooseberry, 28
Gordon, J. S., 281
Government of Ireland Act, 1920, 141, 145, 146, 147
grassland, 27
Grattan, Henry, 93
Gray, Henry J., 118, 159-60, 185-6, 188, 209, 213, 327, 328
   on forestry policy, 203, 204
   on profits, 207
Great Famine, 23, 262
Greece, 289
Greeley, Col., 215
greenhouse effect, 286-7
Greenland, 12
Greenmount Agricultural College, 282
Griffith, Arthur, 132
Griffith, Sir Arthur, 343
Grousset, Paschal, 91
Gwynn, Dr E. J., 335

Hall, Edward, 63
Hamilton, Alexander, 314
Hampton Roads, battle of, 101, 111
hares, 55
Harriman, Averell, 193
Hartig, Georg L., 106
Hartig, Heinrich, 107
Harvey, Maurice, 234, 236, 297
Haughey, C. J., 202
hawthorn, 16, 28, 31
Hayes, Samuel, 122-3
hazel, 15, 16-17, 26, 152
   Celtic classification, 28, 29
   decline in, 72-3, 91
   propagation of, 24
   sacred tree, 23
   use forbidden, 73, 77
   wicker-work, 42-4
heather, 28
Heffernan, Mr, 196
Henry, Augustine, 71, 128, 138-9, 144, 164, 173, 178, 274, 355
   on afforestation, 113, 201
   calls for legislation, 264-5

   on land acquisition, 136
   and new species, 250
   on social forestry, 186, 202
Henry I, 25
Henry II, 37, 39, 41, 42, 44
Henry III, 36, 49
Henry VIII, 49, 50, 66, 71
Hess, Richard, 108
Hickie, David, 368
Hill, H. C., 106
Hinde, Thomas, 25, 138
Hobson, Bulmer, 150, 166, 167, 170, 174, 177, 186, 192, 202, 354
Hodgson, Daniel, 97
Hodgson, Levi, 97
Hoffman, Paul, 193
Hogan, Patrick, 157-8, 159, 329
Holiday (Employee) Act, 1939, 356
Holloway, J. C., 328
holly, 16, 17, 19, 24, 28, 29
holy wells, 19, 20, 21
Home-Grown Timber Committee, 140
Home-Grown Timber Committee, Inter-Departmental, 175
Home-Grown Timber Council, National, 175
Home-Grown Timber Marketing Council, 175
Hood, Robin, 24, 49
Hore, Herbert F., 47, 297, 298, 299, 300
hornbeam, 152
housing, 15, 17, 42-4
   timber demand, 54, 71-2
Howitz, Daniel C., 111, 112, 119-21, 152, 186
Huffel, G., 106-7, 152
Hummel, Dr F. C., 255-6

Ice Age, 9
Iceland, 12
India, 66, 106
Indian Forest Service, 108
industrial development, 51, 54, 108
   timber demand, 54, 68-70, 73-7
   timber shortages, 152-3
Industrial Development Authority (IDA), 229
Industrial Research and Standards, Institute for (IIRS), 214, 271
Industries Committee, 121
Industry, Trade, Commerce and Tourism, Department of, 237
Industry and Commerce, Department of, 269
Innes, Arthur, 84
Interim Forest Authority, 144-5
International Forestry Exhibition, Edinburgh, 106

INDEX

Invermore (vessel), 87-8
Iona, 44
Ireland's Forestry, A Review (UPTCS), 242
Ireland's Long-Term Recovery Programme (White Paper), 192
Irish Agricultural Organization Society (IAOS), 119
Irish-American Cultural Institute, 241
Irish Board Mills Ltd, 363
Irish Co-operative Organization Society (ICOS), 234, 236-7
Irish Creamery Milk Suppliers' Association (ICMSA), 133-4, 236
Irish Farmers' Association (IFA), 133, 242-3
Irish Forest Service, 166-7
Irish Foresters, Society of, 172, 181, 269
Irish Forestry, Committee Report on, 1908, 125-31, 135, 145, 151, 164, 176, 187, 203, 236, 259, 262, 289, 291
    acreage, 260, 263
    private forestry, 148, 174, 233, 267
    on species, 250
    targets, 246
Irish Forestry Board; *see* Coillte Teoranta
Irish Forestry Society, 121-4, 132, 144, 165
Irish Forests, Committee Report on, 1908 proposals, 323-6
Irish Free State; *see* Saorstát Éireann
Irish Free State Agreement Act, 158
Irish Land Commission, 112
Irish Landowners' Convention, 1956, 269
Irish Life Insurance Company, 274
Irish Sugar Company, 365
ironworks, 63, 65, 99
    timber demand, 35, 54, 73, 75-7
Israel, 241
ivy, 28, 29

James, N. D. G., 90
Japanese larch, 114
Joyce, P. W., 44, 243, 319
Judeich, Friederich, 108
juniper, 28, 32

Kalahari desert, 10
Kavanagh, Liam, 330
Keating, Geoffrey, 44
Keating, Mr, 316
Kehoe family, 85
Kelly, Dr Denis, Bishop of Ross, 126
Kelly, Fergus, 30, 32
Kerry, Co., 69, 206
Kilbracken, Lord, 243
Kildare, Co., 64, 222
    tree plantations, 316

Kilkenny, Co., 253
    tree plantations, 315
Killetra, Co. Derry, 71
Kilmartin, 145-6, 350
Kilpatrick, C. S., 281
Kinch family, 85
King and Hermit Dialogue, 31
King's County; *see* Offaly, Co.
Kinsale, battle of, 62
Knockboy Experiment, 111-13, 120, 121, 123, 132, 251, 262, 282
Knocklong, Co. Limerick, 22
Knockmany, Co. Tyrone, 279

La Touche, John, 316
Labour Party, 190
land acquisition, 99, 102-5, 159-61, 164, 187, 200-1, 203, 206, 212-13
    Cameron Report, 198-9
    compulsory, 163
    difficulties, 103, 222, 232-5, 277-8
    foreign investment, 271-2
    history of land ownership, 319-21
    holding of title, 252
    inadequate, 204, 209
    Land Acts, 109-10
    need for Land Use Policy, 289-90
    in Northern Ireland, 282
    price of, 164, 221
    rate of, 206-7, 208, 210
    tenant right, 102-5
    Ulster custom, 103-4
Land Acts, 118, 125, 127-8, 131, 134, 135, 262
    1881, 109-10
    1903, 122, 124
    1909, 134
    effects of, 346-7
Land Commission, 145, 161, 228, 263
Land Purchase Loans, 141-2
Lands, Department of, 182, 197, 212-13, 222, 270, 327-8
    and MacBride proposals, 191-3, 195-6
    opposed to forestry, 162
    and private forestry, 272
    and social forestry, 224
    see also Forestry Division, Dept of Lands
Lands and Agriculture, Department of, 148, 155, 327
Lands and Fisheries, Department of, 161
Laois, Co., 60
    tree plantations, 313
larch, 78, 86, 88, 89, 92, 93, 95, 250, 281
Larne, Co. Antrim, 11
Larnian people, 11
Laudabiliter, 41
le Devenies, William, 338

Le Fanu, T. P., 69, 83, 297, 340
legislation
    Statutes, pre-1800, 301-8
    Statutes since 1920, 322
    see also under forest management
Leinster, Duke of, 103
Leitrim, Co., 75, 234, 253
    hostility to afforestation, 242-3, 277
    tree plantations, 315-16
Lemass, Seán, 216, 223, 360
Lenihan, Brian, 330
Leyland, John, 66
Limerick, 33, 79, 85
Limerick, Co., 55, 103, 253
limes, 26
Littleton Bog, 72
Lloyd George, David, 138
Lodgepole pine, 247, 250-2, 253
London, 98
    Great Fire of, 42, 71, 72
London Afforestation Conference, 185
Londonderry, plantation of, 69, 71
Long, George, 74
Lorentz, Bernard, 107
Lough Neagh, 12
Loughrey, John, 328
Lovat, Lord, 143
Lowery, Martin, 328
Lowther, Mr, 315
Lucas, A. T., 21
Lucey, Dr Cornelius, Bishop of Cork, 200, 201, 359
Luke, Archbishop Thomas, 40-2
Lurgan, Co. Armagh, 63
Lusitania (vessel), 140
Lynch, Marcus Blake, 317
Lyons, Dr R. S., MP, 65, 111

McAlpine, John, 280
MacBride, Seán, 118, 205, 210, 225, 229, 270
    critical of forest policy, 180-1
    forest policy of, 204, 208, 213, 231
    forestry policy of, 190-9
    interest in forestry, 150, 170, 174
    and social forestry, 186
Mac Carthys, 21
McCracken, Eileen, 51, 55, 63, 67, 69, 71, 74, 75
    on iron smelting, 75-6
    on timber merchants, 96-7
McDonagh, Lar, 87
McElligott, J. J., 194
McEvoy, T., 328
McGerty, J. K., 328
McGilligan, Patrick, 191, 194
McGinley, Ciaran, 243
Mc Glynn, 160, 162, 177, 201, 233, 252, 278
    on forest policy, 204-5
Mackey, John, 137, 150, 162, 171, 177, 186, 202
    and Forbes, 148, 156-7, 176
    Forestry in Ireland, 154, 170, 174
    on state neglect, 181-2
    on timber shortages, 152-3
Mac Murrough, Dermot, 37, 41, 42
Mac Neill, Eoin, 21-2, 27, 28, 33, 34, 319-21
McParlan, James, 75
Mac Piarais, Seamus, 209, 327
Macra na Feirme, 234
McRae, Alaister, 146
Madden, Rev. Samuel, 92, 309
Mael Mordha, king of Leinster, 19-20
Magna Carta, 49
Maguire, Hugh, 62
Malachy, King, 21
Manner of Raising, Ordering and Improving Forest Trees, The (Cooke), 78
Manning, T., 327
Manual of Forestry (Schlich), 108
Manwood, John, 37
Maori people, 10, 11-12
Marine, Department of the, 328
marketing, 175-6, 178-9, 180-2, 206-8
    inadequate, 240-1, 276
    planning needs, 225-6, 230-1
    report on, 256-7
Marshall, William, 81
Marshall Plan Funds, 192-5
martens, 55
Martin, Richard, 317
Martin, T. & C., 90
Martyrology of Oengus, 20
Mayo, Co., 206, 234
    tree plantations, 315
M'Donnel, Charles, 316
Meath, Co.
    tree plantations, 315
Meath, Lord, 125
megalithic tombs, 13, 17
Meldrum, J. A. K., 196, 327
mesolithic period, 10-12
M'Evoy, John, 315
Meyrick, F. J., 166, 327, 355
Milton, Mr, 70
mistletoe, 19
Mitchell, Professor Frank, 72
Mitchell, G. M., 12, 15, 30
Moeran, Archibald E., 124
Molloy, Robert, 290, 330
Molony, James, 316
Monaghan, Co., 253
monasteries, 13
Monteagle, Lord, 123, 126, 132
Montgomery, Hugh de Kellenberg, 126

INDEX

Mooney, Oliver, 335
Moran, Michael, 239-40, 272, 329
Morris, James, 109
Moryson, Fynes, 57
Moryson, Sir Richard, 76
Moss, Bridget, 97
Moss, Richard J.
    report on tree plantations, 313-18
Mount Sandel, Co. Down, 11
Mountjoy, Lord, 62
Moylan, Seán, 182, 208, 329
M'Parlan, James, 315
Municipal Councils, Association of, 110
Munster Chipboard Ltd, 363
Munster Plantation, 61-2
Murphy family, 85

Nally, W. F., 327
National Arbor Day, 147
National Development Plan, 1989-93, 245
National Forest Enterprise, 238, 239-40
National Trust for Ireland (An Taisce), 289, 290, 291
Natural History of Ireland (Boate), 76
navy commissioners, 82, 100-1
neolithic period, 10-17, 26, 30
    migration, 14-15
    tools, 16
Nesbitt, Mr, 315
New England, 59
New Forest, 26
New Zealand, 10, 182, 256
Newcastle, Co. Down, 280, 281
Newgrange, Co. Meath, 13
Newry, Co. Down, 20
Niall of the Nine Hostages, 35
Nicall, Capt. Henry, 72
Nine Years' War, 62
Nisbet, Professor John, 123-4, 128
Normans, 38-9, 46, 47-8
    forest law, 24-6, 36-8, 40-2, 58
    and Irish law, 45
    land law, 32, 33, 50
Norse, 13, 22, 33, 44, 45, 79
Northern Forest Service, 284
Northern Ireland, 264
    amenity forestry, 284
    forest acreage, 281-2, 285
    forestry in, 279-85
    and timber tendering, 240-1
Northern Ireland Forestry Act, 1953, 282
Northern Ireland Home Timber Merchants' Association, 282
Norway, 228
Norway spruce, 78, 250
nurseries, 77-8, 146, 222
    premium plantations, 19th c., 313-18
    premiums for, 93-4

RDS incentives, 310-11

oak, 63-4, 67, 76, 92, 93, 95, 100, 103, 113
    Celtic classification, 28, 29
    charcoal from, 75
    conservation of, 55
    distribution of, 26
    dominant, 78, 114
    forests, 87-8, 89, 119, 152
    house-building, 43
    profiteering, 109
    Royal Oak, Co. Antrim, 74
    sacred tree, 19
    and shipbuilding, 80, 82, 86, 89, 101
    survey of, 52-3
    transport of, 96
O'Beirne, Michael, 351
O'Brien, Dr, 254
O'Brien, Sir Edward, 317
O'Brien, Tim, 327, 328
O'Briens, 21
Observations of the Trade and Improvements of Ireland (Dobbs), 89
O'Byrne, Mathew J., 146
O'Carroll, Dr Niall, 328
O'Carroll, Niall, 367
O'Connor, Art, 147, 329
O'Donnell, Hugh Roe, 62
O'Donnell family, 85-6
O'Donovan, Tim, 146
OEEC, 193, 195
Offaly, Co., 60-1
    tree plantations, 314
Ó Fiacháin, M. S., 327, 328
O'Flanagan, Liam, 254
O'Hara, Mr, 315
O'Hara, Robert, 316
O'Keeffe, Paddy, 237
Oliver, Harry, 283
Oliver family, 103
O'Loghlen, P. J., 192
Ó Muirgheasa, Niall, 328
O'Neill, Hugh, 62, 64, 70
O'Neill inauguration site, 21
Operational Forestry Programme, 1989-93, 245, 278, 291
O'Rahilly, Alfred, 154-5, 172, 176, 186
Ó Rathaille, Aodhgán, 300
Ordinance de Melun, 67-8
O'Reilly, D. W., 316
Osmaston, F. C., 35-6, 108-9
O'Sullivan, S. M., 327, 355
O'Sullivan Beare, Donal, 62
O'Sullivan Beare, Phillip, 339
O'Toole, Laurence, 41
O'Toole, Patrick, 237, 330
Ottelt (forester), 107
Ouzel galley, 79

Painter, Sidney, 36
paleolithic period, 10, 11
Palmerston, Lord, 53
paper, demand for, 153, 249
parliament, Irish, 93, 98-9, 261
Parnell, Charles Stewart, 88, 106, 122-3
Parnell, Sir John, 313
Payne, Robert, 261
Payne (undertaker), 51-2
Peacock, Sir Joseph, 316
Penal Laws, 99
Petrie, Mr, 355
Petty, Sir William, 298
Petty family, 63, 76
Philips, Sir T., 76
Piers, John, 314
Pigot, J. L., 127, 130, 324, 348
Pigot, John, 313
pine, 43, 58, 81
    Celtic classification, 28, 29-30
pipe-stave trade, 69-70
piracy, 53
pitch pine, 90
Plantations, 51-3, 53, 60-4, 65-6
    building methods, 71-2
    and deforestation, 299-300
    land improvement, 91-2
plum, wild, 32
Plunkett, Sir Horace, 119, 122, 125, 126, 329, 346
pollen counts, 10, 15, 17, 26, 72
Pomeroy Forestry School, 283
Ponsonby, T. B., 145
Poor Law system, 104-5
poplar, 76, 94, 282
population increase, 104
Portarlington, Lord, 313
Power, Patrick, 330
Practical Treatise on Planting and the Management of Woods and Coppices (Hayes), 122
Price, Henry, 69
Price, Liam, 295
private forestry, 211, 235, 245, 259-78, 264, 270
    acreage, 265, 271
    criticisms of, 241-2, 242-3, 276-7
    decline, 262-3, 265-6
    and EC, 218
    extent of, 246, 259-60
    foreign investment, 271-2
    importance of, 263-4, 269-70
    incentives, 235-7, 266, 270-1, 276
    inventory of, 366
    as investment, 273-4, 275-6
    in Northern Ireland, 280
    taxation of, 269-70

Private Forestry, Memorandum of, 269-70
Private Planting Grant, 269
Prussian Forest Service, 107
Public Expenditure, Dáil Committee on, 276-7
Public Service, Department of the, 237
Public Works, Interdepartmental Committee on, 168
pulpwood industry, 222, 226, 228, 230, 240, 249, 283
Purcell, Thomas J., 366
Pyne, Henry, 52

Queen Charlotte (vessel), 83
Queen's County, 61
    see also Laois, Co.

rabbit warrens, 42
Rahane inn, 44
raised bogs, 10, 30
Raleigh, Sir Walter, 52, 61, 63, 69
Ramblers' Association, 241-2
Rathlin Island, 282-3
Rawson, Thomas, 316
Raymond le Gros, 42
Rea, Thomas, 196, 198, 328
Recess Committee, 119-20, 172, 262
red cedar, 114
red deer, 55
Redmond, William, 126
reeds, 32, 44
Reilly, Mr, 315
Reinhardt, Otto, 169, 172-3, 177-8, 182, 327, 355, 357
Ribbentrop, Berthold, 108
Riverston, Lord, 317
roads, 64, 109
    clearance of, 67
    corduroy roads, 17, 38
Roman Empire, 35
rose, wild, 28
Rostrevor, Co. Down, 281
rowan, 22, 28
Rowley, John, 69
Royal Dublin Society (RDS), 92-5, 97, 98, 106, 261
    fraudulent claims, 312-13
    incentives for planting, 309-18
    report on plantations, 313-18
Royal English Forestry Society, 175-6, 188
Royal Forests, 36, 70
Royal George (vessel), 81
Royal Indian Engineering College, 108
Royal Oak, Conroy Park, 74
Royal William (vessel), 82
Russell, George (AE), 265
Russell, Thomas Wallace, 125-6, 126, 135, 142, 329

Ryan, Dr, 168
Ryan, James, 329

sacred trees, 18-23
   bile, 20-1
St Leger, Sir Warham, 297
St Patrick's College, Maynooth, 102
sallows, 94
sanctuary, 19-20
Saorstát Éireann, 117-18, 147-9, 263, 281
   forest policy, 150-69
   planting scheme, 164
   and private forestry, 264
Sauer, Carl, 16
saw-log industry, 214, 222, 230, 245
sawmilling, 96, 109-10, 118, 124, 240-1, 257-8
   First World War, 140
   increase in, 129
   slump, 228
   Smurfit expansion, 241
Schlich, William, 113, 128-9, 136, 164, 171
   Acland Committee, 141, 143
   and Howitz, 111, 120-1
   influence of, 108, 172-3
Scotch fir, 345
Scotland, 43, 60, 69, 136, 241-2
   forestry schools, 280
   timber imports from, 72
Scots pine, 30-1, 78, 95, 150, 281
   decline, 16, 26
   planting of, 92, 93
Scottish Agricultural Society, 144
Scottish Arboricultural Society, 106
Seagull (vessel), 88
Second World War, 142, 149, 172, 178, 182, 185, 190, 220, 269, 292
   demand for timber, 179-80
   and Northern Ireland forests, 281-2
   outbreak of, 173, 176
   state control of forestry, 177, 181
   timber demand, 217, 266
seeds; *see* silviculture
Selbourne, Earl of, 140
Seskinore Forest Game Farm, 283
Severin, Tim, 14
Shanahan, Cathal, 234
Shannon Hydroelectric Scheme, 192, 194
Shillelagh, Co. Wicklow, 64, 87, 89
shipbuilding, 59
   arrival of iron ships, 100-1
   construction methods, 80-1
   dry rot, 82
   importance of, 79-90, 83
   in Ireland, 79, 83-90
   Norse, 33
   timber demand, 49-50, 54, 68, 70, 73, 261

timber survey, 52-3
Timber Trust, 113
Sidney, Sir Henry, 261
silver fir, 78, 114
silviculture, 50, 106, 137, 344
   definition of, 353
   experimentation, 111-13, 252-3
   lack of seed, 222
   seed supply, 182-3
   species, 140, 149-50, 248
   survey, 256
   see also estate forestry
Sinn Féin, 132, 146-7
Sitka spruce, 109, 114, 219, 221, 250-3
   introduction of, 248-9
   in Northern Ireland, 282
   percentage distribution, 247
Siward, Richard, 49
slavery, 77, 79
Sligo, Co.
   tree plantations, 315
sloe-bush, 28
Smith, H. G., 327
Smith, Michael, 276, 330, 361, 367
Smurfits, 241
social forestry, 142-5, 164, 179, 184-6, 198-9, 200-2, 204, 208, 210-11, 245, 255, 256
   in Cameron Report, 205
   definition of, 224
   employment forecasts, 255
   in Northern Ireland, 283-4
   numbers employed, 256
   and private forestry, 266-7
   wage rates, 353
   workers' conditions, 270-1
soil types, 10
Sophia (vessel), 85-6
Spain, 53, 59-60, 69, 70, 79
Spanish Armada, 80
Spenser, Edmund, 43-4, 57, 61
spindle tree, 28
sport, 25, 35-7, 42, 119, 139
spruce, 81, 92, 153, 281, 345
squirrels, 55
Standing Timber (UK) Order, 1917, 138
Star of Murrisk (vessel), 85-6
state forestry, 92, 110-13, 115, 133, 140, 144, 156, 177, 221, 229, 245-6, 260-1
   acreage, 148, 157-8, 175, 214, 274
   after Second World War, 187-8
   changing species, 250-1
   and civil service, 178-9
   Departmental Committee report, 1908, 126-31, 323-6
   development of, 117-36
   finances, 135-6, 226-7, 238-9
   holding of title, 252
   incentives, 213

income, 217
increased interest in, 1968-76, 215-27
lack of planning, 125
legislation, 176-7
neglect of, 181-2
in Northern Ireland, 281-2
opposition to, 200-2
planting rates, 212, 213, 222, 224
and private investment, 231-3
problems of forecasting, 217-18, 244
rates of investment, 207, 211-12
rotational 'hump', 225
and Saorstát Éireann, 148-9, 151-5
slump, 226-7
surveys, 229-31
valuation, 219-20, 222-4, 254-5, 256
State Forestry Board, 239
State Forests, Inventory of, 1968, 254
State Papers, Calendar of, 71-2, 83
Steward, Sir William, 55
Stewart, David, 146, 279, 280
Stilicho, 35
Stradbally, Co. Laois, 103
Strongbow, 58
Stuart, Henry, 315
Stuarts, 53
Suibhne Geilt, 30
Sullivan, H. J., 328
Supplies, Department of, 356
Survey and Distribution, Books of, 51
Sutherland, Col. John, 280
Svenska Teknolog Foreningens Forlag, Stockholm, 352
Sweden, 257
sweet-blackthorn, 32
Sweetman's Calendar, 42
Swift, Jonathan, 297
Swords, Co. Dublin, 20
sycamore, 95
Synge, George, 314

Taisce, An, 289, 290, 291, 363
Tallow, Co. Waterford, 252
tanning, 73, 74, 77
Tara, Co. Meath, 44
taxation
    and private forestry, 269-70
thatching, 44
Thompson, Robert, 315
Tighe, William, 315
timber industry, 50, 58, 63, 67, 73, 98, 134, 142-3, 167, 202, 222, 225
    carpentry, 45
    Celtic, 24
    corruption in, 109, 113
    cross-border trade, 284
    dominant woods, 78, 249
    EC demand, 229

employment in, 69
export possibilities, 217
and house construction, 42-5
importance of, 66-7, 96
imports, 67, 81, 89-90, 97-8, 137, 139-40, 180
increased demand, 54, 214, 235
industrial unrest, 96
Irish exports, 70-1
and ironworks, 35
kiln-drying facilities, 213
loss of market, 119
neolithic period, 14-17
Norman, 42-5
in Northern Ireland, 283, 284
possible export, 206
processing sector, 235-6, 255
prospects for, 1978, 230-1
quality, 255
under Saorstát Éireann, 149
Second World War, 179-80
and shipbuilding, 59, 80-2
slump, 226, 228
tendering system, 239, 240-1
timber classifications, 29-32, 260
Timber Trust, 113, 119
transport, 82-3, 96
Tudor survey, 52-3
and unemployment, 184-6
worldwide demand, 152-3, 231-2
see also marketing
timber measuring, 97
timber merchants, 96-7
Timber Supplies, Directorate of, 140
Timber Trust, 113, 119
Tipperary, Co., 133, 134, 206
Tokefield, Mr, 76
Tombrickane, Co. Tipperary, 21
Tomkins, Steven, 242
Toome Bay, Lough Neagh, 12
Tourism, Fisheries and Forestry, Department of, 328
training; *see* forestry education
Treatise of the Laws of the Forest, A (Manwood), 37
Tree Council of Ireland, 243
tree-farming, 178, 355, 359
treen, 54
trees
    Celtic classification of, 28-32
    see also sacred trees
Trees for Ireland, 241
Trócaire, 287
Troup, Professor R. S., 354
Tudors, 34, 48, 60
    forest policy, 53-4, 65-6
    and Irish law, 45
    timber requirements, 49-56

see also deforestation
Tullymore Forest Park, Co. Down, 283
Turlough Hill, Co. Wicklow, 13
Turner, J. G., 26, 37
Tyrone, Co., 279, 280, 281
    tree plantations, 315
Tyrrell, Capt. John, 64
Tyrrell, John, Ltd (shipyard), 86-90
Tyrrell's Pass, Co. Westmeath, 64

Uí Dunchada, 42
Uí Faelain, 42
Uí Fiachrach Aidhne, 21
Uí Muiredaig, 42
Ulster
    landlord and tenant, 103-4
    Plantation of, 62
Ulster, Plantation of, 71
Ulster Timber Growers' Association, 282
unemployment; *see* social forestry
unemployment relief, 168, 184-6
Union, Act of, 95, 98-9, 137, 262
Union of Professional and Technical Civil Servants (UPTCS), 242
United Nations Environment Programme (UNEP), 287
United Nations Organization (UNO), 232
United States of America, 70, 81, 102
    Civil War, 100-1, 137, 262
    iron ships, 100-1
    timber shortages, 152
    tree species, 249, 251, 252
Urban and Rural Improvement Campaign (NI), 284

Vandaleur, Boyle, 316
Vavasour, Sir Charles, 63
Viney, Michael, 190, 196, 198, 199, 211

Wakefield, 85
Waldau, Der (Gayer), 107-8
Wallop, Sir Henry, 51
walnut, 76
War of Independence, 146, 148
warfare
    timber demand, 65, 66, 75, 76
water transport, 12, 14
    see also ships
Waterford, 33, 42, 79, 86
Waterford, Co., 52, 84, 206, 222, 252
Waterford, Marquis of, 314
Watson, Dr, Bishop of Llandaf, 98
wattling, 17, 29, 31, 44, 72-3
Weld, Isaac, 313
West Indies, 66
Western Development Programme, 233-4
Western Forestry Co-operative, 234

Western Package Grant scheme, 235, 243, 274-5
Westmeath, Lord, 104
Wexford, 86
Wexford, Co., 58, 133, 134
weymouth pine, 93
Whaley, Jack, 220
whin; *see* furze
Whitaker, T. K., 191, 192, 194-5, 199, 222-3, 224, 244
White family, 63, 76
white hazel, 28
whitebeam, 32
whitethorn, 31
Whooley, Patrick, 328
wicker-work, 42-3
Wicklow, Co., 47, 75, 206, 240
    tree plantations, 314
Wicklow, Lord, 88
Wildlife Act, 1976, 215, 322
wildlife conservation, 215, 224
William III, 71, 76
    forest legislation, 301-2
William the Conqueror, 26
willow, 28, 94
Wilson's Dublin Directories, 96
Wittich, Professor, 355
wolves, 55, 56
Women's Forestry Corps, 138
Wood Quay, Dublin, 74
wood-workers, Celtic, 45
Woodfab, 241
woodkerne, 45, 55, 56
woodlands; *see* forests
Woods, Forests and Land Revenues, Commission on, 109
Woods, Office of, 140
Woods, Surveyor General of, 66
Woods and Forests, Commissioners of, 99-100
woodsmen, role of, 95
Woodstock, Assize of, 25, 39, 40
Woolsey, Theodore, 117
Wright, Richard, 340
Wyndham, George, 122
Wynne, Mr, 315

yew, 15, 45
    Celtic classification, 28, 29
    sacred tree, 19-20
yield class, definition of, 364
Youghal, Co. Cork, 20, 70
Young, Arthur, 64, 298, 299